Engineering Victory

Johns Hopkins Studies in the History of Technology
MERRITT ROE SMITH, SERIES EDITOR

ENGINEERING
VICTORY

How Technology Won the Civil War

———

THOMAS F. ARMY, JR.

Johns Hopkins University Press
Baltimore

© 2016 Johns Hopkins University Press
All rights reserved. Published 2016
Printed in the United States of America on acid-free paper

2 4 6 8 9 7 5 3 1

Johns Hopkins University Press
2715 North Charles Street
Baltimore, Maryland 21218-4363
www.press.jhu.edu

Library of Congress Cataloging-in-Publication Data

Army, Thomas F., Jr., 1954– author.
Engineering victory : how technology won the Civil War /
Thomas F. Army, Jr.
pages cm. — (Johns Hopkins studies in the history of technology)
Includes bibliographical references and index.
ISBN 978-1-4214-1937-4 (hardcover : alk. paper) — ISBN 978-1-4214-1938-1
(electronic) — ISBN 1-4214-1937-8 (hardcover : alk. paper) — ISBN 1-4214-
1938-6 (electronic) 1. United States—History—Civil War, 1861–1865—
Technology. 2. United States—History—Civil War, 1861–1865—
Campaigns. I. Title.
E468.9.A67 2016
973.7′3—dc23   2015026992

A catalog record for this book is available from the British Library.

Special discounts are available for bulk purchases of this book.
For more information, please contact Special Sales at 410-516-6936 or
specialsales@press.jhu.edu.

Johns Hopkins University Press uses environmentally friendly book
materials, including recycled text paper that is composed of at least 30
percent post-consumer waste, whenever possible.

To Virginia, because without your love and support none of this would have been possible

Contents

Maps

Acknowledgments

The word *acknowledgment* strikes me as meaning a cursory nod to those who played a supporting role in the production of this book. The word suggests the obligatory and dutiful recognition of those who have guided and assisted me during the oftentimes lonely and always demanding process of writing this manuscript. To me, the word falls far short of the mark. Instead, the heading should read: "Those who had my back," or "Without the following people I'm not sure I would have started this project, let alone finished it." Constructive comments, encouraging words, small kindnesses, and quiet support from mentors, colleagues, friends, and family truly meant more to me than *acknowledgment* conveys. Throughout the seven years I worked on this project, the unwavering support I received from many remarkable men and women held up both my enthusiasm and my emotions better than any trestle or pontoon bridge sustained the weight of crossing Union or Confederate armies. Countless people went out of their way to offer me words of encouragement or exercised the patience of Job as they listened to me revel in stories about winged dams on the Red River and corduroy roads and pontoon bridges on the Virginia peninsula. To all of you I offer my undying gratitude and most sincere thanks.

This project began in 2002 when I started to wonder aloud about the effects of engineering on the outcome of the Civil War. Stymied by my own efforts to locate resources on the subject, I mentioned my interests and frustrations to friend and Licensed Battlefield Guide at Gettysburg National Military Park, Phil Cole. He had just embarked on a project of his own, and so we spent a fair amount of time standing atop Little Round Top discussing our respective interests. Phil offered to help me in any way possible, suggesting where to look for archival information and encouraging me to follow the breadcrumbs. I did. His practical advice, wisdom, and guidance helped jump-start my journey into the world of Civil War bridge builders.

The research for this book continued with my dissertation at the University of Massachusetts, Amherst, where my advisor, Heather Cox Richardson, helped to sharpen my focus. As my dissertation started to take shape, she read chapter drafts and helped structure and develop my writing. I am thankful for her teaching and guidance.

Leonard L. Richards, professor emeritus of history at UMass, also frequently offered cheerful and smart advice about the structure of my dissertation and championed my topic enthusiastically. I am also deeply indebted to Merritt Roe Smith of the Massachusetts Institute of Technology for agreeing not only to serve on my committee but to shepherd me through the process of converting my dissertation into a manuscript for publication. His knowledge of American technological developments in the nineteenth century and their intersection with the Civil War is remarkable. He was most generous in sharing his knowledge and time with me. Roe Smith also connected me with Senior Editor Robert J. Brugger of Johns Hopkins University Press, who guided me through the process of revising and editing my manuscript. For poring over these pages and offering wisdom and sound advice, I'm grateful to Linda Strange. Spencer Tucker, retired chair of military history at the Virginia Military Institute, was also gracious in answering my emails and in reading several chapters of the book.

Special thanks go to the Reverend Doctor Bennett A. Brockman, a most remarkable scholar and professor of English Literature, Episcopal priest, and good friend, who warmly agreed to read the entire manuscript. Ben offered sound advice, corrected my grammar, and suggested ways to strengthen the manuscript. I am deeply indebted to him for all his assistance.

I am thankful to the many park service people and archivists who generously offered me their time and talent. Park rangers and historians at the Vicksburg, Chickamauga, Chattanooga, and Perryville battlefields were enormously helpful in pointing out important landscape features, especially those that required the attention of Union and Confederate engineers. Several folks with local knowledge also provided directions to the spot where Lee's army crossed the Potomac River after the Battle of Gettysburg, and to the remains of Grant's Canal in northern Louisiana near the Mississippi River. In particular, I want to thank the people at the Flowerdew Hundred Plantation for giving my daughter Priscilla and me a personal tour of the spot where Grant's army crossed the James River in June 1864.

To the archivists and librarians at the National Archives in Washington, DC, the Virginia and New York State libraries, the Museum of the Confederacy in Richmond, the Connecticut Historical Society in Hartford, the Civil War Museum in Atlanta, and the National Park Service Library at Gettysburg, I extend my sincere appreciation.

I would be remiss if I failed to mention Professor Willard M. Wallace of Wesleyan University. He was a role model to me, and one reason I wanted to become a history teacher. He was a quintessential storyteller, scholar, and gentleman. He died in June 2000, about the time I started thinking about Civil War engineering. This book is dedicated, in part, to his memory. I owe a special debt of gratitude to Dr. Richard Griffith, one of the kindest and gentlest souls I have had the privilege to know. Richard and I spent many an evening together enjoying a meal and discussing a range of topics, including our shared interest in the Civil War. A relative of Robert E. Lee, an anesthesiologist, and a PhD in engineering, Richard found my ideas about Civil War engineering exciting and fresh, and his enthusiasm meant a great deal to me. In his spare time, Richard is also a professional videographer, and he graciously offered to assist me in generating a series of maps for my book. Once completed, the maps were given to Chris Robinson, who corrected several of my omissions and made adjustments to the style. To both I offer sincere appreciation.

Next, thanks must go to Roger Franklin, who helped me find the photographs for this book, and to my brother-in-law Dave Sluyter, a computer ninja (or perhaps even a wizard), who prepared the images for publication. Dave also read a draft of the manuscript, offering constructive comments, and was always exceptionally encouraging and expert with technical questions. Kevin Fenner of Toland, Connecticut, took many precious hours away from his own family on Sunday afternoons to assist me with a fair number of computer issues. I also want to thank my buddy Art Mulligan. Art called or emailed me frequently to find out how I was holding up, and when we were together, he always wanted to know how the writing was going. His genuine interest in and enthusiasm for the project helped keep me going.

To my in-laws, Paul and Priscilla Gray, and my parents, Tom and Betty Army, go my abiding admiration, thanks, and love. My parents bought me my first Civil War book when I was ten, encouraged my interests, and always had my back. My parents and in-laws have surrounded me with constant encouragement and support and, as a respite from the daily attention I paid to my keyboard, were always ready to provide a wonderful meal, which usually included ice cream or chocolate cake.

My five daughters, Hannah, Rachel, Priscilla, Elizabeth, and Catherine, have brought joy and fulfillment beyond measure into my life. It has been the greatest privilege of my life to watch them grow into strong, resilient, intelligent, and independent women. Each one of them made contributions to this book, even when it was not apparent to them that they were making those contributions. Surprise phone calls when I was discouraged, teasing about my slow, four-finger typing, or even a guilt trip because their mom and I had not visited them in awhile, all helped to place in perspective the things that really mattered. I am a very fortunate man.

Finally, I am most grateful to my remarkable wife, Virginia. From the very beginning of this project she has been at my side. She never complained about all the time, including family vacations, I spent wed to my computer, and she never wavered in her support and encouragement. She is my companion and soul mate. She is my hero. This project was the result of her faith and love. For all of this and countless other reasons I dedicate this book to her.

Masters and Mechanics

When I learned that Sherman's army was marching through the Salk swamps, making its own corduroy roads at the rate of a dozen miles a day and more, I made up my mind that there had been no such army in existence since the days of Julius Caesar.

General P. T. G. Beauregard, Confederate States Army (CSA), 1865

That engine was made in our shop; I guess I can fit her up and run her.

Private Charles Homans, Company E, 8th Massachusetts Infantry, 1861

In November 1862, Ulysses S. Grant, commanding the Union Army of the Tennessee, considered Vicksburg the key to his western campaign and thus critical to ending the War of the Rebellion. Six times he had tried to reach dry ground east of the city as a staging area for his army. Six times he had failed. He turned now to Major General John A. McClernand to find a way to move the army south of Vicksburg along the western bank of the Mississippi, to a point, as yet undetermined, where the army could cross the river. McClernand ordered the Ninth Division commander, Brigadier General Peter Osterhaus, to take charge of the reconnaissance. In turn, Osterhaus selected Colonel James Keigwin to discover a passable route.

Only the difficulties created by the landscape's semitropical climate, miserable swamps, numerous alligators, countless mosquitoes and mayflies, cottonmouth water moccasins, and occasional quicksand matched the enormity of the task before Colonel Keigwin of the 49th Indiana Volunteer Infantry. He was to determine whether there was a passable road on the western bank of the Mississippi River from Somerset to a point opposite Grand Gulf where troops could cross to the eastern bank of the river. If a staging area opposite Grand Gulf could not be found, the reconnaissance was to extend farther south.[1] The Confederates deemed the terrain impassable. Grant's army was in its fifth month of operations in April 1863, as the general attempted to reach dry ground east of Vicksburg. Once there, he could maneuver his army—45,000 men, 5,500 mules, 2,500 horses, 1,500 wagons, 220 pieces of artillery, and one pontoon train—and attack.[2]

The drawn-out siege at Vicksburg invoked criticism from politicians and Northern citizens. On April 16, a war correspondent for the *New York Times* registered his frustration at the "entire lack of progress that marks the character of this expedition against Vicksburg." As the reporter lamented to his readers, "there is always between us and the enemy a deep, swift river, of a mile in breadth to cross, which, in the face of one hundred and sixty guns which frown upon us from the opposite heights, is beyond reach of human valor, endurance, or ingenuity."[3]

The reporter must have sent his observations on their journey east sometime during the day, because on the night of April 16, and then again on the evening of April 22, the stalled campaign took on a character of frenetic progress. Rear Admiral David Dixon Porter ran his gunships and transports past the Confederate batteries resting on the cliffs overlooking the river at Vicksburg. Using the transports, Grant now had a way to carry his army from the western bank of the Mississippi to the eastern side. An amphibious operation, however, still had two considerable problems. Avoiding the Confederate batteries well-positioned on the river at Grand Gulf, about thirty miles south of Vicksburg, was challenging enough. Even harder was finding among the bayous, bogs, wetlands, and creeks that made up most of the geography along the western bank of the river a staging area for 45,000 men (approximately 22,000 in the first wave), 8,000 horses and mules, artillery, and wagons. This is why Colonel Keigwin was ordered to make a reconnaissance south from Somerset and to possibly extend it beyond Grand Gulf.

On April 25, before Keigwin's patrol left the semidry ground of its camp to begin the reconnaissance, Grant's army had performed yeoman's work in cutting a road forty miles long from Milliken's Bend to New Carthage. Between these two points, Roundaway and Walnut bayous converged. Opposite their mud-colored waters lay the small town of Richmond, Louisiana, occupied by Confederate cavalry under the command of Lieutenant Colonel Isaac F. Harrison. Captain William Patterson's Kentucky Company of Engineers and Mechanics ferried men from the 69th Indiana Infantry across the waterway in yawls, and after the Indianans drove off the rebel cavalry, Patterson and his men were ordered to build a 200-foot-long bridge over Roundaway Bayou. Ripping down barn timbers and sideboards from outbuildings in Richmond, the engineers completed the structure in thirty-six hours, opening the way to New Carthage.[4]

When Grant arrived at New Carthage, however, he discovered that the flooding from the soaking rains of the past three weeks had made the town untenable as a jumping-off point for his three army corps to cross to the east bank of the Mississippi. The one hundred men of the Kentucky Company of Engineers, along with Company I of the 35th Missouri Infantry (the army's pontoon company) and the 34th Indiana Infantry, built four more bridges and two more miles of connecting roads that required, in some places, layers of planking to create sufficient solidity over quicksand to property owned by John Perkins called Somerset Plantation.[5]

With the Missouri engineers attending to the bridges and roadway from Milliken's Bend to New Carthage and the Kentucky engineers opening a passage to Somerset Plantation, Grant decided to extend his line of march to Hard Times and perhaps beyond. His urgent need now was for more engineering troops, so Grant turned to the infantry. He had used infantry regiments before, with great success, when confronted with a shortage of engineers, so Colonel Keigwin's 49th Indiana, the 114th Ohio, a detachment of the 2nd Illinois Cavalry, and a section of gunners from the 7th Michigan battery set out to find a passable route to Hard Times. Cypresses, cottonwoods, several bayous, and five hundred Confederate cavalrymen with light artillery stood in the way of Keigwin's detachment. Lieutenant Francis Tunica of the Corps of Engineers accompanied Keigwin on the march.[6]

In just four days, Keigwin's augmented regiment built four bridges, with a total length of 450 feet. They secured lumber for the bridges by tearing down buildings on nearby plantations, where foraging parties found tools and rope. Three of the bridges were particularly complicated to build because the bayous were wide, the currents strong, and the quicksand abundant. When orders recalled Tunica to headquarters, supervision of these three bridges fell to infantry officers—Lieutenant Colonel John W. Beekman, Captain William H. Peckinpaugh, and Lieutenant James Fullyard. Construction work faced frequent harassment from Major Harrison's Confederate cavalry.[7]

On April 29, Keigwin reported to army headquarters that a practical road from Somerset (just north of Perkins's Plantation) to Hard Times was open for traffic. The story, however, was not over. Grant, believing the Confederate batteries at Grand Gulf were too near to this vicinity, decided to move his Mississippi River staging area another fifteen miles farther south to DeShroon's

Landing. Union soldiers built four more bridges, three requiring "excavation to level the exit and entry points."[8] The combined length of two of the bridges over Bayou Vidal was 600 feet. One of them, reported Herman A. Ulffers, Army of the Tennessee assistant engineer, was "curved upstream and rested on 16 flatboats, 25 to 40 feet long by 12 feet wide."[9]

When the Union army finally crossed the Mississippi on April 30, it had extended the central supply line through seemingly impenetrable terrain. Starting at Milliken's Bend and ending at DeShroon's Landing, through rain, mud, and swamps and in the face of harassment by Southern cavalry, Grant's soldiers had, in thirty days, covered 70 miles and built 2,000 feet of "floating roads" and bridges.

Grant's reputation as stubborn and determined proved itself again and again throughout the war. Once he had decided to find a staging area for his 22,000 men on the west bank of the Mississippi, he pressed his commanders to do so regardless of the obstacles. Three times he observed for himself the potential jumping-off points, at New Carthage, Perkins's Plantation or Somerset, and Hard Times, and three times he ordered reconnaissance farther south. It was clear that Grant believed the men in his army possessed the skill, regardless of difficulty, to deliver him to the necessary strategic advantage point. This convergence between the commander's strategic and tactical objectives and his men's ability to deliver the army to those points gave Grant a significant advantage over his enemy. Grant developed his well-placed confidence in his men's mechanical abilities early in the war. In October 1861, one hundred coal barges were used to build a pontoon bridge at Paducah, Kentucky. A reporter for *Harper's Weekly* wrote that the bridge "surpasses anything of the kind before attempted in the United States."[10]

British Field Marshall Archibald Percival Wavell, a veteran of the Boer War, World War I, and World War II, wrote about this critical relationship between a commander's plan and the reality of its execution. "The more I see of war, the more I realize how it all depends on administration and transportation . . . It takes little skill or imagination to see where you would like your army to be and when; it takes much knowledge and hard work to know where you can place your forces and whether you can maintain them there."[11] Grant understood this gem of military science long before Wavell. In his recollection of the Vicksburg Campaign and the march from Milliken's Bend to Perkins's Plantation, Grant wrote: "Four bridges had to be built across bayous . . . The river falling made the current in these bayous very rapid, increasing the difficulty of building and permanently fastening these bridges; but the ingenuity of the 'Yankee soldier' was equal to any emergency."[12] This and several similar comments in his post-war musings gave credit to his soldiers' abilities and skills to deliver the army to where he wanted it to be and to sustain the supply lines necessary to allow his men to operate against their opponents in favorable positions.

Even so, Grant in his memoirs seldom mentioned the Union army's engineering efforts and the remarkable technical skills his soldiers displayed. This lack of acknowledgment of the role of engineers was true of the participants on both sides and of the writers, journalists, and historians who came after them. The past 150 years have produced many carefully argued reasons for why the North won the Civil War, ranging from overwhelming resources to superior civilian, political, and military leadership, failed Confederate economic planning, a stronger Union navy, emancipation, and according to Robert E. Lee, God's will.

This volume serves as a corrective. The Union's critical advantage over the Confederacy was its ability to engineer victory. The North won because, in the decades before the war, Northerners invested in educational systems that served an industrializing economy. Furthermore, the labor system in the North rewarded mechanical ability, ingenuity, and imagination. The labor system in the South failed to reward these skills. Plantation slavery generated fabulous wealth for a slim percentage of the Southern white population. It fostered a particular style of agriculture and scientific farming that limited land use. It curtailed manufacturing opportunities, and it stifled educational opportunities for the middle and lower classes because those in political power feared that an educated yeomanry would be filled with radical ideas such as women's equality, temperance, and, worst of all, abolition.

In the North, each state, in its own way, attempted to implement basic educational reform by introducing the natural sciences into university curricula and spreading informal educational practices directed at mechanics and artisans. These reforms, coupled with exponential growth in manufacturing, generated a different work-related ethos than that required of the plantation South. Ironically, Southern politicians such as South Carolina's James Henry Hammond and Virginia's Edmund Ruffin viewed the Northern industrial economy as a chimera. They argued that workingmen's associations, immigration, and rampant unemployment destabilized the North and that it was only a matter of time before the North would explode in a class war, creating chaos, collapse, and future uncertainty. It was for this unstated reason, Southern fire-eaters maintained, that Northern politicians—especially Whigs and Republicans—demanded free territory in the west. This was the place, according to Hammond, where

Northerners sent immigrants and the unemployed to escape social unrest. The North would come to see that Southern slavery provided a stable social structure and that, despite price fluctuations in cotton, Southern agriculture was not susceptible to the ebb and flow of production experienced by Northern manufacturers.[13]

The slavocracy missed the larger picture, not because these Southerners were uneducated, but because their system was structured around the brutality of slavery—it was what they knew—and their profoundly conservative, classics-based education could not conceive of an acceptable alternative. As a result of this complete commitment to their "peculiar institution," when the Civil War broke out, few Southern officers were either trained or inclined to solve the massive logistical demands the war presented. Even fewer common soldiers had interest or skill enough to maintain roads, build bridges, repair railroads, or dig fortifications—this was work for slaves.

It could be argued that having few mechanics, civil engineers, and artisans within the ranks of the Confederate army made little difference in the South's ability to fight the war, given that most of the war was fought in the South and Confederate armies in the field did not need extensive logistical support. Southern generals knew the local geography better than their Northern counterparts; they understood the road systems, could count on accurate intelligence from Southern citizens, and often could determine the place of battle.

This cannot be said for the North. Without the unique, novel, and remarkable engineering operations conducted by common laborers, machinists, gun makers, and both common school–educated and West Point–trained engineers, it seems unlikely that the North could have fought the war effectively, much less won it. Union generals were forced to execute a strategy that demanded the control of 750,000 square miles of territory and the defeat of enemy armies, partisan raiders, and cavalry that constantly threatened the North's long and tenuous supply lines. Between 1861 and 1865 the North engineered victory.

This book addresses the general topic of why the North won the Civil War and, specifically why Union soldiers were better than Confederate soldiers at engineering that victory. My purpose is to explain why and how Northern and Southern engineering differed and why the difference mattered.

Perhaps the most common understanding of why the North won the Civil War is its rising capacity in manufacturing and production. The North did possess overwhelming material and manpower resources. General Robert E. Lee recognized this fact. His soldiers, the general believed, were never outfought, nor did they ever display a lack of personal courage or manly skill. After Lee surrendered his tattered and torn Army of Northern Virginia to Union forces at Appomattox Court House, he explained to his men: "After four years of arduous service marked by unsurpassed courage and fortitude, the Army of Northern Virginia has been compelled to yield to overwhelming numbers and resources."[14]

In 1960, historian Richard N. Current reinforced Lee's assessment of why his army surrendered. Current, a revisionist historian influenced by World War II and the Korean War, witnessed firsthand how the industrial might of the United States defeated the Germans and Japanese. He argued that it was improbable that the agrarian South could have defeated its industrial enemy. The North's overwhelming resources included sixty-one percent of the nation's white population, sixty-six percent of railroad mileage, and eighty-one percent of the country's factories. According to Current, these numbers made it only more remarkable that the Confederacy lasted as long as it did. It would have taken a miracle for the South to win. "As usual," Current wrote, "God was on the side of the heaviest battalions."[15]

After the Vietnam War, however, historian James McPherson pointed out that "God was not always on the side of the heaviest battalions."[16] With American withdrawal from Southeast Asia, historians and strategists began the investigation as to what went wrong. United States economic power prolonged the conflict, but it did not bring Ho Chi Minh to his knees. As in our own American Revolution, an undersupplied, underfed, undercapitalized people defeated a wealthier nation. Thus, Civil War historians like McPherson determined that Current's thesis did not adequately answer the question of why the North won the Civil War. Many turned elsewhere for possible answers.

Southern apologists had maintained that adherence to states' rights principles handicapped the central government in Richmond in mobilizing men and supplies for war. Historians suggested that Jefferson Davis was not bold enough to nationalize industries or, more importantly, the railroads, and instead allowed state governors to dictate how the logistical operations of the war were to unfold. Economic historians piled onto the states' rights arguments the suggestion that a lack of economic imagination cost the Confederacy the resources to conduct a prolonged war. They argued that Confederate policies on taxation, cotton, impressments of supplies, and the use of slave labor damaged the South's efforts to win independence. Historian John Solomon Otto added to the criticism of Confederate economic strategy by proposing that the government not only generated poor fiscal policies but also failed both to invest in a deteriorating infrastructure and to focus on feeding its people. By 1864, "Confederate soldiers possessed first-class ordnance but had second class accouterments and ate third class rations."[17]

Using economic analyses to explain Confederate defeat, however, is just the other side of Richard N. Current's coin. For Current, Union economic might and sound fiscal and monetary policies delivered the victory. For some economic historians, the South's poor fiscal and monetary policies led to defeat. Both sides of these economic deterministic arguments fail to address the human equation. Certainly, if the economies of the North and South had been reversed, it would have been almost impossible for the North to conqueror the South. As it was, the South came close to winning its independence. So how did this happen? To come closer to answering this question we must pursue a different line of inquiry. How long would Northerners fight to maintain the Union? What if Union blunders on the battlefield had led to Confederate victories at Gettysburg, Vicksburg, or Chattanooga? These questions appear more important to answering why the Union won or the Confederacy lost the Civil War.

There is a body of work that has survived the test of time and has offered military explanations for the war's outcome. Gary W. Gallagher took up the question of military leadership in his 1992 essay "'Upon Their Success Hang Momentous Interests': Generals" and first paid his respects to historians who explored the "complex and interrelated factors" away from the battlefield as reasons for Confederate defeat. "No shift in civilian morale," he continued, "North or South—and really none of the non-military factors—can be fully understood outside of the context of the military ebb and flow."[18] Gallagher agreed with James McPherson "that defeat caused demoralization" and thus major military events during the course of the conflict altered the direction of the struggle. For McPherson, this was the role of contingency, the moments when "victory or defeat hung in the balance, and the issue might easily have been resolved either way."[19] The "turning points" were McClellan's failure to capture Richmond, the failure of Confederate offensives at Antietam and Perryville, Union victories at Gettysburg, Vicksburg, and Chattanooga, Sheridan's campaign in the Shenandoah Valley, and Sherman's capture of Atlanta, ensuring Lincoln's election.[20] Thus, for Gallagher, "generals made a very great difference in determining the outcome of the war."[21] Following closely upon this logic, he concluded that the two generals who needed the most careful scrutiny were Grant and Lee.

Many historians have taken up the cause of the two generals. Revisionist interpretations of why the North won the Civil War, however, developed around the thesis that it was Grant—unencumbered by the theories of Antoine Henri

Jomini, the French intellectual of the early nineteenth century who first estab-lished the principles of modern warfare—who was responsible for the Union's victory. Jomini analyzed the campaigns of Napoleon, pointing out why the em-peror's campaigns needed to be studied and emulated. Grant later admitted that he never paid much attention to these lectures on Jomini in his West Point classrooms. Grant's strategy of annihilation, historian Russell Weigley argued, was more like the warfare of the future than the warfare of the past. Brooks Simpson, in his 2001 essay "Facilitating Defeat: The Union High Command and the Collapse of the Confederacy," credited Grant not only with creating and then implementing a strategy of annihilation but also with winning the peace, which helped to avoid a Southern guerrilla war in the years following Appomat-tox. "In the end, of course," Simpson writes, "it was Grant and Lincoln who suc-ceeded in devising a way to remove Virginians from the Confederate army. The general believed that the Appomattox terms, embodying Lincoln's notion of a lenient peace, would facilitate the end of the war."[22]

Some historians, however, have challenged the notion that Grant was the most intrepid and skilled general the war produced. Albert Castel wrote in 2011 that Robert E. Lee was the war's most brilliant general.[23] Not surprisingly, Cas-tel is in good company. Beginning immediately after the war, former Confeder-ate generals started looking for answers as to why they lost the war. (Lee never commented on this.) Several scapegoats emerged from the battlefield detritus and an aura grew around the stately, honorable, courageous, and stoic Robert E. Lee. Lee became the son to the father George Washington. For five score years Lee was sacrosanct.

In a provocative reappraisal of the Confederacy's strategic decision mak-ing, Thomas Lawrence Connelly and Archer Jones argued that Lee's central concern about the Virginia theater of operations influenced Davis's judgment concerning the crucial western theater. Western generals such as Joseph E. Johnston, Braxton Bragg, and P. G. T. Beauregard, according to Connelly and Jones, formed a "western bloc" that challenged the strategic ideas of Lee. Davis was left to decide between the two factions, and he chose Lee's suggestions. Lee's tunnel-vision strategy of the war then led to the loss of western battles, territory, and ultimately the war itself.[24] Other historians argue that Lee "squan-dered Confederate blood and resources in campaigns and battles that did not resolve anything and, worse, precluded victory."[25]

We should keep in mind, however, that despite its limited resources, key battlefield losses, and perhaps some questionable leadership, the South al-most won its independence. To all the actors in this wretched drama, for most of the war the outcome was never inevitable. As the war continued through the spring and summer of 1864, and Grant was stopped at Petersburg and Sherman frustrated in Georgia, the Northern home front grew tired of the bloodshed. Citizens North and South were well aware that if Lincoln were reelected in November he would continue to prosecute the war. They also knew that if the president's Democratic opponent, former General George McClellan, were elected he might continue the struggle, or he might negotiate a peace for an independent Confederacy or, more likely, a peace with slavery intact in a re-United States, meeting all the demands of the South. We can only speculate what McClellan would have done as president, but in July 1864 hopes sprang eternal in the hearts and minds of Southerners.

Using a counterfactual, we can also deduce that the outcome of the war would have been different had the South won, specifically, at Antietam, Gettysburg, Vicksburg, Chattanooga, or Atlanta. James McPherson believes this is so because, with a Confederate victory at Antietam or Gettysburg, Peace Democrats in the North would have applied unrelenting pressure on the Lincoln administration to negotiate a settlement with the Davis government. A successful invasion of the North would have had a severely damaging impact on Northern public opinion. Furthermore, in the case of Antietam, the Preliminary Emancipation Proclamation would not have been issued, and those political voices in the North, and in the Union army itself, that subscribed to the notion that the war should not be about slavery would have become louder than they already were.

Yet, even after crucial Union victories at Antietam (1862) and Gettysburg (1863), the South's geographical advantage loomed large, and the strategic possibility that the North could conquer the South remained remote. In addition, political problems remained for the Lincoln government. The idea of placing weapons in the hands of Northern blacks would be experimental at best and infuriated many Democrats, including those that supported the war effort. And the Northern public would not tolerate and support a long war.

Union army operations against specific geographical points and the Southern armies that guarded them were the key elements in destroying the Confederacy. Controlling the Mississippi River and extending supply lines and operations into the Confederate heartland, along with destruction of the South's army defending Virginia and avoidance of Union military disasters, defined the Union's strategic goals.

Consequently, a study of campaigns must reconsider why the Northerners were allowed to further encroach upon Southern territory and to devastate

Rebel morale. The campaigns of momentous consequence—Fort Henry and Fort Donelson, Island No. 10, Vicksburg, Chattanooga, Atlanta, Petersburg, and the Carolina Campaign—maintained the Union's strategic initiative, improved Northern morale, and weakened the South's strategic defenses. All of these have one thing in common: complex engineering operations were essential to their success. Seldom highlighted in the Civil War literature, Northern engineers, over an enormous geographical space and with limited time, kept supply lines open, built roads to move tens of thousands of men through impassable terrain, and repaired roadways, bridges, tunnels, and railroad lines almost as quickly as enemy soldiers and partisan raiders destroyed them. Engineering emerged as a critical factor in the Union war effort.

Several studies both provide a useful backdrop for and suggest the timeliness of this book. Among those exploring the North's scientific and technological advantage over the South before the war, *Military Enterprise and Technological Change: Perspectives on the America Experience*, edited by Merritt Roe Smith, challenges the notion that the rise of nineteenth-century American "industrial capitalism" was solely linked to and influenced by private individuals and firms operating in the marketplace. Instead, the military enterprise played a major role in the United States' rise as an industrial power. Specifically, there were three areas of technological change in the nineteenth century that were tied to the military: interchangeable manufacturing, machine tools, and railroads.[26] Dirk J. Struik's *Yankee Science in the Making: Science and Engineering in New England from Colonial Times to the Civil War* explores how New England engineers began to transform the American landscape, how doctors contributed to advances in medicine, and how communities established educational institutions devoted to the study of science and engineering.[27]

Several Civil War studies describe the operations of Northern engineering regiments. These works offer insight into the formation and composition of such units and describe how bridges were built, canals dug, and road systems created and how engineers provided vital support for battlefield operations and campaigns. Historian Earl J. Hess has written a compelling series of books that explore the development of field fortifications on both sides during the war. Pointing out the increasing sophistication of the earthworks, Hess pays particular attention to the impact of these dirt fortresses and trenches on the outcome of the Petersburg Campaign.[28]

The other books on Northern engineering during the war are Phillip Thienel's *Mr. Lincoln's Bridge Builders: The Right Hand of American Genius* and Mark Hoffman's *"My Brave Mechanic": The First Michigan Engineers and Their*

Civil War. These two books make a case for the skills and courage of Union engineers. Thienel's focus is on bridge building, and Hoffman explores the versatility of engineers who not only repaired bridges, railroad tracks, and telegraph lines but also fought in a number of engagements.[29]

The few books devoted to Confederate engineering, such as James L. Nichols's *Confederate Engineers* and Harry L. Jackson's *First Regiment Engineer Troops P. A. C. S.: Robert E. Lee's Combat Engineers*, fail to answer the question of why the Confederacy had so little and the Union so much in the way of engineering support.[30] To answer this, I argue that there is a direct correlation between labor, manufacturing, agriculture, and educational systems in the North and South before the war and the effectiveness or ineffectiveness of engineering operations during the war. Looking critically at these relationships and their outcomes, I ask three primary questions: How did the economic, social, and educational patterns (as they related to engineering and technology) in the North differ from those in the South before 1860? How were Union and Confederate senior officers recruited and engineer regiments formed, and how did they differ, if at all, from each other? What role did engineering play in the critical campaigns of the war?

The outcome of the Civil War depended on the Union army's ability to improvise and take the war to the South. Northern armies operated on unfamiliar terrain, which included mountain ranges, swamps and wetlands, alluvial plains, forests, and rugged hills, all of which were difficult to cross because of dismal road systems and poorly mapped landscapes. Grant and other Union generals came to expect engineers and their men to subscribe to the theory that there are no fixed principles in war.[31]

So, as the calendar turned to May 1, 1863, Colonel James Keigwin of the 49th Indiana Infantry and approximately 22,000 soldiers crossed to the eastern bank of the Mississippi River. General Grant now began the task of chasing Confederate General John Pemberton's army across southwestern Mississippi, to move within striking distance of Vicksburg. It took Grant six months to get into this position. Now, to invest or besiege the city's earthworks, the army faced a final challenge. On the morning of May 17, the Union offensive against Southern forces stalled. The muddy, sluggish, deep, seventy-five-yard-wide Big Black River stood between the Confederate fortifications surrounding Vicksburg and Grant's army. Retreating Confederate forces had destroyed the existing bridges over the Big Black. Grant wanted to cross the river immediately to maintain the initiative and prevent giving the Southerners time to plan an escape across the Mississippi, but the equipment to build floating bridges, known as pontoons, sat fifteen miles from the river. Moving the essential material quickly forward proved herculean. Heavy traffic of men, horses, munitions, and quartermaster wagons clogged the roadways between the river and the pontoons. To bring the train forward would require time—at least eight to ten hours.

The pontoon train, with 34 pontoon wagons, 22 chess wagons, 4 tool wagons, 2 forge wagons, and approximately 300 horses and 100 to 500 men, lumbered to its destination. The pontoon wagons carried multiple small, flat-bottomed boats, very heavy and hard to transport. The chess wagons delivered the boards used to hold the pontoons together, and once soldiers built the roadway, the engineers used the remaining boards to build side rails. The tool wagon consisted of shovels and carpenter tools, and the forge wagons contained anchors and horseshoes.

Bridling under the pressure of time, Grant asked his engineers to find solutions to the bridging problem without waiting for the pontoon train. Under considerable pressure and lacking equipment, three different teams of engineers came up with three different solutions to the problem of crossing the Big Black River. Each built a bridge using scant resources and a large dose of Yankee ingenuity. One group, led by Captain Andrew Hickenlooper (a non–West Point civil engineer and an artillery officer when the campaign began), used cotton bales found in an old warehouse. Directed by the officers, men recruited from the infantry regiments closest to the river fastened these 500-pound bales of cotton, which would float, to frameworks that the men had built out of salvaged timber, creating a substitute for pontoons. The men built three sections of frames, fastened them together, and then tied them to guy lines. Planks from the warehouse nailed to the top of the cotton bales made a roadway that completed the bridge.

The bridge was finished by 3:00 a.m. on May 18. In the next nine hours, 15,000 men, horses, wagons, and artillery pieces crossed to the west side of the river. Two other bridges built downstream helped the rest of Grant's army to cross the river and besiege the city. With no obvious escape route for the 30,000 Confederate soldiers inside the Vicksburg fortifications, Confederate officials surrendered the city forty days later. The Mississippi River was, once again, open to Union river traffic, inspiring President Lincoln's ageless statement: "The Father of Waters again goes unvexed to the sea."

Why was it that Grant's engineers and his infantry, turned engineers, were able to tackle these unusual problems with such ingenious solutions? The answer rests in the textile mills, railroad yards, small farms, and mechanics' shops of antebellum America.

Part One

The Education and Management Gap

*Schooling, Business, and Culture in
Mid-Nineteenth-Century America*

CHAPTER ONE

Common School Reform and Science Education

A Yankee mixes a certain number of wooden nutmegs, which cost him
¼ cent apiece, with a quantity of real nutmegs, worth 4 cents apiece,
and sells the whole assortment for $44; and gains $3.75 by the fraud. How
many wooden nutmegs were there?

Southern algebra textbook by Daniel Harvey Hill, 1857

As Colonel Keigwin's men swatted mosquitoes and kept a keen eye out for alligators, perhaps some soldiers' thoughts turned to life before the war. Standing in knee-high water and feeling the bone-chilling dampness, these soldiers may have dreamed of the warmth of hearth and home and of daily work that did not require dodging Confederate bullets and artillery shells.

The division commander, Brigadier General Peter Osterhaus, had been chosen to lead the reconnaissance because both at Boonville, Missouri, in 1861 and again at Pea Ridge, Arkansas, in 1862, he had successfully served as the point man during the Union armies' operations. Osterhaus also had demonstrated an appreciation for the importance of logistics.[1] He ordered Keigwin's regiment and the 114th Ohio to conduct the advance scouting presumably because the men of these regiments had been recruited from areas of the country involved in the early construction of the railroads and had mechanical ability, in case it was needed. As it turned out, the soldiers' mechanical abilities were needed. The men under Keigwin's command not only discovered an acceptable route to extend Grant's supply line but also were able to build the necessary bridges in unconventional ways using the limited raw materials and tools available. The example of Keigwin's reconnaissance points to the significant advantage that Union generals held over their Confederate counterparts: they could send forward a regiment of soldiers with both the skills to build bridges and the combat experience to fight off raiding parties and guard the bridges once built.

Why was it that so many of these Union soldiers had some mechanical skill? A look back at the country's common schools and scientific advancements before the war provides some answers. Understanding the educational opportunities of soldiers of the North and South and their exposure to new inventions and mechanical ideas helps explain the different approaches and practices of the Union and Confederate armies. An investigation into conventional attitudes in the North, South, and West about education, economic development, societal structures, and cultural formations makes clear the reasons for the disparity in military engineering between the Union and the Confederacy and explains why, thanks to its engineering prowess, the Union prevailed at places like Vicksburg and Chattanooga and ultimately won the Civil War.

At the beginning of the nineteenth century, American schools reflected the national ethos. Wealthy merchants, plantation and business owners, attorneys, ministers, and scholars sent their sons to the small number of regional private schools and then on to the great universities, where these young men studied Latin, Greek, classical literature, and philosophy. Knowledge in these areas of study defined what it meant to be a gentleman in American society. Young men were trained to quote Shakespeare, command respect, and fight to defend family and honor. Upper-class women were trained to look for these traits in a man.

For the sons and daughters of artisans, craftsmen, yeoman farmers, and storekeepers, learning republican virtues, the Christian faith, basic literacy, and arithmetic required a rudimentary education. Some local communities operated a small school, yet many children learned their lessons at home because boys spent most of their time assisting on the farm and girls assisted with rearing the younger children and running the household.

By the 1830s, common or district schools began to appear throughout the North. Local communities controlled the funding, maintained the cold and dreary one-room schoolhouses, and provided the teachers—who were often incompetent. William Watts Folwell, born in Seneca County, New York, in 1833, and a future Union army engineer, started attending school at five years old. Folwell remembered the "green hardwood lumber [floor], which had shrunk so as to leave wide cracks between the boards." He recalled that most of each day was spent learning to spell, read, and write. Sometimes an ambitious teacher would furtively introduce geography, grammar, and algebra, although "any proposal to introduce advanced studies was steadfastly opposed."[2] District school had harsh discipline, sporadic attendance, and no grading system. The 1840 Census revealed that in Massachusetts and New York, of the total population of five- to sixteen-year-olds, only twenty-three percent attended some kind of common school.[3]

During the 1830s, while inconsistency marked efforts to better organize and run common schools, the Northeast began to show signs of structural change in its economic system that provided an irresistible impetus to educational reform and innovation. With the rapid growth of textile manufacturing and the parallel development of machine tool industries, a transformation occurred in the workforce. Commerce, transportation technology, finance, and innovation highlighted the importance of reading, writing, and arithmetic. New roads, canals, and railroads opened access to urban markets, and population growth forced young people to search for occupations other than farming.

Many New Englanders in the 1830s believed that home schooling and the common schools were adequate to meet the future needs of children. Yet, as early American industrialism developed, formal and better education outside the home became important, and reformers in Massachusetts such as Horace Mann called attention to the decrepit state of common schools. In 1838 Mann wrote: "It is commonly believed down to the present hour . . . that in all the world there is nothing to be compared to the common school system of New England . . . Till dreamings and gloryings as contemptible as these are banished from our firesides and legislative halls, nothing will be accomplished, and our common schools will continue in the process of deterioration."[4]

Mann's concerns alone, however, were not enough to launch a widespread educational reform movement throughout New England, Pennsylvania, New York, Ohio, and the Northwest. This meant that families wishing to see their children become productive workers in the emerging commercial/industrial economy and wanting their children to receive more than a "three R's" education had to seek alternatives beyond the common school. William Folwell's father chose to send his son approximately one hundred and fifty miles away to Livingston County, New York. There the boy stayed with his uncle and aunt and attended a private academy, the Nunda Literary Institute, where he was instructed in English literature, geology, chemistry, and Latin.[5]

As young Folwell continued to demonstrate a penchant for things academic, he went on to attend more advanced private schools, including the Geneva Union School and Ovid Academy. Serendipitously, when studying at Ovid, he stayed in the same boarding house as his science teacher, William H. Brewer. The future Dr. Brewer, who was to become a professor at Yale College, taught German to Folwell and encouraged him to study at the Yale Scientific School in New Haven, Connecticut.[6]

Will Folwell was fortunate. The circumstances of his birth, combined with hard work, a demonstrated aptitude, his father's support, and some good luck, opened the doors to a fine education both in classical studies and in mathematics and science. Yet, at the same time that Folwell was reciting Virgil and solving problems in plane trigonometry, Northern educational reformers were advocating the transformation of common schools into institutions that, as grassroots reformers such as the Philadelphia Workingmen's Association pointed out, "will be an instrument of unlimited good to the great mass of the people when they shall posses that degree of intelligence which will enable them to direct it *for their own benefit*; but at present this very blessing is suffered, through our want of information, to be directed against our prosperity and welfare by individuals whose interest is at variance with ours."[7] Education would develop talented citizens from families of the poor as well as the rich. It would produce informed citizens who, in turn, could influence party platforms and legislation within the framework of a broadening popular democracy, launched in the mid-1820s by the Democratic-Republicans. Education would also provide moral training and discipline—essential skills for Americans, unique among the people of the world.

One year later, a "Report of a Committee of Philadelphia Workingmen" advocated for a strengthening of common schools to strike at "an aristocracy of talent . . . in the hands of a privileged few" and for opportunities for an education, orchestrated by the state legislature, to ensure that liberty and "equal prosperity and happiness" were attainable for everyone, not just those in cities or towns committed to educational reform.[8] This request of the Philadelphia Workingmen that the state government should pass legislation to establish and regulate public education throughout the state was a radical departure from the standard practice of allowing local government to determine the extent of support for its own common school without interference from the state. In 1831, Pennsylvania passed a law making state funds available for schools. It was a first step.

And it was a step vulnerable to reversal. Four years later, legislators drafted a bill to repeal the Public School Act of 1834 because it "removed instruction from religious control, and it forced the whole group to pay for instruction to benefit only those who had children in school."[9] It took a resolute Thaddeus Stevens to convince his colleagues to pocket their proposed repeal. "In New England free schools plant the seeds and desire of knowledge in every mind, without regard to the wealth of the parent, or the texture of the pupil's garments," he insisted.[10]

In Connecticut, the impetus behind educational reform was the result of the urbanization that accompanied industrialization. For example, in mill towns and cities, unemployed adolescent boys, "transient mill workers and immigrants" who worked on jobs like the Farmington canal project and new railroads, were seen as threats to social stability and as contributors to prostitution, drunkenness, poverty, and crime.[11] Henry Barnard, the first secretary of the Board of Commissioners of Common Schools in Connecticut, reported his concerns to the legislature in 1840, arguing that the state could not combat the social problems of recalcitrant, uneducated youths running amuck if the government would not commit to increased funding and improved teacher training. "That our schools are not as good as they should be, or as they can be made with or without a school fund, that there are defects, and great defects . . . is painfully evident." Furthermore, "the source of much if not all of the inefficiency of common schools everywhere [is] the want of a suitable number of well qualified teachers."[12]

As a result of Barnard's passionate persuasion, the state passed legislation in 1841 to revamp the common school system by granting district school societies the power to levy school taxes and "prescribe rules and regulations for the management, studies, books, classification and discipline of the schools in the society."[13] Each district school organized its own curriculum, and all common schools taught reading, arithmetic, spelling, writing, and history. Some schools also began to teach rhetoric, geography, geometry, algebra, chemistry, surveying, and bookkeeping.[14] Yet, like other Northern states before the Civil War, the Connecticut system of public schools looked better on paper than in reality. Significant problems remained.

The Connecticut system was haphazard. In 1852, 217 school societies supervised 1,642 districts.[15] Fifty districts held school for two hundred days or more. The Fairfield North School District in 1857–58 kept school for two hundred days, and during the winter term of that year, twenty-two students were enrolled. The average daily attendance was eighteen.[16] For the decade of the 1850s, Connecticut common schools enrolled only fifty-seven percent of the school-age population. As the Fairfield District example illustrates, gender and age determined the actual number of days a student was in school. When they were in school, students learned to read and spell using the *McGuffey Reader*, younger students memorized multiplication tables and did "mental arithmetic," and older students solved arithmetic word problems.[17]

New England and New York advocates of common school reform occasionally moved to the Midwest and attempted to share their knowledge with members of developing school systems in, for example, Ohio, Michigan, and Wisconsin. All of Wisconsin's superintendents of public instruction between 1849 and 1868 were born in New York or New England, and one, Josiah Pickard, went on to become president of the University of Iowa.[18]

Midwestern educators learned valuable lessons about curricula, administration, state support, and teacher training from eastern reformers such as Horace Mann, Henry Barnard, and Catherine Beecher. The New York State school system, however, garnered the most attention and received the highest accolades from western reformers. The praise was well deserved.

After New York State developed a system of common schools in 1812, the responsibilities of the superintendents and regents overlapped until 1842 when the superintendent of common schools became a member of the Board of Regents.[19] The organizational structure of the state school system provided significant advantages over other school systems in the North. The state monitored teacher training, local library development, curriculum development, textbook use, and attendance, in addition to serving as a powerful voice within the halls of the state legislature and governor's office. On January 4, 1831, using data supplied by the regents, Governor Enos T. Throop, in an address to the legislature, stated clearly that "there is no one of our public institutions of more importance, or which has better fulfilled public expectations, than that providing for instruction in common schools."[20]

After reporting an increase in the number of children enrolled in school from the previous year and accounting for state money apportioned among school districts, the governor challenged lawmakers and educators to consider more change: "For mere purpose of reading and arithmetic, selections may be made among various books extant, of such as are perfectly adapted to the purpose. But I feel confident that, under proper regulation, a vast amount of knowledge in the arts and sciences, connected with agriculture and handicraft, which are simple in their principles . . . might be taught to children during those years which are usually spent at common schools."[21] Thirteen years later, the state encouraged school districts to teach rudimentary farming principles, which included a basic understanding of chemistry. In 1856, New York attempted to establish, with limited success, two-acre farms near common schools to be cultivated by the male pupils, under the supervision of a teacher. "Young ladies" were encouraged to tend an ornamental garden.[22]

Abstracts of reports submitted to the superintendent of public instruction (the title of this position was changed in 1854) paint an extraordinary canvas. The ratios of number of children taught to total number of children in each county were impressive. Furthermore, some, such as Oneida County, maintained libraries that boasted 54,000 volumes. Of course, these numbers varied, and the small Maine County had only 1,613 volumes in its library. The state apportioned money for the purchase of new books.[23]

Problems remained, however. In a letter to Superintendent V. M. Rice in 1856, School Commissioner for Oswego Tioga County E. Powell reported that "the obstacles that lie in the way of the progress of our schools ... are, the change of teachers twice a year, a want of uniformity as regards a system of instruction, and of text books, inexperience of teachers and a dire want of interest among the inhabitants."[24]

For those in the North, regardless of their common school experiences and their states' problems in fixing them, significant educational opportunities were available once a student started working full time. Foremost among these was the lyceum movement, a fixed component of adult learning as it disseminated knowledge and useful information to a broad audience, both rural and urban, and contributed to greater farm and manufacturing production, opening the way to potential upward mobility for men interested in inventing new and more efficient tools and machines that contributed to the overall wealth of a region.

The idea for the lyceum, as developed first in the North, probably began in the United Kingdom at the turn of the century. After George Birkbeck graduated from the University of Edinburgh Medical School in 1799, he began giving popular lectures on scientific topics. Birkbeck's lectures gained such a following among local artisans, smiths, and carpenters that he decided to establish the Glasgow Mechanics' Institution in 1821. Birkbeck moved his medical practice to London in 1823 and in an effort to continue to provide technical education to the working classes, he founded, with several colleagues, the London Mechanics' Institution.[25]

The concept of this mechanics' organization crossed the Atlantic, and in 1826, a former laboratory assistant to Benjamin Silliman at Yale, Josiah Holbrook, started the Millbury (Massachusetts) Branch Number One of the American Lyceum.[26] For Holbrook, the lyceum was an educational and social institution and a crusade to diffuse knowledge to thousands who might otherwise develop habits "which will lead to their ruin."[27] Holbrook believed that the lyceum would extend to include everyone in the community, "old and young, the male and female, the learned and illiterate," that it would bring together ideas for improving common schools, and that for a young person growing up "under the advantages and influence of an Association well conducted," the lyceum would provide "more useful, practical information than he would be likely to obtain in a College course."[28]

The Millbury Lyceum was the first of many, and as the idea spread, so did the breadth and depth of the presentations and lectures. In 1836–37, the Salem Lyceum offered lectures entitled "Popular Knowledge as Applied to Scientific Improvements," by Daniel Webster, "Application of Science to Common Life," by Elisha Bartlett, and "Electro Magnetism," by Charles G. Page.[29] Other speakers included Charles Francis Adams, George Bancroft, Ralph Waldo Emerson, Horace Mann, George Catlin, and John Quincy Adams, and topics ranged from "Common School Education" to "The Legal Rights of Women" and "Life and Times of Oliver Cromwell."[30] The Lincoln, Massachusetts, Lyceum recorded that on January 19, 1847, the speaker was Henry David Thoreau, and after his lecture a discussion question was posited: "Are the present customs of society in this country calculated to develop the mental and physical powers of its young men?" The lyceum selected four men: two argued in the affirmative and two argued in the negative. After hearing the debate, the "house voted in the affirmative."[31]

In the North, the common school and lyceum movements provided citizens with an opportunity to expand their knowledge, collect new ideas, and compete in a growing industrial and mechanical environment. To the contrary, in the South, the opportunity for commoners to receive a basic education was so inadequate that by 1860, only eighteen percent of children between five and nineteen years old attended a common school.[32] This inadequacy was not just the result of a Southern economic system centered around agriculture and slave labor; men in power, with hierarchical notions of how society was controlled, chose to reject educational reform. Furthermore, when Southerners encouraged by Northern industrial success sought to open new manufacturing operations, the dominant planter class that controlled state legislatures often blocked these efforts.

Wealthy planters could afford to hire private tutors to instruct their children and, along with artisans, military officers, doctors, lawyers, and clergy, had the additional option of sending their children to one of the large number of academies and church schools. By 1860, the number of academies in the South as a whole (2,435) was almost double the number in the Middle Atlantic region (643) or New England (988).[33] Episcopalians, Methodists, Baptists, and Presbyterians all established academies. Finally, some communities built log cabin schoolhouses in old fallow fields, hired teachers, charged tuition, and provided a basic elementary education for anyone who could pay. These "old-field schools" operated at the whim of the community and often recruited men who were looking to make money rather than to teach. Governor George Gilmer of Georgia recalled his old-field teachers as a "loud, violent Irishman, an impoverished Virginia gentleman who drank too much, and a well-qualified, sober Georgian." Benevolent societies, as well as states such as Virginia that had a small fund for teaching poor children, would reimburse old-field schools for pupils whose families could not afford the tuition.[34]

An example of someone of middle-class lineage who was educated at a private academy was future Confederate engineer John Morris Wampler, born in Baltimore in 1830. John's father, Thomas Jefferson Wampler, a tin plate maker, died fourteen months after the boy's birth. The child's mother, Ann B. Johannes Wampler, was fortunate enough to pay off her husband's debts and, through her connection with the Episcopal Church, befriended Margaret Mercer, a Baltimore socialite.[35]

Miss Mercer's father, John Francis Mercer, served as a Maryland delegate to the Constitutional Convention, was a member of the House of Representatives and a two-term governor, and saw fit to provide his daughter with a superior education that included Hebrew, Latin, and French literature. Margaret put this education to good use. She ran several schools for girls and was actively involved in freeing slaves and colonizing freed slaves in Africa. When she met Ann Wampler and her son John, Mercer was about to open a new school "on one of the largest tracts in eastern Loudoun County [Virginia]."[36] The Belmont estate, albeit in disrepair, was the perfect site to start her new school, and she brought along Ann Wampler as her assistant. As a result, young John became the only boy to attend a girls' boarding school, Belmont Academy.

For ten years John Wampler learned chemistry, mathematics, Latin, history, English literature, and music. He read Charles Lyell's *Principles of Geology* and Sir John Frederick William Herschel's *Treatise on Astronomy*, although the school did not have a telescope or operate a laboratory. Consequently, the next phase of his education needed to include practical experience, and at seventeen years old Wampler left the comfortable surroundings of Belmont and landed a position with the US Coast Survey.

Under the tutelage of Superintendent Alexander Dallas Bache, Wampler learned the science behind topographical engineering. Working under the direction of senior members of the Coast Survey, including Montgomery Meigs, future quartermaster general of the Union army, Wampler found that mapping the coastline of the United States "involved the observation and calculation of an interlocking series of triangles." Wampler resigned his Coast Survey position in April 1853, tried working for various railroad companies, and eventually became chief engineer for the Baltimore City Water Works.[37]

Wampler's childhood circumstances presented him with an educational opportunity that would not have been available had he grown up on a farm or been part of the laboring class in the South. Social hierarchy and the institution of slavery represented the sine qua non of conservative social thought and were the most universal obstacle to the formation of common schools below the Mason-Dixon Line. Southerners argued against the North's competitive, free-market, wage-labor system. According to Southern nationalist Edmund Ruffin, the struggle between capitalists and laborers in the North did not exist in the South. Under slavery, Ruffin wrote, "there is no possibility of the occurrence of the sufferings of the laboring classes" that characterize the "class-slavery of labor to capital."[38]

For conservative thinkers, hierarchy, aristocracy, and social control were the fundamental principles of God's divine plan in which rich and poor, intelligent and illiterate, leaders and followers could coexist in harmony and happiness. Education could upset the social balance. Public education could be dangerous. South Carolina attorney William Harper, an extreme proslavery ideologue, argued that "men of no great power of intellect, and of imperfect and superficial knowledge, are the most mischievous of all . . . Of all communities, one of the least desirable, would be that in which imperfect, superficial half-education should be universal."[39] The free exchange of ideas and the ability of the entire population to read could produce revolutionary ideas among slaves, as it did with Frederick Douglas, and radical thinking among Southerners who might sympathize with Northern abolitionists. An 1852 article in the *Southern Quarterly Review* succinctly stated the case: "Throughout the whole country, from the Hudson to the Bay of Fundy, a settled determination exists to abolish slavery at the South . . . The diffusion of education in New England is likely to effect a dissolution of the Union . . . men whose lives are spent in humble toil, have little time for reflection."[40]

Western Virginia, in the three decades before the Civil War, did clamor for better public education, based on a belief that its section of the South had the natural resources in coal and other minerals to forge an industrial center similar to those in the North. Business interests in western Virginia were not afraid of Northerners' ideas about slavery and, instead, sought Northern advice and guidance on manufacturing and commerce. Henry A. Wise, a Whig congressman from Virginia, challenged his constituency to "*Educate your children, all your children—every one of them!*" After attacking the Southern educational system as one based on charity instead of opportunity and the privilege of citizenship, Wise continued: "Does anyone suppose that if education had been diffused universally among our people . . . that her [Virginia's] agriculture and mechanic arts would be in the low state they are now in? That the rich bowels of her inexhaustible mountain mines of iron and coal would be undug and almost unexplored? That her manufactures would have languished as they do?"[41] Wise reminded his audience that the axe helve, plow handle, handspike, and ox chain were all "levers of knowledge."[42]

Yet, eastern plantation Democrats made it exceedingly difficult to create common schools, and in 1860, when western Virginia voted against secession, the *Wheeling Intelligencer* reported that one grievance against eastern Virginia was that tidewater plantation owners denied western Virginians common schools.[43] In May 1861, the First Wheeling Convention, organized to reject secession, resolved that "in view of . . . social, commercial and industrial interests in Northwestern Virginia . . . the State of Virginia have not only acted unwisely . . . but have adopted a policy utterly ruinous to all the material interests of our section . . . drying up all the channels of our . . . prosperity."[44]

In western North Carolina, where Quakers and churchmen talked of anti-slavery feelings and where the Whig Party was strong, the common school movement enjoyed more success than in any other place in the South, although opponents of public education were vocal. In 1839, the North Carolina legislature passed a law establishing public education, but in theory only. Many towns refused to pay for local schools. Eventually the state recognized the need for an educational leader with the passion and administrative skills to turn the concept of the common school into the reality of the common school. In 1852, the government established the office of superintendent and selected Calvin H. Wiley for the job.

Superintendent Wiley's first annual report to the North Carolina legislature in 1854 demonstrated the progress the state was making in public education reform. He documented the growth in the number of common schools and of children who attended them. The 1840 census, he reported, showed the state with 632 common or subscription schools and 14,937 pupils enrolled. By 1850 that number had jumped to 2,131 common schools and 83,873 pupils in attendance.[45] Whereas Virginia distributed about eight cents per capita for white education, North Carolina spent fifty cents. This compared favorably with Connecticut and Pennsylvania at ninety-five cents per capita and Ohio and New York at one dollar.[46] North Carolina had three times as many children in school as South Carolina and six hundred more common schools than Virginia, and Virginia's population was 340,000 more than that of the Tar Heel State. "Upon a calm review of the entire facts," Wiley wrote, "it is neither immodest nor unjust to assert that North Carolina is clearly ahead of all slave-holding states with their system of public instruction, while she compares favorably in several respects with some of the New England and Northwestern states."[47] The superintendent predicted that in ten to twenty years the state would produce as well-educated a citizenry as that of Massachusetts. The 1860 Census confirmed that North Carolina was on the right educational path. In Massachusetts, fifty-seven percent of children between five and nineteen years old were in common schools, and in North Carolina, forty-two percent. These numbers were in stark contrast to Virginia's twenty-one percent and South Carolina's dismal eighteen percent.

Yet, North Carolina, like the entire South, had significant cultural obstacles to overcome before it educated all its children. Many men believed "book learn'n" was useless, and as a result, many male teachers had to travel North to receive proper training and common schools often had to rely on Northern instructors. Again, the institution of slavery was written into the equation. At the Louisiana Constitutional Convention of 1845, the "Report of the Committee on Education" stated that "Southern men should have southern heads and hearts, with sentiments untarnished by doctrines at war with our rights and liberties. It is of the first importance that correct impressions be made upon the minds of children; for it is difficult to unlearn what has been learned amiss."[48] The report noted that when young men returned home from their Northern school experience, they often had to be "re-acclimated" to the Southern way of life.

At the outbreak of Civil War, the new Southern Confederacy took stock of its institutions to determine which ones had been tainted by excessive Northern influence and then offered the proper prescription to correct the illness. According to Edward Pollard, associate editor of the *Daily Richmond Examiner*, education was wounded because of the "curse of New England Society, and the great revolutionary element of the North . . . We believe that the education of the New England common school is carried out to that point where learning is dangerous."[49] Other writers also believed that common school education in the North had produced an indoctrination and radicalism among the masses. "Nothing is more to be dreaded than a community half-educated, and who consider themselves learned."[50]

Poor public school systems, an inadequate number of textbooks, a dearth of teachers, and an ideological belief that education for the masses was a repulsive Northern concept combined to limit basic literacy, mechanical aptitude, and inventiveness in the antebellum South. The region, nonetheless, did witness the rise of an emerging middle class, with nonagricultural professionals making up about ten percent of urban populations.[51] Before the war, the sons of these men received a solid education at an institution that would make its mark on Southern society in the antebellum period and in the years that followed—the Southern military academy.

In her careful analysis of schools such as the Virginia Military Institute, historian Jennifer R. Green saw a link between middle-class alumni of military schools and Southern elite, although the former possessed "a separate status." According to Green, by 1850 Virginia Military Institute alumni made up twelve percent of all students who attended colleges in the state that year. By 1860, these young men served as instructors or superintendents in the ninety-six military secondary schools or military colleges in the South.[52]

The superintendent of the South Carolina Military Academy (now The Citadel) reported that of eighty-seven men who graduated in 1854, ten percent became agriculturally employed and the rest served as teachers, businessmen, attorneys, or doctors. No one became a civil engineer.[53] Nonetheless, unlike nonmilitary private academies and colleges, some private and all state military schools received public funding, and unlike civilian schools, military academies taught advanced mathematics, English, chemistry, physics, and French.

Just as the Southern upper class felt no compunction to lead in the common school movement, it felt no responsibility to lead the lyceum movement.[54] This meant that lyceums could form only in more densely populated areas where middle-class artisans and mechanics had a vested interest in joining a society that supported continuing education. It is not surprising, then, that the few cities in the South that did have lyceums included large urban areas with a middle class, such as Richmond, Savannah, Macon, and Nashville. Thomas Grimké attempted to start a lyceum in Charleston in 1834, but nothing came of it. A "Yankee enclave" in New Orleans established the Library and Lyceum Society in 1844.[55]

In addition to population density, there were other factors that worked against education in the South. First, with easy transportation available only to cities, it was difficult to bring in outsiders to lecture, especially from the North. Second, after Nat Turner's Rebellion, a culture of caution informed decisions about inviting Yankees to speak. The fear of abolitionist viewpoints led to a pervasive anti-intellectualism. Third, as historian John Majewski observed in *Modernizing a Slave Economy*, "A region composed of isolated farms and plantations generated fewer subscribers for periodicals and newspapers, had fewer potential members for mechanics' institutes or literary associations, and provided fewer students for schools and colleges."[56]

The foundation of Southern life was built on an intellectual defense of slav-

ery and on maintaining a hierarchical social structure. As Northern abolitionists attacked the institution of slavery and called the men who practiced it indo-
lent and immoral, Southern politicians and intellectuals grew more determined
to defend and justify their institution, their society, and themselves. Thomas R.
Dew, an intellectual disciple of Thomas Jefferson, analyzed the proceedings
of the Virginia State Constitutional Convention in 1829–30 and, as a result,
published the first systematic defense of slavery in *Review of the Debates of the
Virginia Legislature.* At the center of the debate was the relationship between
slavery and the apportionment of political power in the state. The slaveholder
faction won and, consequently, controlled the legislature for the next thirty
years. Dew argued that the South had inherited slavery from the colonial fathers
and, once introduced, it could not be abandoned because blacks would face un-
bearable misery; to abandon them would be immoral.[57]

A colleague of Dew's at William & Mary College was William Barton Rogers.
A Philadelphian by birth, Rogers went to school at William & Mary and then
taught chemistry and natural philosophy at the college beginning in 1828. His
papers on the mineral deposits of eastern Virginia grabbed the attention of the
Virginia legislature, and in 1835 the lawmakers placed Rogers in charge of a geological survey of Virginia. Concurrently, he took a position on the University of
Virginia faculty as professor of natural philosophy. The scholarly Rogers found
the results of the geological survey to be a tale of two regions. Whereas the
eastern part of the state provided the natural resources essential for the state's
plantation owners and small farmers, the western half provided a plethora of
raw materials to meet westerners' interests in developing a center for Southern
industry. Eastern planters, who made up the majority interest in the legislature,
"associated coal and other emerging industries with Yankee interests." Their
skepticism and fear and their commitment to slavery led them to cut off funding for the survey.[58]

Rogers discovered from newspaper reports that his funding had been cut. In
correspondence with Judge J. F. May of the Virginia House of Delegates, Rogers
wrote: "Indeed, without the assistance of the corps in the laboratory and at the
drawing-table, and in arranging the cabinet . . . I should feel incompetent to perform the task of drawing up my Report without great additional delay."[59] May
dismissed his concern, and the report was not finished.

In the same year, the university hired mathematician James Joseph Sylvester, who upon his arrival was greeted by students with verbal insults and physical assault because he was an English Jew. The *Watchman of the South,* a period-
ical of the Presbyterian Church, wrote of Sylvester's hire: "This is the heaviest

blow the University has ever received. The great body of the people of this Commonwealth are by profession Christians and not heathen, nor musselmen [Muslims], nor Jews, nor Atheists, nor Infidels."[60] Sylvester was gone from the university in six months, prompting Rogers to write to his brother Robert: "I have been unable to shut out the contrast between the region in which I live and the highly cultivated nature and society of glorious New England . . . Would you believe it, that a series of essays has been published condemning the Visitors for the appointments of a Jew and a Catholic [Kraitzer], and sweeping charges at the same time made against the character . . . of the University!"[61]

Whether the events of 1841 had anything to do with the next crisis for Rogers at the university is speculative, but three years later the Committee on Schools and Colleges of the House of Delegates was instructed to investigate "the past history and present condition and influences of the University of Virginia, with a view of forming their opinion upon the question of repealing the Act of Assembly granting an annuity of $15,000 to that Institution."[62] Rogers responded with a lengthy report answering the charges of Jewish influence on the university and the appropriations remained, but not without scaring the embattled professor.

He had begun to imagine the model for a new type of scientific school unlike anything that had been attempted at universities across the country. Rogers saw relationships between developments in technology and developments in science and envisioned a polytechnic institution that connected science to its practical applications. He tendered his resignation at the university and shortly after wrote to his brother Henry about a place to establish the school he imagined. "The occupations and interests of the great mass of the people [in Boston] are immediately connected with the applications of physical science, and their quick intelligence has already impressed them with just ideas of the value of scientific teaching in their daily pursuits."[63] Rogers challenged conventional wisdom by insisting that the tools of science be placed in students' hands.

In the fifteen years before the Civil War, colleges and universities in the North, following Harvard's and Yale's lead, expanded offerings in the applied sciences, yet programs received no financial support from their institution and lacked laboratory facilities for students. Professors ran the experiments and students observed.

Other schools such as West Point "came close to Roger's ideal," but by 1855 the course of study turned away from the theoretical to just the practical. The Rensselaer School used laboratories for instruction, yet by 1848 students could complete their course of study in one year. Other schools demonstrated interest in educational reform, but none entirely resembled Rogers's goals. These schools included The Citadel (1843), the US Naval Academy (1845), Polytechnic College of Pennsylvania (1853), Brooklyn Polytechnic Institute (1855), Glenmore School (1859), and Cooper Union (1859).[64]

By 1860, Rogers's opportunity came knocking. The city of Boston was looking to place an educational institution in its Back Bay development. A conservatory was formed, and at a meeting on October 5, 1860, Rogers presented his plan to establish the Massachusetts Institute of Technology. The following year, an act of the Massachusetts State Legislature incorporated MIT, and Rogers was named the institute's first president.

For most college students in the mid-nineteenth century, developing a knowledge and passion for science was a hit-or-miss proposition. This was William Watts Folwell's experience at Hobart Free College in Geneva, New York. After attending the Geneva Union School, Folwell enrolled at Hobart in 1854 as a sophomore.[65] Over the next three years, he studied Greek, Latin, English literature, calculus, and chemistry. The latter subject, as was typical of all colleges in the 1850s, was taught by lecturing, but Folwell was fortunate that his teacher, John Towler, was also a practicing chemist. Folwell recalled that "to a few of us he gave a short course in laboratory work, which was worth while because it showed us what real chemistry might be."[66]

After graduating in the spring of 1857, Folwell taught at Ovid Academy and then accepted a position at Hobart as a mathematics instructor. During his two years at the college as a teacher, he attended a literary and scientific club established in Geneva by the Reverend Doctor William Dexter Wilson, a philosophy professor at Hobart and, according to Folwell, the leading intellect on the faculty. It was during this time that Folwell had an epiphany: he would abandon mathematics and pursue linguistics and language.[67]

The examples of William Watts Folwell and John Morris Wampler demonstrate, in general terms, how science and engineering came to be applied to solving the practical mechanical problems of the mid-nineteenth century. Both men developed a passion for and greater understanding of applied science after they left school. There was no doubt that Folwell received advanced studies in mathematics, but it was not until—purely by good fortune—he took a brief laboratory class in chemistry at Hobart and, after graduating, started attending lectures at the Geneva Literary and Science Club that he became interested in applied science. For Wampler's part, it was only after working for the US Coast Survey that he gained enough experience to develop and patent an automatic mechanism for operating the surveyor's graphodometer.[68]

Mid-nineteenth-century colleges, universities, and private academies did not teach the principles and practical application of science in the real world. This is why William Rogers's concept of a new scientific school was unique and met with such opposition among the more traditional thinkers within the academic community. Men who worked with their hands and were constantly seeking better and more efficient ways of doing business were doing the work of innovation and invention. For example, Charles Goodyear, inventor of vulcanized rubber, Joseph Henry, who helped develop practical devices using the electromagnet (including Samuel Morse's telegraph), George Henry Corliss, inventor of the Corliss steam engine, and Elias Howe, developer of America's first patented sewing machine, all were self-taught scientists and engineers.[69] Besides being self-taught, they were all educated at common schools.

Common schools, administered by state agencies, started to emerge in the 1840s and 1850s, especially in the North. Although they provided a rudimentary education, their real effect was to establish a cultural ethos that made it clear to families who enrolled their children in the one-room schoolhouses that knowledge was important. Learning was critical to improving standards of living for everyone. New inventions made work more productive and made inventors, possibly, wealthy.

Knowledge and information were exchanged at colleges and universities in mid-nineteenth-century America, but for most people, practical ideas, better practices, and innovations were discovered elsewhere. As we shall see, it was at mechanics' institutes and agricultural fairs that people were exposed to useful solutions to everyday problems. In 1836, thousands of people attended the fair of the American Institute in New York City. Inventions from all over the Northeast were on display, and the editors of *Mechanics' Magazine* noted that "we found [there] some of the most ingenious and celebrated mechanics of this country or of any country."[70] Fairs were places where ideas intersected with practicality, where knowledge and imagination were celebrated. Ingenuity was cultivated, and it would become an important and useful weapon in the war that was coming.

CHAPTER TWO

Mechanics' Institutes and Agricultural Fairs

*Transmitting Knowledge and Information
in Antebellum America*

It's well known among engineers that the most important inventions
in a particular field are often made by people who are new to that
field—people who are too naïve and ignorant to know all the reasons
why something can't be done, and who are therefore able to think more
freely about seemingly intractable problems.

The Contrarian's Guide to Leadership *by Steven B. Sample*

Throughout the antebellum period, patterns developed in the economic fabric of the three regions of the country. Geography played a role in establishing favorable conditions for particular economic development: fertile soil and favorable growing conditions in the South, important river systems in the West, and running streams and woodlands in the North. Geography alone, however, was not the only reason for differences in regional economies. How each section of the country defined democracy, community, labor, education, and upward mobility determined how resources were spent, laws were made, schools were formed, and labor was regarded. These were choices: to encourage common school education or to warn of the dangers of learning; to establish institutions to improve workers' skills or to regard labor as nothing more than a necessary cog in the social hierarchy. They were the conscious decisions that political leadership wove into each region's fabric.

The development of manufacturing, agriculture, science, technology, and higher education was also a choice for each region of the United States. How these endeavors were adopted and utilized was important and, consequently, became important to the individuals living in those regions. On the eve of the Civil War, more than one million Northerners worked in large and small factories scattered throughout the states. Technological innovations increased industrial production. Mechanics such as Christopher Spencer, Francis Pratt, and Amos Whitney worked out ideas in their years at the Colt Armory in Hartford, Connecticut, and then built their own businesses. Henry and Clement Studebaker learned the trade of blacksmithing from their father in South Bend, Indiana, and in 1860 turned a small carriage and wagon shop into a major manufacturing operation.

Accompanying the North's industrial growth was the West's economic development. People's willingness to move into the new states of Ohio, Indiana, Illinois, Michigan, Wisconsin, Minnesota, and Iowa demonstrated their natural entrepreneurial spirit and replicated and advanced what was happening in the Northeast with skill and tenacity. Capital, land, and education were available. And the obstacle of slavery was absent.

In the South, before the war, farming generated great wealth, and slavery served as the foundation of that wealth. Any attempt to question the value or integrity of the "Peculiar Institution" met with harsh rebuttals on the grounds that this labor system provided the essential component of the South's continued social stability and wealth. The South's hierarchical structure and aristocratic character created an ethos unique to the region. As the country moved toward civil war, little did it appreciate or understand how these regional differences would influence the fighting and affect the outcome.

The stories of David Ross of Campbell County, Virginia, and Chauncey Jerome of Canaan, Connecticut, illustrate how regional differences in the South and North affected the conduct of business, technological expansion, and information exchange during the antebellum period. In 1811, Ross was approximately seventy-two years old and Jerome was eighteen. Ross had emigrated from Scotland to the American colonies sometime in the mid-1750s, established himself as a tobacco merchant and ship owner in Richmond and Petersburg, and by 1788, according to the tax lists, owned four hundred slaves and more than one hundred thousand acres of property scattered across twelve Virginia counties.[1] Sometime around 1777, Ross opened the Oxford Iron Works, south of the James River and eight miles from Lynchburg. By 1811, he employed 220 black bondsmen, ninety percent of his entire labor force. The slave labor force in other industrial operations in Virginia told a similar story. By 1850, the Buffalo Forge and the Etna Furnace (later combined to form the Bath Iron Works), both in Rockbridge County, together had 100 to 120 bondsmen. The Tredegar Company in Richmond, by 1860, had approximately 800 slave laborers.[2]

At Oxford Iron Works, slaves were taught every job and so worked as carpenters, blacksmiths, forgemen, miners, and laborers. The small percentage of whites employed served as supervisors and bosses. The red hot glow of the furnaces did nothing to ignite Ross's profit margins. By 1812, profits were significantly dampened and his iron business faced insolvency. Increased competition, mismanagement, and poor workmanship led Ross to chide that "the ruin of the estate is founded in the management of it."[3] He believed that his manager and clerks were "a very trifling people and incapable of making use of the means in our power." Oxford had "Mechanicks & labourers adequate to every purpose," yet his nephew, Robert Richardson, "has not improved in the Smallest degree." With venom directed at his nephew, Ross continued: "those that ought to represent the master are not infrequently inferior to the Servant."[4]

Ross recognized that his slaves had little incentive to increase productivity and improve the quality of their work, yet good supervision, he believed, would have corrected the bondsmen's poor and lazy performances. Forever hopeful, Ross developed plans to move his operation nearer the James River, but severe credit problems prevented him from acquiring the necessary loans and his business continued to spiral downward. Plagued by continuous health problems, he died on May 4, 1817. His accrued debts were finally paid off by 1819 with liquidation of the Oxford slave force.[5]

The paradox of this story is that Ross understood how the system of slavery limited his ability to improve his workers' performance, not only because bondsmen lacked incentive, but also because he would not allow free ironworkers traveling South to have access his slaves, for fear that free workers would spread incendiary ideas. He recognized that his slaves lacked fresh technological knowledge, but he was willing to deprive them of this knowledge because he believed it was more important to protect the institution of slavery than to educate his workers. He admired and wanted to emulate the technological growth taking shape in the North, but he did not want his labor force to have contact with those who might share the information necessary to spawn that growth.

Historian Charles B. Dew, in *Bond of Iron: Master and Slave at Buffalo Forge*, pointed out that in contrast to Ross, William Weaver, owner of Buffalo Forge, paid slaves for "overwork." Slaves such as Phill Easton built up considerable earnings on the overwork ledgers or "Negro Books," as they were called; Easton was thus able to improve his family's quality of life and, as Dew suggested, "to stake out some precious independence . . . in the midst of a system that theoretically held him totally bound to the will of his owner."[6]

Weaver, like Ross, recognized the cost-effectiveness of using slaves instead of free labor, but unlike Ross, by adopting an overwork system, Weaver provided his slaves with some incentives and curtailed incidents of manufacturing sabo-

tage. To discipline slaves by beating or by threatening to sell them only destabilized the business and slowed or significantly damaged production. Weaver
believed in a system that trained a group of skilled slave artisans and provided
them with some monetary incentive, and it worked. The result was that for four
decades, Weaver emphasized stability in the workforce. Yet stability did not
imply innovation.[7] Weaver ignored the technological innovations that had rev-
olutionized Northern industry, and during the Civil War, when the Confederacy
need wrought iron, Weaver's forge and others like it could produce only a limited amount. Furthermore, by virtue of their social position, slaves skilled in industrial arts could not or, given the opportunity, would not contribute to Confederate engineering operations during the war.[8] Therefore, the slave economy
was itself a central reason for the shortage of skilled engineer soldiers in the
Confederate army. During the antebellum period, most artisans and mechanics
who worked for Southern manufacturing businesses were slaves.[9]

Chauncey Jerome was born in Canaan, Connecticut, in 1793. At the time of
Ross's death, Jerome's life had been as painful as the crude wooden shoes he
walked in. When he was eleven years old his father died of the "black colic," and
young Jerome became the man of the family. His mother sent him off to be a
farm hand until, at fifteen, he became a carpenter's apprentice. Neither enam-
ored with his work nor pleased with his lack of boots and warm clothing, Jerome
arranged to take four months during the winter to find other work by which to
earn the means to purchase a coat. He found work in Waterbury, Connecticut,
making dials for old-fashioned long clocks. Jerome continued as a carpenter
and part-time clock-dial maker until after the War of 1812, when by chance he
landed a job in Plymouth, Connecticut, helping clockmaker Eli Terry set up his
new business in manufacturing his patented shelf clock.[10]

Jerome studied Terry's operation and the machinery used to produce his
clocks, including "arbors" and "mandrels," prototypes of lathes and jigs, as well
as Terry's secret weapon, a circular saw, a novelty only Terry possessed. After
learning from Terry, the master of his craft, Jerome went into business for himself as a "jobber" assembling clocks, casing them, or manufacturing cases. Four
years later things started to get interesting for Jerome. Relying on his training as
a craftsman and his desire to improve his station in life, Jerome sold his house
to Terry for six hundred dollars, one hundred of it in wooden-clock movements,
"with dials, tablets, glass, and weights."[11]

In Bristol, Connecticut, in 1821, Jerome bought a two-story house with seventeen acres of property. The price asked for by the original owner was two
hundred and fourteen Terry Patent Clocks. "I told him I would give it," Jerome

wrote, "and closed the bargain at once. I finished up the one hundred parts which I had got from Mr. Terry, exchanged cases with him for more, obtained some credit, and in this way made out the quantity for Mitchell [the seller]."[12]

Working with his brother, Jerome began to make the Bronze Looking-Glass Clock to compete with Terry's Patent Shelf Clocks. To manufacture his clocks, Jerome installed the first circular saw ever seen in Bristol—although quite familiar to Plymouth, Connecticut, only ten miles away, and to Mr. Terry. Fifteen years later Jerome introduced a one-day brass clock, which he mass-produced cheaply and sold for between $1.50 and $2.00.[13] By comparison, wooden one-day clocks sold for $14 and brass eight-day clocks for $20.

Business boomed as Jerome sold his inexpensive timepieces domestically and overseas. As a result, he moved his entire operation to New Haven, Connecticut, partnered with Benedict & Burnham, brass manufacturers of Waterbury, Connecticut, and became the New Haven Clock Company. Business boomed. In 1853 he sold 444,000 clocks.[14]

Yet, like David Ross in Virginia, Jerome made poor business decisions that, by 1855, led to bankruptcy. In a buyout of a clock company in Bridgeport, Connecticut, owned by the great bamboozler of the nineteenth century, P. T. Barnum, Jerome found Barnum had sold off important assets and left Jerome with considerable debt. He never recovered from the burden, and he died penniless in 1868.[15]

David Ross and Chauncey Jerome shared parallel experiences in the course of operating their nineteenth-century American businesses. Both enjoyed success, and both suffered from the vicissitudes of doing business. Both made good and bad investment decisions, and both took risks. Each man died broke. Yet, they differed in ways that reveal the influence of their surroundings on each business and of each business on its surroundings.

When Ross arrived in Virginia, he immediately understood the value of shipping and commerce with Great Britain. Tobacco was in demand, and he shipped enough of it to purchase an enormous plantation and more than two hundred slaves. Having imagination enough to see the potential for mining operations and iron production, he started a business venture that generated a huge profit. Using some of his own slaves along with bondsmen rented from other slave owners, Ross put his black labor force to work in his companies' mines and forges. He recognized the mechanical abilities and skills of white miners and ironmen, many from the North, and the need for skilled whites to train his workers in areas of new technical developments. He also recognized, as his fellow slaveholders did, the dangers of allowing whites, who might have radical ideas, to come in contact with his bondsmen. Also, white workers' attempt to gain economic security was viewed as a threat to the status quo. In 1851, South Carolinian James Henry Hammond wrote that "in all other countries, and particularly manufacturing states, labor and capital are assuming an antagonistical position. Here it cannot be the case; capital will be able to control labor, even in manufactures with whites, for blacks can always be resorted to in case of need."[16]

Ross understood, albeit superficially, that his black workers had little incentive to work hard, minimize the waste of resources, and carefully follow directions. He expected his management team to compensate for all these problems by remaining current with recent technological changes, better supervising his workers, and demanding more productivity from them. After his death, Ross's slaves were sold to other plantations, and the mining and forging business ended. The slaves who had learned the skills necessary to work in the iron business had no choice about where to work, so unless they were bought or leased for forge work, their skills simply disappeared from manufacturing.

Chauncey Jerome started out with nothing, working as a farm hand and carpenter's apprentice, neither of which he liked. In the surrounding towns, however, new mechanical inventions were cropping up, and he was fortunate enough to find work with clockmaker Eli Terry. Terry's mechanical skills and imagination rubbed off on Jerome, and the young man designed and built his own clocks using Terry's work as his model. Related inventions and businesses contributed to Jerome's success—circular saws, lathes and jigs, planing machines, a sandpaper wheel, and the brass industry. Laborers, at first paid by the piece, had incentive to produce more parts and perhaps invent a new machine or process, and then could leave and start their own business. Jerome was an example of this.

In *A New Nation of Goods: The Material Culture of Early America*, historian David Jaffee argued that men such as Terry and Jerome represented the "face of a manufacturing and market revolution that drew in a broad range of consumers throughout the United States." It was in rural communities where innovative artisans and Yankee peddlers transformed American material culture: "The flexible and decentralized rural system of production—pioneered by clockmakers and soon to be found elsewhere—relied on the labor of countless young men and women in the countryside willing to work for wages and on the desire of middle-class families across the nation to fill up their parlors with consumer goods."[17]

Terry's and Jerome's technical information and mechanical know-how were passed on to many men who worked with them. Craftsmen working the floors of the factory were given the freedom to suggest improvements and even to design their own tools. All this created a unique and evolving work environment that gradually created a socioeconomic phenomenon difficult to pinpoint, but real nonetheless. The job of laboring in mills or factories was a completely different experience from working in the fields. Thus, it was not just the type of employment that changed; men also had to make psychological and emotional adjustments to this new kind of work.

The new machinery and technology made Americans in the North redefine democracy and freedom. Inventor Robert Fulton believed that technology assisted republicanism because, in providing economic opportunity, it led to "equality in class status" and promoted "the welfare of the many as opposed to the special privileges of the few." "Every order of things, which has a tendency to remove oppression and meliorate the condition of man directing his ambition to useful industry, is, in effect, republican."[18] At the 1851 Crystal Palace Exhibition in London, Edward Riddle, reporting on the exhibition to the commissioner of US patents, noted that "the Russian exhibition was a proof of the wealth, power, enterprise, and intelligence of Czar Nicholas; that of the United States [was] an evidence of the ingenuity, industry, and capacity of a free and educated people."[19]

Manufacturing changed the idea of working alone to the idea of working as a group, as part of the wheel of production. Everyone had to do his or her job to build the finished product. In the case of Jerome's clocks, the carpenters built the case, others built the dial, glass tablet, machinery, and weights. Eventually, Jerome's business faltered, but his workers moved on with marketable skills, and they spread their knowledge to others within the region, perhaps improving upon machinery in new and evolving manufacturing endeavors.

The clock business was only one of hundreds of new industries emerging in the North and Northwest in the decades before the Civil War. The growth in textile, iron, lead, lumber, glass, paint, shoe, and tobacco manufacturing over the first six decades of the nineteenth century was staggering. Between 1810 and 1860, the manufacturing value of these products grew by a factor of ten to $1,885,861,676.[20] Furthermore, comparing numbers from the three regions of the country between 1840 and 1860 revealed remarkable evidence of the differences in economic development during the antebellum period.

The Southern states experienced unprecedented economic development in the two decades before the Civil War, earning the sobriquets "The Cotton South" and "King Cotton," but the region's focus was on producing the raw material not refining it. As the North and West demanded raw materials from the South, and the South demanded finished products from the North and West, a symbiotic relationship developed between the various sections of the country. This led to distinct labor patterns and business and management systems, along with logistical networks to support and develop the different priorities in each section. These distinctions were prevalent on the eve of the Civil War.

The cotton plant, from which extraordinary wealth derived, was commercially known as short-staple (black seed), green seed, and Upland cotton, and the amount of cotton grown was remarkable. The cotton crop in 1830 was 350,000,000 pounds. By 1840, the crop had grown to 790,000,000 pounds, and by 1860, 2,154,820,800 pounds were picked and ginned by four million African American slaves. They produced 5,387,052 bales of 400 pounds each.[21]

Rice, sugar, and tobacco also generated considerable wealth. All of these staple crops were refined and produced by the hands of slaves. Iron manufacturing and coal mining operations also hired slaves to do most of the work, much to the grievance of white artisans or mechanics. Plantation owners hired out their slaves to work as masons, coopers, and carpenters. White mechanics in the Tredegar Iron Works of Richmond, angered because the owner hired mechanics from the North to teach slaves "the skilled processes of puddling and rolling," went on strike in 1847. The owner ignored the strikers. The Richmond *Times and Compiler* declared that such strikes "attacked the foundations of slavery by maintaining the principle that the employer may be prevented from making use of slave labor."[22]

Where white operators did outnumber slaves was in the small and modestly successful textile industry in the South. Of course, some mills employed slaves as workers, such as the Rocky Mount Manufacturing Company, located at the falls of the Tar River in North Carolina, and the Salem Cotton Manufacturing Company in the Piedmont section of North Carolina.[23] William Gregg of South Carolina, however, advocated the establishment of textile industries to diversify the economy of the South and proposed using poor whites as the labor force. Many in the state legislature, controlled by the planter class, were hostile to Gregg's ideas.[24] He continued to work toward his goal of building a mill and employing white operatives, and in 1846 he finally constructed a factory in Graniteville, South Carolina. This factory, combined with mills in Macon, Eatonton, Columbus, and Athens, Georgia, gave rise to a stable textile industry. The 1840 Census reported 1,581 hands employed in textile factories in Georgia, 2,122 in South Carolina, 1,830 in North Carolina, and 6,081 in Virginia.[25]

Twenty years later, the textile industry in the South had flat-lined. Thirty-three mills operated in Georgia with approximately 83 workers per mill, for a total of 2,813 employees. South Carolina listed 891 factory hands in seventeen mills; North Carolina, 1,755 workers in thirty-nine mills; and Virginia, a decline to 1,631 workers in just seventeen mills. Also, by the 1850s, bondsmen were supplied to factories by slave owners to reduce operating costs.

Other forms of manufacturing with more than a million dollars of capital investment in these four Southern states included tobacco manufacturing, coal mining, and iron production (bar, sheet, and railroad) in Virginia; turpentine distilleries in North Carolina; and steam engine manufacturing in Fulton County, Georgia, the county that included Atlanta. All four states invested in lumber mills and the manufacturing of flour and meal. The total capital investment in manufacturing in these states came to $54,451,894. Yet, this number fell short of the capital investments made in just the counties of Middlesex, Essex, and Worcester in Massachusetts.[26]

As the demand for cotton to feed the nascent Northern industry grew, mills built on the Samuel Slater model sprang up in the Quinebaug-Shetucket and Blackstone River valleys in Massachusetts, Connecticut, and Rhode Island. Census figures showed unprecedented expansion in textile manufacturing. For example, Massachusetts employed 26,000 people in the textile business in 1840, with a capital investment of $17,414,099. By 1860 those numbers had grown to 37,145 workers and $32,685,514 in capital investments.[27] In addition, unlike the period before the War of 1812, when most textile mills made their own machinery, machine building became more sophisticated, and specialized companies emerged.[28]

In the early nineteenth century, there was no distinction between mill and machine. The waterwheel was built with the mill, and each operation within the mill had its own wheel, controlled by its own water gate. Therefore, increased improvements in machinery demanded improvements in "water wheels, speed controls and power distribution."[29] Power was generated for the machines through a system of shafts, gears, and pulleys. Shafts were first made of wood then of iron, and water power initiated by the total fall of a river often passed through so many machine works, gristmills, lumber mills, and textile factories that a mill's waterwheel was rendered inefficient as a result of the water's overuse. The replacement of heavy gearing with lightweight leather belting and improvements in metallurgy and iron water-turbine technology, first developed in France in 1832, improved the efficient use of rivers and canals. To cope with the complexity of a mill operation, traditional millwrights gave way to men who specialized in engineering the construction of mills and their power sources.

These mechanical engineers were trained through an apprentice system. Like Chauncey Jerome of clock-making fame, Paul Moody, a mechanic with the Merrimack Manufacturing Company in Lowell, Massachusetts, first learned his engineering skills as a machine shop laborer. Other workers left the textile mill to design and make their own metal products, machine tools, locomotives, sheet-iron ware, carpenter's tools, and brass clocks. Between 1840 and 1860, the Northern labor force dramatically changed. Approximately seven times more carpenters, civil and mechanical engineers, and factory hands worked in the North than in the South. New York and Pennsylvania together employed 67,000 carpenters, 8,300 engineers, and 5,000 ironworkers in 1860. Virginia and North Carolina together employed 12,000 carpenters and 1,000 engineers. Only 600 ironworkers worked in those two Southern states.[30]

English observers reported on the quality of the mills and skills of the workforce throughout the Northeast. Joseph Whitworth presented his report on New England manufacturing to the House of Commons in February 1854. Whitworth was no stranger to machine operations and engineering. He pointed out that industrial development "on a national scale demanded that nuts and bolts of the same diameter be interchangeable."[31] Whitworth had proposed such a system in 1841, and by 1854 it was in common use in England. He was considered the world's foremost manufacturer of machine tools. At the Crystal Palace Exhibition in London in 1851, "he had displayed his lathes, his machinery for planing, shaping, slotting, drilling, boring, punching, and shearing, and his standard gauges and measuring apparatus."[32] He also devised a rifled breech-loading cannon and a rifled gun with a hexagonal bore, both of which were used by the Confederacy during the Civil War.

It was not until 1857, however, that the United States would begin to develop standards for screw threads. Other measurement standards had already been developed in the clock and firearms industries.[33] William Sellers of Philadelphia was the first to develop a gear-cutting machine in 1857 that introduced commercially interchangeable nuts and bolts.[34] Before Sellers's gear-cutting device, machinists had developed their own system to meet particular construction needs, and consequently, the use of special threads prevented outside repairs on their machinery. This explains, at least in part, the use of the various track gauges by American railroad companies, especially in the South.

Whitworth commented extensively on both the manufacturing companies and the labor force. For example, he noted how the Hadley Falls Company was formed in 1847 for the purpose "of turning to account the water power supplied by the river Connecticut, buying water privileges, and purchasing land to form the site of a manufacturing town." The town laid out streets and constructed a thousand-foot-wide dam across the river. Whitworth also applauded town officials for appropriating $3.72 per student for the education of each child. Within seven years, Holyoke, Massachusetts, had two cotton mills employing 1,100 workers, a machine shop employing 365 workers, and a paper mill.[35]

Whitworth concluded his report by saying that "the intelligent and educated artisan is left . . . free to earn all that he can, by making the best use of his hands, without hindrance by his fellows. It may be that the working classes exhibit an unusual independence of manner, but the same feeling insures the due performance of what they consider to be their duty with less supervision than is required where dependence is to be placed upon uneducated hands." He also found that any workman who developed "peculiar skill" could rise to become a superintendent of the mill. This was so, Whitworth wrote, because common schools were "placed within the reach of each individual, and all classes avail themselves of the opportunities afforded."[36]

Although he did not visit the South on his tour of the United States, Whitworth no doubt heard conversations that raised questions in his mind about the effectiveness of slave labor. His fellow commissioner, George Wallis, reported that "in some middle States . . . attempts are being made to bring slave labour to bear upon manufacturers, to the reduction of prices, and, consequently, of the remuneration of free labour." "The prevailing idea is," Wallis continued, "that slave labour can never by any possibility be made to compete, or to pay in comparison with free labour. This is held equally by the opponents and advocates of the institution of slavery."[37]

In the two decades before the Civil War, skilled labor and new machinery were at a premium and were sought by owners of factories and mills, especially in the North. Unskilled workers—and there were many of them—were not as effusive in their assessment of how new tools and inventions offered remunerative opportunities to laborers. Many people were impoverished, and machinery threatened to take away the finite number of jobs available. In addition, with immigration burgeoning in the 1840s and 1850s, there seemed to be an infinite number of unskilled laborers ready to fill those jobs.

This was particularly true in the South, where even skilled labor felt threatened by new machinery. Inventions could displace workers, but compared with the North, there were fewer factories and thus fewer opportunities to work elsewhere. Owners of iron forges, coaling operations, and textile mills further weakened the white labor market by using slave labor for both skilled and unskilled jobs. The difference in attitude toward machinery among skilled workers North and South is nowhere better illustrated than in the manufacturing of guns, cannons, and ordnance in the twenty years before the Civil War. The operations of the US armories in Springfield, Massachusetts, and Harpers Ferry, Virginia, demonstrate how to one side technology was progress, and to the other side, anathema.

Sometime before 1819, Thomas Blanchard, a New England inventor, designed an apparatus for manufacturing gunstocks. After revising his patent specification in 1820, he presented to the superintendents at Springfield and Harpers Ferry his "engine for turning or cutting irregular forms out of wood, iron, brass, or other material or substance, which can be cut by ordinary tools."[38] Springfield's superintendent Roswell Lee embraced the new machinery; Superintendent James Stubblefield at Harpers Ferry did not. Other technological improvements in waterwheels, trip hammers, and machines for milling, drilling, and trimming iron components were also slow to take hold at Harpers Ferry. As historian Merritt Roe Smith pointed out, traditional artisans were not receptive to mechanization because it threatened their job security and their traditional way of life. "Combined with close-knit local kinship groups that held outsiders suspect and outside ideas alien to customary practices, these inbred feelings fostered curious technological conservatism at Harpers Ferry."[39]

The relationship between Stubblefield and his workers at Harpers Ferry grew from a culture of patriarchy and patronage. Stubblefield was a member of the town's leading family. His brothers-in-law operated the major businesses in town, and Stubblefield was friends with Virginia senator James Barbour and Kentucky's Henry Clay. To maintain his business interests, quell class antagonism, and hold his political power, Stubblefield protected worker's jobs at the armory, and in return, workers voted as he desired and accepted his paternalism.[40] Change felt threatening to everyone, so change, and anyone associated with it, was a pariah. The Clock Strike of 1842 embodied this prejudice against new ways of operating at the Harpers Ferry factory.

The origins of the Clock Strike dated to the appointment of Major Henry K. Craig as the first military superintendent at Harpers Ferry. He inherited from Stubblefield a workforce that still believed in the idealized craftsman, a skilled laborer who took pride in his work by applying individual attention and artistry to each piece he made. Not only did new machines devalue a laborer's expertise, but they also punctured an artisan's identity as a person. Technology challenged workers' belief in human agency. Skilled laborers found anonymity rather than satisfaction in the results of their work. For several decades, innovations in arms manufacturing crept into their lives, lightened their physical labor, increased their output, at least theoretically, and altered their culture. These men lived in two worlds: one inside the armory, where industrialization brought rapid change to the work environment, and one outside the factory, defined by a lifetime of habits, traditions, and values unique to their region. Furthermore, the men at Harpers Ferry, like many people in the South, might enjoy the material goods industrialization produced, yet their culture and customs were still bound by farming, plantation life, paternalism, and slavery.

For years, James Stubblefield understood that to maintain harmony among his workers in what they considered a threatening environment, he had to allow for certain customs such as twelve-hour workdays so that workers had time to complete piecework at their own pace and, in the same twelve hours, had time to read a newspaper, argue and debate, drink alcohol, and take long breaks in the middle of the day. Unfortunately for the workers at Harpers Ferry, Major Craig did not see things their way.

Craig determined that written rules and regulations were needed to operate the armory efficiently. Violations of the rules needed to be accompanied by stiff fines. One regulation that antagonized workers was the new ten-hour workday. No drinking or debating, no trips home at lunchtime, all work to be finished in ten hours, and work to be of high quality. The craftsmen in the shop believed time should not be governed in this way because it degraded the workers and exacerbated their feelings about being replaced by machines. Craig dismissed their complaints. Consequently, on March 21, 1842, the entire labor force, led by the pieceworkers, walked off the job. This began the Clock Strike. A select handful of workers boarded a boat headed to Washington to present their petition for the removal of Craig to President John Tyler.

The president told the petitioners to "go home and hammer out their own salvation." Tyler's answer did not satisfy the workers. According to Smith, Major Craig's regulations "not only struck at the craft ethos but also threatened basic norms within the community. These factors awakened the armorers to a spirited defense of the old order."[41] That defense was so spirited that by December 1858, Secretary of War John B. Floyd removed Craig because the armory had fallen into disarray. The quality of the work was poor and expenditures had exceeded the value of production.

Even as Southern politicians and business leaders in the 1850s encouraged some industrial development, production quality suffered for reasons of limited access to improved materials, limited machine tools and training, and general unease about how the new technology would alter people's way of life. Industrial development required a skilled labor force, engineers who understood the latest technology, and competent managers who would adapt to new ways of doing business—all resources much less available in the South than in the North.

In the North, in addition to the way information was shared among workers on the shop floor, science and mechanics' institutes and agricultural fairs contributed to the free flow of information among owners and laborers in the mid-nineteenth century. When the *Saturday Evening Post* reported on February 14, 1824, that the Franklin Institute in Philadelphia was formed to "advance the general interests of Manufactures and Mechanics," this news ushered in a wave of new organizations. By 1861, these organizations would produce large numbers of skilled mechanics and artisans and encourage entrepreneurs and inventors.

Under the leadership of John F. Frazer, a member of the Board of Managers and a leading citizen in Philadelphia, the Franklin Institute recruited speakers with superb reputations to lecture adult members and guests on topics such as hydraulics, civil engineering, and practical chemistry. Frazer successfully enlisted specialists to submit articles for a new journal.[42] For example, John C. Trautwine wrote an article addressing the problem of designing a ship canal through the Panamanian isthmus.[43] Frazer also encouraged Trautwine and others, such as Colonel Joseph Totten of the Corps of Engineers, to publish their research in book form, which they did. The institute became the copyright proprietor of Trautwine's and Totten's work. These publications were a significant resource for entrepreneurs in the city and their manufacturing of heavy machinery, including large lathes, railroad turntables, and large boring machines. As a result of this information exchange, men such as William Bement, who had apprenticed in the Amoskeag Machine Shop in New Hampshire and the Lowell Machine Shop in Massachusetts, moved to Philadelphia to establish machine tool businesses.[44]

In addition, in 1841, the Franklin Institute, under Frazer's leadership, made the decision to hold annual fairs that provided a fivefold benefit. First, the prize money at the fairs encouraged new inventions and improvements to machines and tools; these innovations then buoyed domestic manufacturing throughout the country. Second, the fairs provided craftsmen, artisans, and working men with an opportunity to study the new machinery, gain valuable information about repairs to their equipment, and immerse themselves in the culture of

America's emerging technological revolution. Third, the fairs brought in essential revenue for the Franklin Institute. Fourth, the fairs reinforced the notion that education, scientific study, and mechanical aptitude were appropriate masculine pursuits. Finally, not only did the fairs inspire the creation of other fairs throughout the North, but they also led to collaborative efforts with the Massachusetts Charitable Mechanics' Association "for proper American representation" at the 1851 Crystal Palace Exhibition in London. By 1854, approximately one hundred thousand people were visitors to the institute's exhibitions.[45]

One year before the Crystal Palace Exhibition in London, the Massachusetts Charitable Mechanics' Association opened its sixth exhibition in Boston, on September 11, 1850. Groups from Bangor, Portland, Salem, Lowell, Worcester County, Providence, the New Bedford Mechanics Association, and the Franklin Institute were present. Notes of support were received from the American Institutes, the New York Agricultural and Mechanic Institution, and the Montreal Mechanic Institute. Chairman of the association's Board of Managers, Henry H. Hooper, remarked on the contributions the exhibition received from around the country but lamented the paucity of contributions from the West and South. "It was highly gratifying to find in the Exhibition, specimens of the skill of Machinists and Manufacturers of the West and South. We regret that we had not more of this honorable competition." The managers had made a concerted effort to attract artisans and inventors from different sections of the country. Given the simmering sectional tension, including the free soil movement and the Compromise of 1850, the managers wished it "to be distinctly understood that the Exhibitions held under the sanction of our Association bear no sectional character. They are designed to promote the interests of our whole country." The managers implored every section of the country to "unite with us on this occasion, and thereby strengthen the cords which unite us as one body striving to promote the best interests of the whole."[46]

The absence of Southern and Western participation in the exhibitions could be explained partially by distance. Time and money were needed to transport new inventions or machinery to Boston, and the requisite stay of nearly three months was onerous for many. There was the cost of getting participants/exhibitors to the city, finding lodging and food for the required two months of residency before the fair and for the seventeen days the fair operated, and then getting home. Another reason for lack of participation from the South was the limited interest in manufacturing and the modest and reluctant financial support from powerful Southern businessmen. Manufacturing accounted for a small percentage of the South's wealth in the years before the Civil War, and the industrial goods sought after and purchased were imported from the North. Southerners' ability to purchase manufactured goods from the North depressed the demand for Southern artisans and mechanics. The limited number of artisans meant hardly any opportunities for continuing education because few were interested in investing in mechanics' institutes. The mechanics in the South, therefore, were unable to stay abreast of the latest developments in technology.

Conversely, textile mills in Rhode Island, Massachusetts, and Connecticut provided opportunities for the creation of industries in machine building. "Under urban influences," wrote historian Victor Clark, "a cabinetmaker's shop became a furniture factory and a smithy expanded into engine works."[47] Cheap coal made steam available in Philadelphia. Expanding rural settlements in New York and Pennsylvania produced more lumber than any other states in the Union, but demand exceeded supply. Maine was a small consumer of lumber, so by 1832 its annual output was 38,000,000 board feet.[48] Bangor was the greatest sawmill center in the country, yet by the 1850s, Saginaw, Green Bay, and sections of Minnesota and Wisconsin rivaled Bangor. After introduction of the steamboat, Cincinnati, Buffalo, Cleveland, Pittsburg, and Louisville became the centers of the river steamboat industry. By the 1840s the industry had attracted capital and workmen from the East, "encouraged local manufacturers of cordage and naval stores," increased the number of sawmills, led to the creation of machine shops and engine works, and "disseminated a knowledge of mechanical arts and science."[49]

The growth of industry in the West coincided with development of the region's common school movement and formation of mechanics' institutes. The Ohio Mechanics' Institute founded in Cincinnati in 1828 was established to facilitate the "diffusion of useful knowledge" to "ingenious artisans and mechanics"; the Akron Mechanics' Association founded in 1846 was established to do the same.[50] By 1860, Illinois, Indiana, and Ohio had approximately 5,000 civil and mechanical engineers, 43,000 carpenters, and 4,000 ironworkers. At the same time, Virginia, North Carolina, and South Carolina reported a combined total of 1,300 civil and mechanical engineers, 14,000 carpenters, and 880 ironworkers.[51]

Joseph Whitworth's 1854 report to the House of Commons identified the "energy" and "peculiar aptitude" of Northern laborers to capitalize on their natural resources and their ingenuity in producing machinery in "almost every department of industry." Whitworth keenly observed that the intelligent and educated artisan was free to earn all he could "by making the best use of his hands" and that, particularly in the North, "education is, by means of the common schools, placed within the reach of each individual, and all classes avail themselves of the opportunities afforded."[52]

Northern agriculture also contributed to technological developments, and farmers, by necessity, learned how to repair equipment, make tools, and exchange ideas about the importance of education for progressive farming practices. To take the place of crude winnowing machines, which separated wheat from chaff, dirt, and weed seeds, inventors designed and built threshing machines. In the 1830s, businesses in Worchester, Massachusetts, manufactured cast-iron plows by the thousands each year, with replaceable parts.[53]

In the North, Massachusetts and New York took the lead in promoting agricultural societies and efforts to introduce basic farm studies into common schools. In New York State's *Second Report of the Special Committee for Promoting the Introduction of Agricultural Books in Schools and Libraries*, the authors noted that the Massachusetts legislature passed an act in 1819 suggesting that agricultural societies should raise $1,000 for the improvement of agriculture and receive a subsidy from the state of between $200 and $600 per year. The report went on to recommend that scientific developments in farming should be spread "among the mass of practical and laboring farmers . . . It is believed that no more effectual instrument can be employed for reaching both the rising generations and the adult population than the school district libraries."[54]

The New York State Agricultural Society, established in 1832, had begun to receive state appropriations for its work in the 1840s, and the society's business was published as documents of the New York Legislature. The annual fair hosted by the society offered handsome cash prizes (premiums) for practical inventions that would assist the farmer in his work and make life more comfortable and efficient for his wife. As the fair attracted people from all over the United States, it assumed the character of a competition anyone could enter and became an idea factory, an outdoor classroom, and a gathering place—in other words, a sort of nineteenth-century Home Depot. The list of premium winners included C. H. McCormick (Chicago) for a grain reaper, William Hovey (Worcester, MA) for a hay and straw cutter, and Reuben Daniels (Woodstock, VT) for a self-sharpening straw cutter. Others' agricultural inventions included an excavating scraper, new cradles, scythes, and snaths, and a seed sower built by H. L. Emery and Company of Albany, New York.[55]

Since many Northern farmers spent the bleak winter months working on machinery they sold for cash, the 1850 fair also exposed rural communities to other technologies developed to assist in winter jobs or aid in the repair of equipment. Wood-planing tongue-and-groove machines with stationary cutters, a large engine lathe, waterwheels, a machine for salting meat, a rotary fire engine pump, and a hot- and cold-air furnace for making pig iron—all were on display.[56] By 1856, there were 912 local and state agricultural organizations in the United States, of which all but 165 were in the North and West.[57]

Contributors to farm journals began to write about the need for American colleges to teach agricultural science. An editorial in September 1847 in Chicago's *Western Prairie Farmer* called for common schools, private schools, high schools, and colleges to introduce the study of agriculture into their curricula. "There is no rational doubt, that a general introduction . . . of such studies as bear upon agriculture, is demanded by the general interests of society, and will be demanded by society itself, so soon as the case is made clear to its apprehension."[58]

In Connecticut, a leading pioneer in agricultural science, Samuel William Johnson, after two years abroad, returned to the Yale Scientific School where he became professor of analytical chemistry, analyzing fertilizers for crops. By 1859 he was professor of agricultural chemistry.[59] Johnson's work caught the attention of politicians who hoped to establish agricultural colleges in their own states. Michigan and Pennsylvania passed legislation in 1855 for establishment of the Michigan Agricultural College and the Farmers High School, the latter to be renamed Pennsylvania State College.[60] New York Governor Washington Hunt, in his 1857 annual message, proclaimed that "it cannot be doubted that an institution of the character proposed (an institution for advancement of agricultural science and of knowledge in the mechanic arts) would promote the dissemination of agricultural knowledge and elevate the condition of people." Whether the governor's motives were economic, social, political, or altruistic, he spoke about the importance of education and how "the elevation of the laboring classes . . . is worthy of the highest ambition of the statesman and patriot."[61]

Thirty-three months later, another politician spoke to the importance of agricultural education, both in schools and through the vehicle of the state fair. In an address to the Wisconsin Agricultural Society's fair held in Milwaukee, Abraham Lincoln told the audience—eager to get on with the awarding of the premiums and not necessarily interested in what the speaker had to say—that "the chief use of agricultural fairs is to aid in improving the great calling of *agriculture* . . . to make mutual exchange of agricultural discovery, information, and knowledge." In general, education enabled the farmer to study soils and seeds, crop diseases, implements and machinery, and a "thousand things of which these are specimens." Lincoln concluded by saying, "A capacity, and taste, for reading, gives access to whatever has already been discovered by others. It is the key . . . to the already solved problem . . . It gives relish, and facility, for success-fully pursuing the unsolved ones."[62]

Reading did open up a world of information to Northern and Western farm-ers, but in the South, where illiteracy was common and fewer than a third of those between five and nineteen years of age attended school, the ability to diffuse agricultural information to yeomen and poor farmers was increasingly problematic. Philip St. George Cocke, president of the Virginia State Agricul-tural Society, complained in 1857 that "seventy thousand of our adult popula-tion can neither read nor write!" He implored his fellow Virginians to open their "hearts and means until every child within the limits of our broad Common-wealth shall, at least, have the advantage of a Free School education."[63]

Improving conditions in the South for the middle and lower classes, es-pecially small farmers, would require disengagement from the institution to which wealthy and powerful Southerners were inexorably committed—slavery. Plantation agriculture, as John Majewski pointed out in *Modernizing a Slave Economy*, had come with a high price tag. The acidity of the Southern soil made it difficult to supply fallow fields with the proper nutrients to allow them to re-cover quickly. Hay and clover, used to fix nitrogen in Northern fields, "failed to thrive in the warm, humid climate." Cattle were fed with low-nutritional grasses and peas, and erosion from heavy rains (cotton and corn were row crops) "created channels that carried away topsoil."[64]

Instead of relying on crop rotation, farmers burned forests and used the ash as fertilizer. But it took approximately twenty years for a forest to regenerate so it could be burned again. This meant that plantation owners would grow crops on a plot of land for five years and then move to another plot for the next five years. At any one time, only a quarter of their property was under cultivation. Furthermore, areas with isolated farms and plantations had only a small market for agricultural periodicals, a small interest in agricultural fairs, and a small sup-ply of people interested in attending agricultural schools and colleges.[65]

By the mid-1850s, some Southern nationalists perceived the South's disad-vantage and envisioned an independent South that combined slavery, technol-ogy, and a strong manufacturing base.[66] One planter was bold enough to suggest that Southerners needed more "Yankee ingenuity" to design and build new ma-chines that would lead the South into the future.[67] The South's attempt at an agricultural reform movement called for improvements in farming that would contribute to the growth and wealth of plantations and yeomen's farms. Infor-mation about agricultural experiments was disseminated in periodicals such as Edmund Ruffin's *Farmers' Register*, and reformers at the Southern Agricultural

Convention in 1854 proposed establishment of the Southern Central Agricultural College.[68] The reform movement, however, remained tied to the institution of slavery. For example, Ruffin declared at the Virginia Agricultural Society fair in 1852 that "slavery was of divine origin and promoted the industry, civilization, refinement and general well-being of mankind."[69]

The great fortunes generated by slavery and cotton made it improbable that Southern leaders would see the economic forest for the trees. As they imagined an independent Confederacy, they understood the importance of establishing their own manufacturing and commercial capacity, but like farming, it was to be carried out on the backs of African American slaves. This meant a continuation of social controls to buttress the established social hierarchy, but it would also require educational controls. Information would need to be gained but disseminated carefully. The proposal for an agricultural college, open to the sons of the planter class, was an example of this model in which the upper classes would remain the society's intellectuals, acquire new knowledge, and dispense it as necessary.

Information exchange in the South contrasted sharply with that in the North, especially when it came to ideas about new technology. Mechanics' institutes, agricultural fairs, and manufacturing exhibitions introduced average Northerners to new tools and machines, which were adopted for everyday use. This exposure also had the effect, in general terms, of making "everyman his own mechanic."[70] A young William Watts Folwell recalled that a "revolution took place in the tools of the carpenter and blacksmith as in the implements of the farmer." There were also "examples of light and beautiful implements by individual artisans."[71]

In addition to these educational and practical innovations, a revolution taking place in transportation made possible the rapid movement of goods and services from one section of the country to the other. At the center of this marvelous transformation was the railroad.

CHAPTER THREE

Building the Railroads

Early Development of the Modern Management System

> By an arrangement now perfected, the superintendent can tell at any
> hour in the day, the precise location of every car and engine on the line of
> the road, and the duty it is performing.
>
> *Henry Varnum Poor, on the general superintendent of the*
> *New York & Erie Railroad, Daniel C. McCallum*

Advances in technology and improvements in machinery and in the exchange of information were closely aligned with railroad development in the United States in the antebellum period. For example, the use of coal for smelting led to enhanced iron production, while improvements in metal-working tools and the availability of interchangeable parts created more sophisticated machinery. With an understanding of steam engines in the development of water transportation, American designers began to build new locomotives and the tracks on which they operated. By the time the Civil War began in 1861, the United States had approximately 31,500 miles of track, 22,000 miles in the North and 9,500 miles in the South. In addition, the railroad industry produced tens of thousands of machinists, toolmakers, engine builders, and repairmen whose skills contributed to America's transportation revolution and the expansion of businesses and markets throughout the country. After shots were fired in Charleston, South Carolina, on April 12, 1861, "in the dawn's early light," many of these artisans and craftsmen would take up arms with the Union or Confederacy. Little did they or their leaders imagine that their prewar skills would become essential elements in attempting to bring the war to a successful conclusion.[1]

To build the network of railroads to move commerce and people across mountains, rivers, and plains, civil engineers were needed. Yet, in the 1830s, there were only a few trained civil engineers, most of them educated at West

Point, and there were no railroad engineers. Ingenuity, creativity, determination, and patience were required not only to lay track over a multifaceted landscape but also to build the locomotives and rolling stock necessary to develop this transportation system. Growth came quickly. On March 27, 1839, the people of Boston watched for the first time a locomotive in operation. The *Boston Evening Transcript* reported: "We noted it as marking the accomplishment of one of the mighty projects of the age, and the mind, casting its eye back upon the past, as it was borne irresistibly onward, lost itself in contemplation of the probable future."[2]

The crowd on Tremont Street gazed in awe at the futuristic locomotive, which was in all likelihood built in Lowell, Massachusetts. The Lowell Machine Shop became the early center of the railroad industry, and many of the mechanics who learned their craft in Lowell eventually took their knowledge of design and construction with them to join or start new companies. Locomotive construction created new opportunities and occupations for machinists, mechanics, and ironworkers. Furthermore, trial-and-error experiments in new engine technology generated ideas in other areas such as hydraulics, textile machinery, and bridge building.

The locomotive industry in Lowell developed as a result of two West Point engineers and skilled mechanics employed by the city's Locks and Canal Works. George Washington Whistler (West Point, 1819) and his friend and mentor William Gibbs McNeill (West Point, 1817) were army engineers before they started to work in the embryonic New England railroad business. Whistler, while still in the army, had received permission to serve with the newly established Baltimore & Ohio Railroad Company and traveled to England in 1829 to study and meet with George Stephenson, designer of the first steam-powered locomotive. After Whistler's return he continued to work with the B&O and, in 1833, went to the Northeast to work as a surveyor for the proposed Providence to Stonington, Connecticut, section of the Boston to New York railroad.[3]

McNeill's railroad experience was more extensive than his protégé's. He superintended the survey for the Baltimore & Susquehanna, the Patterson to Hudson River, the Boston to Providence, and the Taunton to New Bedford railroads. He had traveled to England with Whistler and encouraged the latter's interest in railroad engineering. No doubt the two consulted before Whistler resigned his commission in 1833 to accept the position of engineer for the Proprietors of Locks and Canals at Lowell.[4]

Whistler soon had the machine shops in Lowell building their own locomotives. At around the same time, another West Point–trained engineer, William H. Swift, joined Whistler and McNeill, and the three men embarked on extending the Boston & Worcester Railroad through the Berkshires to Albany. This feat required solving some unique engineering problems. For example, in Lowell, toolmakers, machinists, and mechanics had to build locomotives to traverse the mountainous spine that separated western Massachusetts from eastern New York, overcoming a steep incline while pulling heavy freight.[5]

The men in the machine shops met the construction challenges that the engineers placed before them. The mechanics demonstrated such remarkable skills that soon many of them left Lowell to share their expertise with other railroad engine manufacturers throughout the country. Most were actively recruited. Others founded their own companies. James and Nathan Ames, mechanics in the Lowell shops, founded the Ames Manufacturing Company in Chicopee, Massachusetts, and Elias Howe received his training in Lowell before setting to work on his sewing machine. Other mechanics who worked on locomotives took advantage of the wealth of engineering knowledge shared in the mechanic shops in the city and left to develop other technical innovations. James Bicheno Francis, at twenty-two years old, went from working as an engine mechanic to chief engineer of the Locks and Canals. In 1857 he published *Lowell Hydraulic Experiments*, which became the standard reference in the field.[6]

Lowell became a spawning ground for America's locomotive industry and the dissemination of technical knowledge. By 1860, at least twenty-nine American companies built 470 engines worth $4,866,900. Eighteen of the companies were in the North, five in the border states, and six in the South. Two shops in Philadelphia alone constructed 172 locomotives, averaged 600 workers, and purchased most of their machinists' tools from Pennsylvania or New England companies.[7] In Massachusetts, the two largest locomotive shops were in Taunton: one, with 175 workers, built twenty-three engines; the other, with 425 men, built cotton machinery and fourteen engines.[8]

Although most locomotives and rails were manufactured in the North, a small number of engines were made at companies such as the Nashville Manufacturing Company, and railroad iron bar and car axles were made at the Etowah Iron Works in Georgia. In addition, by 1861, a number of companies had started to make their own rolling stock and passenger cars. Yet, according to the leading historian of Southern railroads, Robert C. Black III, "manufacturing . . . was scarcely adequate for the proper maintenance of its [the South's] railroad system."[9] The largest collection of locomotives belonged to the South Carolina Railroad, with sixty-two, and a close second was the Georgia Central, with fifty-nine.

The South Carolina Railroad also led the way with 849 cars of all kinds.[10] These numbers compare favorably with those of companies in the North and West, but most Southern railroads averaged fewer than fifteen locomotives and 235 cars, whereas the New York & New Haven Railroad alone had twenty-nine locomotives and 440 cars, and the Boston & Worcester had thirty engines and 725 cars.[11]

Locomotives, cars, tracks, bridges, and repair shops were the tools of the railroad trade, but most important to the operation was the labor force and the way it was managed. As John E. Clark Jr. argues in *Railroads in the Civil War*, railroad management involved constant inspection, superintendents, mechanics, track repair supervisors, and stationmasters all along the line. This was especially true for railroads that operated over 200 miles of track. "Decentralized management required unprecedented delegation of authority. Size and distance dictated that local managers had to master every aspect of their responsibility. They alone would have to resolve whatever problems arose, to find solutions and make decisions regardless of the problem's nature or complexity."[12]

In "The Corps of Engineers and the Rise of Modern Management, 1827–1856," historian Charles F. O'Connell Jr. argued that early military management procedures had a significant influence on the development of railroad management. Between the efforts of Secretary of War John C. Calhoun (October 1817 to March 1825) and Brigadier General Winfield Scott's *General Regulations for the Army* (completed in 1821 and revised in 1825), the army built a comprehensive organizational hierarchy, which included personal and, for officers, financial accountability, especially in the quartermaster, ordnance, and engineer departments.[13]

With passage of the General Survey Act of 1824, the army became involved in private domestic transportation improvement projects, and by 1827, eleven military engineers, including George Whistler and William McNeill, were assigned to the new Baltimore & Ohio Railroad Company. Captain McNeill was asked by the company's president to draft an official code of regulations to govern the operations of the B&O's Engineering Department. McNeill drew upon Scott's *General Regulations for the Army* to establish a management structure governing all railroad employees.[14]

Working with the Pennsylvania Railroad in the early 1850s, Herman Haupt (West Point, 1835) modified the B&O management model, devising an organizational plan that divided responsibility for the operation of the railroad among four departments: Transportation, Maintenance of Way, Motive Power, and Maintenance of Cars. Haupt also created a General Transportation Office whose responsibility consisted of maintaining regulations and uniformity.[15] Haupt's innovations were shared among companies through periodicals, the most prominent of which was the *American Railroad Journal*.

The journal had been in existence for seventeen years when Henry Varnum Poor became the editor in 1849. Poor was born in Andover, Maine, in 1812 and attended the local common school until he was twelve. Young Henry frequently borrowed books from the town's library, and after being tutored by the local minister, he attended Bridgeton Academy for two years then matriculated to Bowdoin College in 1831. As a practicing lawyer in Bangor, Maine, Poor used his legal work and business experience (controlling several lumber operations) to become involved in railroads. His brother, John Alfred Poor, was responsible for creation of the Maine railroad system in the late 1840s.[16]

Poor understood business and railroad operations. He printed in the *American Railroad Journal* information on brakes, axles, rails, locomotives, and bridge building. Articles that discussed mechanical engineering, waterworks, and river drainage were reprinted from European papers. Under Poor's direction, the journal provided information for railroad investors and advertising space expanded to include engineering books, new instruments, and brokerage house securities for sale. Poor pioneered advertising placed by merchants supplying railroad equipment.[17]

It was in the journal that Poor first commented on the new management system established by the general superintendent of the Erie Railroad, Daniel C. McCallum: "Mr. McCallum's strong point lies in his power to arrange and systematize, and in his ambition to perfect his systems."[18] McCallum designed a management chart marking the five operating divisions and the various subdivisions under them. Orders had to follow a chain of command within the proper branches of the company. "All subordinates," McCallum declared, "should be accountable to, and *be directed by their own immediate superior only*; as obedience cannot be enforced where the foreman in immediate charge is interfered with by a superior officer giving orders directly to his subordinates."[19]

McCallum attended common school in Rochester, New York. He became a carpenter and then a design expert before working with and later becoming president of the Erie Railroad. Poor admired the work of John Edgar Thomson of the Pennsylvania Railroad. According to James A. Ward, Thomson took McCallum's ideas to create the first line-and-staff managerial organization in American corporate history: "Thomson's scheme allowed men at the lower levels enough authority to demonstrate their talents, although always under the oversight of higher line officers." Of equal importance, Ward continues, "The new corporate structure . . . freed Thomson from the necessity of overseeing his road's daily operations."[20] This new system would become the model for the United States Military Railroad in 1862.

Southern railroad superintendents before the war were aware of Thomson's ideas, but they had little time to implement the new management system before hostilities broke out between North and South. Moreover, given the South's traditional views about centralized authority and the relationship between master and slave, it was hard to imagine the Confederacy establishing an administrative structure beyond its society's hierarchical construct. Although Southern railroads tripled their track mileage in the decade before the war, a shortage of industrial managers, skilled technicians, and engineers inhibited creation of an efficient railroad system, and the quality of the work left a lot to be desired.

The records of the men who led Southern railroads in the antebellum period are difficult to locate, but some provide a glimpse of the type of people who worked on the railroad below the Mason-Dixon Line. Some were Southern born and reared in families that owned and operated small businesses. Many of these sons of the South's middle class were educated at private academies and studied civil engineering at colleges such as Virginia Military Institute and the University of North Carolina. After graduating they went to work for newly formed railroad companies as engineers. Others were born into the planter class, were educated at the best schools, trained as lawyers, owned their own plantations, and eventually expanded their business interests by building railroads. A fair number were born and educated in the North then migrated South to find their wealth and happiness.

William Mahone was born in the little town of Monroe, Virginia, in December 1826. His father, Fielding Mahone, was instrumental in bringing an end to Nat Turner's Rebellion five years later. The Mahones moved to Jerusalem, Virginia, on the Nottoway River when William was thirteen, and the father opened a tavern and hotel business, with the help of his family and three slaves. William was described as a "sandy-haired, freckled-face little imp, who hung around the stores in Jerusalem . . . He was the leader in all deviltry, and the terror of all good country mothers whose boys occasionally went to town."[21]

Unlike many of his mischievous companions whose fathers paid little heed to education, Mahone was forced to attend a private school in Rosedale, Virginia, held at the home of Captain William J. Sebrell and conducted by Hannah and Sarah Armstrong, natives of Maine. At fifteen he was enrolled at the Littletown Academy, where prominent families sent their sons, and two years later he received an appointment to the Virginia Military Institute as a state cadet and, consequently, was given free board and tuition.[22]

Whereas his upper-class classmates excelled in literature and foreign language, Mahone was at the top of his class in mathematics, engineering, and chemistry. He graduated eighth in a class of thirteen. Mahone may have been described as a tobacco-chewing, prolific-swearing gambler of the coarse, unrefined middle class, but he dedicated himself to becoming a civil engineer. He worked on several railroad projects, on surveying expeditions, and on building plank roads, which many Southern leaders believed were superior to and much less expensive than railroads.[23]

By 1850, in the State of Virginia, businessmen began to question what they considered the shortsighted nature of the state's transportation policy. Plank roads were fine, but the commercial future of cities like Norfolk clearly depended on the railroad. The railroads that did exist had established short-line roads that failed to connect to the major cities. In particular, the city of Norfolk had failed to keep pace with other seaports along the Atlantic coast. Building a railroad linking Norfolk to Petersburg, Virginia, and connecting there with other railroads extending into the interior of the South was essential for future economic development.

Mahone was elected the chief engineer of the proposed Norfolk & Petersburg Railroad in 1853, and within five years the line was completed. The entire project was well engineered, including Mahone's twelve-mile roadbed designed to skirt the edge of the Great Dismal Swamp between South Norfolk and Suffolk, Virginia. By 1861, Mahone was both president and chief engineer of the railroad.[24] The phlegmatic Confederate leader who stopped the Union advance at the Battle of the Crater in 1864, which earned him a place of honor in Confederate folklore, built one of the great railroads of the South.

In contrast to Mahone, David Levy Yulee, John Motley Morehead, and William Sheppard Ashe were men of refinement, of the plantation and slave-owning elite of the "Old South," who became involved in railroad building as a horizontal expansion of their lucrative sugar, cotton, and rice businesses. All had studied law and promoted Southern railroad expansion. Yulee, known as the father of Florida's railroads, was the first Southerner to use state grant money. With it he successfully built a line from the Atlantic to the Gulf of Mexico.[25]

Yulee was born in the West Indies, and at nine years old was sent to a private school in Norfolk, Virginia. He eventually moved to St. Augustine, Florida

Territory, where he was admitted to the bar in 1836. He devoted considerable efforts toward Florida statehood with slavery, and when it became the Union's twenty-seventh state in 1845, Yulee was elected as its first Democratic senator. The following year he purchased a five-thousand-acre sugar cane plantation along the Homosassa River in the western part of the state. To provide transportation to deepwater ports on the Gulf of Mexico and Atlantic Ocean, Yulee started a plan to build a railroad. Using federal and state land grants, in 1853 he chartered the Florida Railroad. Construction began in 1855 and the railroad was opened in 1861, running from Fernandina on the Atlantic coast to Cedar Key on the Gulf of Mexico.[26] The Florida Railroad was Yulee's private enterprise. In 1862, when the Union navy captured both Fernandina and Cedar Key, the Confederate government hoped to dismantle the railroad's track so it could be used elsewhere. Yulee argued for an injunction against the state and won. The track remained untouched for the rest of the war.[27]

John Motley Morehead was a keen proponent of internal improvements in North Carolina in the 1840s and 1850s. As a boy he attended private schools, and he went on to graduate from the University of North Carolina in 1817. As an attorney, politician, and businessman, Morehead called for constitutional reform to recognize the growing population in the western half of the state. He promoted the common school movement and the construction of railroads. He served four years as governor, and shortly after his tenure became invested in a cotton mill. Recognizing the need to move textiles from plantation to mill to consumer, he started the North Carolina Railroad (NCRR) and became its first president.[28]

Henry Varnum Poor, editor of the *American Railroad Journal*, preached that the most successful railroads were the ones with chief executives who knew the business from top to bottom. This was not the case with Morehead. Yet, his financial and political skills made up for his lack of technical knowledge. Between 1850 and 1855, when the road was under construction, Morehead viewed his job as hiring a good engineer and letting that man worry about the details of building the line.[29]

The man Morehead hired as chief engineer was Walter Gwynn (West Point, 1822), an accomplished civil engineer. Born in western Virginia, Gwynn had worked on several railroad projects, and he managed construction of the NCRR with professionalism and skill. The line was completed in early 1856. Morehead also hired a New Englander, Thomas E. Roberts, to serve as master mechanic, a position he held until 1860.[30]

Unlike the Florida Railroad that operated as David Yulee's own business, three-quarters of the NCRR was owned by the state. As a result, state representatives on the board of directors kept a tight grip on the purse strings. NCRR salaries for the president, superintendent, road master, and treasurer were near the bottom of those for comparably sized railroads. In addition, there were frequent disagreements over how much money should be spent for company shops.[31]

Like Morehead, William Sheppard Ashe was not a railroad man. Born in North Carolina, he studied at Trinity College in Hartford, Connecticut, became a lawyer, and owned a plantation that cultivated rice. His stature in the local community and the state qualified him to become president of the Wilmington & Weldon Railroad. Ashe's chief engineer, however, was from the North. Sewall L. Fremont was born in Vermont, appointed to West Point from New Hampshire, and graduated in 1841. After serving in various postings, he resigned from the army and in 1854 became assistant engineer in the service of the United States on improvement of the Cape Fear River, North Carolina. In the same year he was appointed chief engineer and superintendent of the Wilmington & Weldon.[32]

William M. Wadley, another New England native, moved to Savannah, Georgia, at twenty years old after learning the blacksmith trade. Known for building the railroad bridge over the Oconee River, he served as superintendent of the Georgia Central Railroad (1849–52), the Western & Atlantic (1852–56), and the New Orleans, Jackson & Great Northern Railroad (1858).[33]

The president of the New Orleans, Jackson & Great Northern Railroad was also a Northerner. Henry Joseph Ranney was born in Middletown, Connecticut, and educated at the Partridge Military Academy, from which he graduated as a civil engineer in 1828.[34] Working first as an assistant engineer for the railroad, Ranney became its president in 1861. Like their Northern counterparts, Ranney and his Southern railroad presidents tried to establish a management system with a chain of command that included a general superintendent, master machinist, journeymen machinists, and carpenters.[35] Unlike many Northern railroads that required the addition of separate divisions of personnel to operate over several hundred miles of the company's roads, Southern railroads did not develop the divisional system, for a couple of reasons: most companies had less than 200 miles of track, and a decentralized management system required a skilled workforce that management trusted.[36]

For most of the backbreaking track repair, and in some cases the skilled work of carpenters, brakemen, and firemen, Southern railroad companies used slaves. Again, as was the case with blacks working in iron foundries or textile mills, many of them learned and developed unique mechanical ability but were not allowed to pass on their skills to white workers or share them with other companies. When it came to investing in "cognitive and aspirational" capital, as was done in Lowell, Massachusetts, in the 1830s, the South earned low marks during the antebellum period.[37] Many lines spent as much as $125,000 per year to purchase slaves. Remarkably, the twenty-nine-mile, five-foot-six-inch-gauged Baton Rouge, Gross Tete & Opelousas invested $115,000 in slaves.[38] The March 22, 1861, issue of the *Richmond Dispatch* reported that the contractor, presumably of the Wilmington, Charlotte & Rutherford Railroad, looked to hire 100 "Negroes (men and Boys)" for nine months.[39] An article in the December 16, 1861, *Daily Richmond Examiner* reported that the Virginia & Tennessee Railroad was looking to hire bondsmen as laborers, carpenters, train hands, and blacksmiths to work on the roads and in the repair shops.[40]

Major railroads in the North succeeded in building and operating trunk lines—for example, between Albany and Buffalo, the New York Central operated 236 miles of primary line and 314 miles of secondary track. But the largest lines in the South—and few were longer than 200 miles—had to rely on 30- to 50-mile roads built by local investors to serve local needs.[41] In the South, the longest line operated by a single railroad company was the Mobile & Ohio, which ran from Columbus, Ohio, to Mobile, Alabama, covering an impressive 469 miles. The directors first proposed that the railroad should run through the towns of Columbus and Aberdeen, Mississippi, but the towns refused. After several months of deliberation Columbus decided it did want the railroad, but by then the owners had moved the trace of the line west and the line cut through Artesia, Tupelo, and Corinth, Mississippi. Eventually, a trunk line was built, the only one on the Columbus & Mobile line, about 20 miles from Artesia east to Columbus, Mississippi, located beside the Tombigbee River.

There were few junction points along these main lines of operation, and the lines that formed junctions with the main lines were often part of a series of tracks managed by different railroad companies, with different gauges, designed to serve different purposes. A line such as the Southern Railroad of Mississippi, from Meridian to Vicksburg, carried local freight to the Mississippi River and so was built of material that could withstand only a limited amount of tonnage. Before the war there was no line that extended beyond Meridian east to the Tombigbee River, about 50 miles. Along the 469 miles of north-south track there were only three junctions along the Columbus & Mobile.[42]

In addition, serious gaps existed between lines. As trunk and small railroad company lines reached toward other lines, they often did not connect. There was no direct line from Wilmington and Charlotte, North Carolina, to Atlanta, Georgia. In Nashville, no track was built west of the city, and the central line from Nashville to Chattanooga was broken off before reaching Tullahoma. Competing interests between railroad men and teamsters, along with short-sightedness and parochialism among city fathers, prevented the creation of an integrated Southern railroad system. The teamsters had lobbied the Virginia legislature to pass a law that prohibited railroads from laying track in the streets of the city without permission of the local authorities, and "as late as 1861 local liverymen had prevented the intersection of *any* of the five railroads entering Richmond."[43] Because of bridging difficulties, the line between Savannah, Georgia, and Charleston, South Carolina, ended on the riverbank opposite Charleston, where passengers and freight were unloaded and reloaded onto ferry boats to complete the final 500 yards over the Ashley River.[44]

Next to the tracks themselves, the spewing, roaring, rattling, and hissing locomotives were the central ingredients of the railroad. In the South, the lack of locomotive repair facilities and trusted skilled workmen to maintain them hampered railroad development. Southern railroads had fewer engines than their Northern counterparts, and Southern companies often lacked repair parts and the mechanics to fix the trains.

No one in the country could anticipate the scope of the war to come. Yet, the different states of the railroad industry in the South and North foreshadowed the eventual outcome. Track mileage, number of locomotives, proximity and number of foundries and machine shops, and availability of men skilled enough to keep the trains operating—all overwhelmingly favored the North. As in the development of educational reforms discussed in chapter 1, technological innovations, labor formations, and new management systems stimulated by railroad operations exposed an Achilles heel in the South's future ability to conduct a war over a vast and complex geographical landscape.[45] The South had to rely too much on Northern engineers—and slave labor—to build, maintain, and manage its railroads, to the detriment of its logistical operations in the war to come.

So, as the sun rose on April 11, 1861, planters, farmers, and slaves headed to the fields. Men went to work in textile mills, clock factories, iron forges, coalmines, armories, machine shops, and railroad yards. Apprentices learned trades, children learned to read and write, university students learned Latin and Greek, and some tried to learn chemistry and biology. William Folwell studied philology in Berlin, John Morris Wampler served in the Virginia militia, David Ross was dead, Chauncey Jerome was bankrupt, William Barton Rogers prepared to open MIT, and Edmund Ruffin prepared to fire on the Federal fort sitting in the Charleston, South Carolina, harbor. In a short time, these farmers, planters, slaves, metal workers, toolmakers, engine builders, carpenters, engineers, laborers, factory hands, draftsmen, professors, and students would become soldiers and be organized into standing armies, the size and scope of which not even the professionals from West Point could foresee. The armies would need cannons, rifles, and ordnance. They would need infantry to fight and die, artillerymen to kill and maim, cavalry to scout and raid, and officers to lead and sacrifice. And, yes, they would need engineers: engineers to assist in getting at the enemy, engineers to deliver food and equipment through dangerous and often impassable terrain, engineers to build roads and bridges over distances unimaginable in April 1861. It was now time for the engineers to go to war.

Part Two

Skills Go To War

CHAPTER FOUR

Wanted: Volunteer Engineers

If mere volunteers can wear them [engineer insignia] they are no longer badges of distinction.

Captain James C. Duane, United States Army, Corps of Engineers, 1861

Just three months after the fall of Fort Sumter, new and untested Union and Confederate armies first drew blood thirty miles southwest of Washington, DC, near a railroad crossroads called Manassas Junction and a river called Bull Run. Both sides believed this clash of arms would settle the question of an independent Confederacy. It did not. Instead, both sides came to the harsh realization that the war could be a bloody and protracted conflict, requiring more men and materials than first imagined, as well as the development of complex strategic and logistical plans. Part of this logistical planning required some consideration of the role of military engineers. At the time of the First Battle of Bull Run, each side had engineer officers attached to army headquarters, a model from the United States' previous wars. But would this small cadre of officers be enough to meet the armies' growing needs? As the war began, only the Federal army had a company of engineer soldiers, and some questioned whether this would be enough. Both sides, therefore, had to decide where to invest their manpower resources. Should men recruited as infantry be converted to engineer troops? Would making engineer soldiers from infantry recruits weaken each army's ability to fight? Would these men need special skills? Who would lead them?

The formation of the 50th New York Volunteer Engineers provides insight into how the Union army tried to address these questions. Twenty-six days after Bull Run, on August 16, 1861, an article headed "A New Company from Rome" appeared in the *Rome (New York) Citizen*. Captain Wesley Brainerd, a member of the New York State Militia's Company "A" 46th Regiment, the Gansevoort

Light Guard, was recruiting a company of engineers. Charles B. Stuart of Geneva, New York, had promised a captaincy to Brainerd if he could enlist approximately one hundred men. Brainerd's father and Stuart had been in the railroad business together and were old acquaintances, so Wesley Brainerd received a favorable response from Stuart regarding the attempt to raise a company of soldiers. The newspaper article stated: "The Regiment is to be armed and equipped as a rifle regiment . . . But its main labors . . . will be the rebuilding of railroads and bridges destroyed by the enemy, the running of trains, or the performance of any other kind of mechanical and engineering work that may be necessary."[1]

Brainerd, like many men living in Rome, New York, had engineering skills. With only a common school background, he learned his mechanical skills while operating a saw and turning lumber, first for the manufacturing of railroad cars and then for the manufacturing of bedsteads and fanning mills. Eventually, he converted a building into a gristmill "making a fair living when the war broke upon us."[2]

Brainerd believed Rome was the perfect place to collect volunteers for an engineer company. "The style of men required are able-bodied and experienced military, civil and mechanical engineers, mechanics, boatmen, lumbermen and carpenters, farmers, and laborers. There are a considerable number of young men in this village and the vicinity, of the classes above indicated who should be glad to join the Engineer Corps."[3]

By mid-September 1861, Brainerd and the entire regiment of the 50th New York Infantry were encamped at Camp Lesley in Elmira, New York. Within a month, the unit would be converted into an engineering regiment and would see its first major action during General George B. McClellan's Peninsula Campaign. Before that happened, however, Union (and Confederate) military and civilian leaders would need to determine how they wanted to structure their forces based on the high commands' assessment and development of their strategic plan. Stuart's men had a role to play. At first this was a minor role, but it soon became a major one, a role essential to the outcome. Engineers had served a vital part in the military campaigns of the past and they were about to do so again. Northern war planners would figure out the mechanical and technical capacity of their men. Between 1861 and 1865, the engineers' role would expand and become indispensable to Northern victory.

This transformation from a limited to a larger role required generals and politicians to recognize the importance of engineering skills and technology; consequently, it required time to develop. One hurdle to overcome was to increase the number of engineers. This meant overcoming resistance to the idea that civilian engineers could perform well as military engineers. A few professional soldiers had the wisdom, by the time the Civil War began, to challenge the well-established notion that only West Pointers made good engineers. These same men understood that with expansion of the war, engineering operations required skilled laborers. It was not enough just to have trained officers. It was incumbent upon the army to enlist personnel with the mechanical skills and in-genuity to solve problems, often without adequate resources and in trying situ-ations.

At first, the War Department and Congress saw no need to expand the Union army's engineering forces. Most senior officers believed that the war would be over by the summer of 1861 and that the size of the Corps of Engineers was ad-equate for the great battle ahead. To everyone—the government, the army, the public—a clash of arms would decide the fate of the Union and Confederacy. President Lincoln had called for 75,000 ninety-day volunteers, based on the Militia Act of 1795, and then expanded the number on May 3 to an additional 42,000 three-year volunteers. Then he called for an increase of 23,000 men in the regular army, to which the Corps of Engineers and Corps of Topographical Engineers were attached. No additional engineer units, however, were autho-rized. The public's expectation and the administration's belief was that the Fed-eral army would march into Richmond and crush the rebellion. An undersized engineer corps did not matter as long as engineers could skillfully maintain har-bor defenses, as they did at Fort Pickens, which guarded the passage of ships to Pensacola Bay, and at Union-held forts along the southern Atlantic, Gulf Coast, and Great Lakes region.[4]

After May 23, 1861, when the people of Virginia adopted that state's ordi-nance of secession, the Corps of Engineers and Corps of Topographical Engi-neers became even smaller as engineers from Virginia went with their state, but no one in the Federal government or military appeared worried. Between May and July 1861, only forty engineers and thirty-seven topographical engineers were active in the service of the Federal army. Nine engineers and seven topo-graphical engineers had resigned their commissions and taken positions in the new Confederate army.[5] These small numbers, although not initially significant, eventually revealed problems that both the North and South would need to ad-dress: a significant shortage of engineer officers and troops. This shortage espe-cially taxed the South because there were just not enough civilian engineers or mechanics, artisans, and railroad workers to meet the ever-growing demands of the Confederate army for fortifications, bridges, and roadways.

The expansion of both armies was staggering. On Sunday, July 21, 1861, the

Union's General Irvin McDowell had approximately 28,000 men under his command from the Department of Northern Virginia and Washington as he prepared to attack approximately 21,000 Confederates at Bull Run. At the time of the attack, the Union army overall had approximately 187,000 soldiers spread throughout departments in Ohio, Kentucky, Missouri, and Florida. By January 1, 1862, the army had grown to 600,000, and by 1863, there were 900,000 men divided into fourteen army groups, ranging in size from 40,000 to 120,000 men, scattered across the Midwest, West, and South.[6]

Such vast concentrations of men and equipment required comparable logistical support. To move an infantry regiment of 1,000 men on a three-day march of fifteen miles a day, for example, required approximately 20 wagons carrying 3,000 pounds of camp equipage, ordnance, quartermaster stores, and medical supplies. An army of 100,000 required approximately 19,000 wagons for a march of twenty-five days. This included feed for about 33,000 animals.[7] Moving these armies over poor roads and rivers without bridges presented a daunting task. Forty engineers, thirty-seven topographical engineers, and one hundred engineer soldiers were a paltry sum to accomplish the strategic and tactical movements of the Union army over the 750,000 square miles of the Confederacy.

The Confederate army also grew in size and complexity. In July 1861, the South, from the Shenandoah Valley to central Kentucky to Pensacola, Florida, fielded seven small armies totaling 60,000 men. By 1863, the army had expanded to approximately 465,000 men in eighteen armies functioning as independent commands. Like its Northern enemy at the time of the First Battle of Bull Run, the South had only a handful of engineer officers. Unlike the North, however, the South had no engineer soldiers and little in the way of mechanics, carpenters, machinists, and factory workers to fill the ranks of the engineers.

The United States Military Academy, in the spring of 1861, had a monopoly on producing professional engineers, whether for the army or for civilian life. While West Point graduates of 1802 had served in the regular army as officers in the infantry, cavalry, and artillery, many cadets aspired to be assigned to the engineers. The Corps of Engineers was considered the most elite branch of the service. Beginning with the class of 1819, cadets received a class rank, and only the top three or four in the class qualified to enter the engineers. Between 1819 and 1860, eighty-seven graduates had earned the distinction of being placed immediately into the engineers. Sixty-eight of them came from a free state and nineteen from a slave state.[8] By the time of the Civil War, forty-eight officers were serving in the Corps of Engineers. Of the forty who remained in the Union army, twenty-nine served as engineers during the Civil War and thirteen served as field commanders. Of the eight engineers who joined the Confederacy, five served as field commanders.[9]

More remarkable was where these cadets were born and from which states they were appointed. Between 1802 and 1860, of the 108 graduates who served in the Corps of Engineers, 75 were from New York, Massachusetts, Pennsylvania, Ohio, and Vermont. Four of these states had adequate common school programs during the antebellum period, and one, New York, which produced 27 engineers, had developed an exceptional common school program. Furthermore, as industry and mechanization altered the economic landscape in the free states, Northern colleges and universities introduced more science into the curriculum. Northern boys of all classes, who had more access to education than their Southern counterparts, would be exposed to the benefits of studying natural philosophy, mathematics, chemistry, and physics. Increasingly, Northern educators and politicians would encourage and promote these courses in schools. It was not a coincidence that the number of Northern graduates and Southern graduates entering the engineers differed by a factor of close to four to one.[10]

Within the Corps of Topographical Engineers, of the seven men who resigned their commissions to join the Confederate army, only two held a rank above second lieutenant. Experience remained in the Federal army. This experience included West Point graduates who had served in other branches of the service before the war. These men amassed a wealth of engineering knowledge as a result of being transferred into the engineers or, after resigning their commissions, serving as civilian engineers or teaching science and engineering at their alma mater. For example, Montgomery C. Meigs and Daniel P. Woodbury (class of 1836) both entered the artillery after graduation and were transferred to the engineers. John M. Wilson (class of 1860) was another artillery officer transferred into the engineers. William S. Rosecrans (class of 1842) attended common school in Utica, New York, and then moved to Mansfield, Ohio, where he was appointed to the military academy. After serving as a professor of engineering, he opened a mining business in western Virginia around the time that eastern planter interests feared industrial competition.

It was another West Point engineer that the government turned to after the North's devastating defeat at Bull Run on July 21, 1861, to quell the panic, defend the city, and assume command of the Military Division of the Potomac.[11] Upon arriving in the capital, Major General George B. McClellan wrote to his wife Ellen: "I find myself in a new and strange position here—Presdt, Cabinet, Genl

Scott & all deferring to me—by some strange operation of magic I seem to have become the power of the land . . . I almost think that were I to win some small success now I could become Dictator or anything else that might please me— but nothing of that kind would please me—*therefore I won't* be Dictator. Admirable self-denial!"[12]

The Federal government could breathe a sigh of relief that General McClellan would not become dictator, but it did fear a Southern invasion of the capital. Manassas Junction was only thirty miles southwest of Washington. The door was wide open. Yet, in that moment, the right man had been called. McClellan would fortify the capital, expand the Corps of Engineers, reorganize the army, and serendipitously introduce volunteer engineer officers and soldiers into the Union army.

McClellan's understanding of military science was significantly influenced by the writings of Simon François Gay de Vernon, Antoine-Henri Jomini, Dennis Hart Mahan, and Henry Wager Halleck. Mahan, a student of both Vernon and Jomini, taught both Halleck (1839) and McClellan (1846) at West Point. As historian Edward Hagerman points out in *The American Civil War and the Origins of Modern Warfare*, McClellan accepted "the two innovations in a basically eighteenth-century view of warfare that responded to the realities of the mid-nineteenth century: Mahan's rejection of the open frontal assault for an offensive tactical organization that emphasized the primary role of field fortifications; and the general awareness of the strategic potential of railroads."[13]

McClellan was also influenced by his experience in the Mexican War and by what he had observed in the Crimean War. Both conflicts had supported his predisposition for field fortifications and turning movements rather than frontal assaults. In Mexico, McClellan was assigned to Company A, Corps of Engineers. On May 15, 1846, Congress had finally passed legislation to create an engineer component from enlisted personnel of the regular army.[14] McClellan witnessed the important work of engineer soldiers and saw the tactical effectiveness of turning movements, especially at Cerro Gordo and Contreras. He also learned the destructive effects on attackers in a frontal assault on an entrenched position. He concluded that one of the reasons Americans succeeded in that war was that the Mexican fortifications were poorly constructed.[15]

In 1855, with Major Alfred Mordecai of the Ordnance Department and Major Richard Delafield of the engineers, Captain McClellan had traveled to Europe and was a military observer at the siege of Sevastopol.[16] Here McClellan noted the significance of rifled weapons in a frontal assault and, consequently, was able to see with crystal clarity the central principles he would adopt as he was handed command in Washington seven years later: defense wins wars, solid entrenchments and fortifications are the critical elements of defense, frontal attacks upon the enemy are to be avoided at all costs, and excellent artillery must accompany successful siege operations. Engineer troops were essential to the army in order to conduct the type of warfare to which McClellan was philosophically bound.

McClellan's vision of a war dominated by field fortifications and entrenchments, planned and guided by professional military engineers, no doubt was on his mind when he received word that on August 3, 1861, Congress had authorized three additional engineer companies, which brought the strength of the Corps of Engineers to 49 officers and 550 enlisted men.[17] As a result, Company B, recruited in Portland, Maine, and Company C, recruited in Boston, Massachusetts, became part of the provisional Engineer Battalion under the command of Captain James C. Duane. It was at this time that Duane began writing his *Manual for Engineer Troops*, which would be used along with extensive field training to prepare the new battalion for the campaigns ahead.[18]

Infantry commanders were skeptical of the engineers because the infantry wanted to take the offensive and fight, whereas the perception was that engineers wanted to take the defensive and dig. The administration, Congress, and members of the press believed McClellan's focus on the engineers might foreshadow a less aggressive approach to waging war against the South rather than emphasizing a hell-for-leather charge to the front. McClellan's Peninsula Campaign in May–June 1862 did reveal that Lincoln's concerns about McClellan's recalcitrance were well founded. The administration and Congress by then wanted action. Eager to take the war to the rebels and deliver a crushing victory, Lincoln and Congress were reluctant to sacrifice manpower resources for engineering troops. This desire to take the war to the Confederacy helps explain why Lincoln continuously encouraged the general to go on the offensive, and why McClellan, the professionally trained military engineer soldier, considered these overtures anathema. He wrote to his wife on October 11, 1861: "I can't tell you how disgusted I am becoming with these wretched politicians—they are a most despicable set of men & I think Seward is the meanest of them all—a meddling, officious, incompetent little puppy—he has done more than any other one man to bring all this misery upon the country & is one of the least competent to get us out of the scrape. The Presdt is nothing more than a well meaning baboon."[19]

Although an invidious narcissist and a self-styled savior of the Union, McClellan did understand the need for more engineer officers and troops, and it was here that he contributed most to the outcome of the war. In September 1861, the army had lost fifteen officers to the Confederacy (both engineers and topographical engineers), and the remaining officers were assigned to various staffs, coastal defenses, or prewar civilian internal improvement projects.[20] In addition, at least seven men left the engineers to accept line commands. So by October 1861, McClellan found he lacked engineer officers and men to expand the project to fortify Washington, begun in May and made critical by the results of Bull Run and the belief that a Confederate attack was impending—from at least 150,000 rebels, McClellan believed. Eventually, sixty-eight forts, ninety-three batteries, and twenty miles of rifle entrenchments were built to protect Washington.[21]

Major John G. Barnard, superintendent of West Point in April 1861 and an apostle of Dennis Hart Mahan, was given the responsibility of designing and overseeing the building of the first forts around Washington. The results were Fort Corcoran, which commanded the approaches to the Aqueduct Bridge; Fort Runyon, built at the northern end of the Chesapeake and Ohio Canal and guarding the approaches to the Long Bridge; and a smaller Fort Albany, about one mile west on the Columbia Turnpike.[22] Also, a chain of lunettes was built, known as the Arlington Line. These forts—De Kalb, Woodbury, Cass, Tillinghast, and Craig—faced southwest and connected forts Corcoran and Albany.[23]

Pleased with Barnard's work, General McClellan promoted the major to brigadier general of United States Volunteers and chief engineer of the newly formed Army of the Potomac.[24] "Little Mac" (the men's affectionate nickname for their beloved general) then handed Barnard an additional task beyond the continued construction of the Washington defenses: to build portable bridge trains. The army had no bridging equipment except for India-rubber pontoons left over from the Mexican War, which McClellan deemed unsatisfactory.[25] So, with no bridging equipment or tools, with forts under construction, and with an inadequate number of engineer troops for the job, McClellan ordered Barnard to construct the new-model French bateau pontoons and the wagon trains to haul them. Barnard, in his turn, assigned Lieutenant Colonel Barton Stone Alexander to prepare the necessary equipment for the Army of the Potomac, an assignment that was to prove profoundly important.[26]

At the same time, Secretary of War Simon Cameron was informed that although Congress had authorized two additional companies of engineer soldiers for the Engineer Battalion, they had yet to arrive in Washington. They would not arrive until mid-December. McClellan needed manpower, and he boldly suggested to the secretary that volunteers should be assigned to the engineers.[27]

This suggestion crossed an important line, whether Little Mac knew it at the time or not. Military engineers were the elite of the army because their technical skills were highly prized by commanding generals, but also because, unlike other branches of the service, engineers had performed valuable assistance to the nation in working on critical internal improvement projects throughout the United States and the Territories. Now McClellan was requesting engineer troops from volunteers, which he would get, but along with these new soldiers he would get volunteer engineer officers as well. The army's Corps of Engineers, Hagerman noted, would need to "give way to the practical need for broader perspectives" in a war that was to demand "ad hoc improvisation."[28]

The corps' unique and elite role within the army was understandably accompanied by a large amount of self-aggrandizement. When Captain Duane told his men not to wear the insignia of the engineers because, if volunteers could wear it, it was no longer a badge of distinction, he was expressing a feeling held by many West Point engineers. Regular army engineer officers and men were better than any other soldiers. "Why," said Duane, "President Lincoln can make a brigadier general in five minutes, but it has taken five years to make . . . an engineer soldier."[29]

Fortunately for the North, in the years before the war, new ideas in technology and industrialization developed by men from all walks of life had altered the notion that ideas came from one class of society and laborers from another. McClellan was willing, under the circumstances and out of necessity, to expand the engineer battalion with men who had enlisted as infantry but who the general believed had developed, as civilians, the basic skills and intelligence necessary to perform engineering tasks. The man who would bring the value of the volunteer engineer into sharper focus and establish the military doctrine that the army would follow to the successful conclusion of the war was Lieutenant Colonel Alexander.

One day after McClellan wrote to the secretary of war suggesting volunteers as engineers, Alexander sat down at his desk and penned a visionary letter. Writing to Barnard, Alexander noted that "we have as yet no bridge equipment, no engineer trains, and no instructed engineer troops . . . What, then, are we to do? . . . The answer must be, however, we must make them. Our country is full of practical bridge-builders. We must secure their services." Alexander was convinced that men who possessed mechanical talent could be found within the ranks of volunteers originally recruited as infantry regiments. With roads and railroads to build and repair, telegraph lines to install, and bridges to construct and destroy, Alexander argued that men "with previous pursuits" that made them mechanically inclined would make excellent military engineers. The India-rubber pontoon bridge could be used and perfected "if a proper proportion [of volunteers] are sailors," and skilled carpenters could rapidly make canvas pontoon boats. The Quartermaster's Department had a hundred corrugated-iron wagons that Alexander believed could be converted into a bridge. All of this work, Alexander concluded, would give the engineers several types of bridges that could be used, depending upon the circumstance.[30]

The following day, General McClellan wrote to Assistant Secretary of War Thomas A. Scott to say that after considering the matter more fully, he concluded that the time and manpower to construct bridge trains was limited. Therefore, McClellan continued, "It is necessary to avail ourselves at once of all the resources which the mechanical skill and ingenuity of the country can furnish in this matter." He concluded his message to Scott with his second request to the War Department to secure the services of "such regiments of volunteers or such portions of regiments as may prove best adapted to the duty."[31]

Scott proved to be an excellent choice as assistant secretary of war. As vice president of the Pennsylvania Railroad in 1860, Scott had worked alongside the company president, J. Edgar Thomson, and witnessed the development of Thomson's innovative management system, and Scott proved to be a skillful administrator. He responded to McClellan the following day, October 15, with an authorization reflecting the management practices of the railroad: "You have full authority to detail the whole or parts of volunteer regiments for engineer service, and will exercise your own discretion in relation thereto."[32]

Eleven days later, the 15th and 50th New York Volunteer Infantry received orders to report to Washington to become the 15th and 50th New York Volunteer Engineer Regiments. Upon his arrival at the capital, Captain Wesley Brainerd of the 50th wrote: "We crossed the Long Bridge for the first time and marched to our ground near the Navy Yard on the East Branch of the Potomac, sometimes called the Anacostia River. Here we commenced the establishment of our permanent camp."[33] Both of these regiments were selected because New York State had a considerable number of skilled mechanics, carpenters, blacksmiths, masons, and civil engineers within their ranks. For example, in the 50th New York, Private Horace Herion was a mechanic; Isaac J. Cox, a carriage maker; James B. McGregor, a blacksmith; and Francis S. Newton, a carpenter.[34] The outfit enrolled men from areas affected by the railroad and Erie Canal, including the counties of Elmira, Geneva, Fulton, Oswego, Potsdam, Rochester, Albany, and Buffalo. New York also had a tradition of establishing engineer troops.

Eight militia regiments that enlisted for ninety days in April 1861 all had engineer companies attached to them.[35]

The 15th Volunteers were organized as sappers and miners at New York City, Long Island, and Newark, New Jersey, under the command of Colonel John McLeod Murphy. Murphy had served as a midshipman during the Mexican War and then, in civilian life, worked as chief engineer for the Brooklyn Navy Yard. At the time the Civil War began, he was also in the New York State Senate. In the late summer of 1861, the Federal government mustered Murphy's regiment into the Army of the Potomac as an infantry regiment. The average age of these men was twenty-six, with the youngest being eighteen and the oldest forty-two. Many were teamsters, dock builders, boilermakers, masons, and mechanics.[36]

Before the two regiments reported to Lieutenant Colonel Alexander in Washington to begin their training, another volunteer engineer regiment from the East was already in the field. Edward Wellman Serrell had offered to organize an engineer regiment to supplement the engineers in the regular army. Serrell had considerable political connections and the reputation of being a first-class civil engineer. Before the war he had served as assistant engineer to the commission on the Erie Canal. As assistant to the chief of the Corps of Topographical Engineers, he was involved in the Hoosac Tunnel project in western Massachusetts, and he had planned and supervised the building of a suspension bridge across the Niagara River at Lewiston and one at St. John, New Brunswick. Serrell spoke with Secretary of War Cameron about recruiting an engineer regiment, which Cameron approved, as did the governor of New York, Edwin D. Morgan. Mustered in on October 11, the newly minted Lieutenant Colonel Serrell found himself the commanding officer of the 1st New York Volunteer Engineers, also known as Serrell's Engineers. Most of the men in the first five companies were from New York City (all the members of Company K, mustered in on December 3, were from northern New Jersey), and on November 8 they began operations to capture Port Royal, South Carolina, near Hilton Head Island.[37]

While Serrell's regiment sailed from Fort Monroe, Virginia, to Hilton Head Island, South Carolina, Colonel Stuart's (50th) and Colonel Murphy's (15th) soldiers and officers were learning their trade under the watchful eyes of regular army engineers. They built wooden French-type and canvas Russian-type pontoons, learned how to construct corduroy roadways, and built field fortifications.[38] With repeated practice they learned how to build a pontoon bridge over a river.[39] Finally, the engineers practiced the art of making fascines and gabions for siege operations.[40]

For the remaining months of 1861, the New York men trained hard in anticipation of General McClellan's great campaign, expected to commence in the early spring. Many members of Congress demanded action. In late December 1861, when Barnard asked for an additional $150,000 to complete the fortifications surrounding the capital, several politicians rebuked him. Senator Preston King said: "I would not expend an additional cent on the fortifications of Washington. In my opinion, the best defense for Washington is the destruction of our enemies where they can be found—at a distance from Washington."[41]

Lincoln and his advisers understood, however, that any military effort to defeat the Confederacy required more than concentration in the Virginia Theater of operations. Union strategy had to include controlling the border states of Missouri and Kentucky and then. from those launch points, attacking south into the heartland of the Confederacy. The president's strategic thinking was at an embryonic stage in the fall of 1861, yet he recognized four important objectives: (1) the concentration of troops from Ohio and Kentucky for a movement on east Tennessee (a pro-Union area); (2) consideration of simultaneous advances along the Mississippi, Tennessee, and Cumberland rivers, making it difficult for the Confederacy to defend two or three places at once; (3) efforts to establish areas where men and supplies could be moved quickly to an extended battlefront and to prevent the Confederacy from doing the same; (4) direct action to pacify Missouri.[42] Lincoln's generals—Henry Halleck in the west and George McClellan in the east—agreed with this overall strategy. To accomplish it required moving men and material over great distances, employing railroad and water transportation. Engineers would be required to build bridges and repair roads, tracks, rolling stock, and engines. McClellan had adopted the regular army's four engineer companies and converted two volunteer infantry regiments into volunteer engineering regiments. For the armies in the West, the engineer soldiers would all be volunteers, as would many of their officers, and when these men were not enough to complete the tasks assigned to them, infantry regiments would be drafted for the work and formed into detached companies known as pioneers.

A strike at the Confederate heartland west of the Appalachian Mountains required the consideration of four possible routes of invasion. Three of these approaches, the Mississippi, Tennessee, and Cumberland rivers, provided the North with lengthy supply lines protected from partisan raiders and cavalry. The fourth potential access point was along the Louisville & Nashville Railroad, with a line in Nashville connecting to Chattanooga. From there the army could advance northeast toward Virginia, southwest into Alabama, and south and southeast into Georgia.[43]

In the South, President Jefferson Davis assigned to General Albert Sidney Johnston the task of preventing the Yankees from punching through the Confederacy's imagined defensive line. Johnston was the quintessential soldier— handsome, brave, intelligent, and experienced. He had served in the Black Hawk War and the war for Texan independence and had led the Utah invasion against the Mormons.[44] He was the highest-ranking officer in the United States army to resign his commission and join the Confederacy. He became the second-ranked full general in the new Southern army.[45]

Johnston's strategic defensive line included building forts along the major rivers—at Columbus, Kentucky, on the Mississippi, 175 miles south of St. Louis; Fort Henry on the Tennessee; and Fort Donelson on the Cumberland. Finally, he placed Brigadier General Simon Bolivar Buckner and approximately 25,000 men at Bowling Green where the Louisville & Nashville Railroad met the Memphis & Ohio Railroad. Johnston's plan to cover the entire Kentucky and Tennessee border with an undersized army was textbook Jomini. Johnston operated on interior lines of supply and communication, which meant that from Bowling Green he could quickly shift resources and manpower along the Memphis & Ohio to forts Henry and Donelson, and using two additional lines, he could send men and material to Columbus. All the critical points that blocked an invasion force were linked together by the railroads over a distance of 163 miles.[46]

Missouri was not among the four fingers that the North could use to reach into the center of the Confederacy, yet it was crucial that the state remain in the Federal government's hands. The Union controlled the Missouri River and the important railway hub of St. Louis. The lines extended west, east, and southwest. In the spring of 1861, however, Missouri Governor Claiborne Jackson offered assistance to the newly formed Davis government, as Confederate soldiers pressed along the Arkansas and Missouri border and occupied most of the southeastern portion of the "Show Me State." Francis Preston Blair Jr., brother of Lincoln cabinet member Montgomery Blair, worked closely with Brigadier General Nathaniel Lyon to maintain control of St. Louis and stall for time as Blair asked Washington for help. Help arrived in the person of Major General John C. Frémont, the new military leader of the Department of the West.[47]

Frémont had no engineer troops to repair railroad tracks, cut new roads, or build fortifications.[48] In July, Colonel Josiah Wolcott Bissell proposed to Frémont the establishment of an engineering regiment with men, "either mechanics, artisans, or persons accustomed to work as laborers under mechanics," recruited from Missouri, Iowa, and Illinois. Before the war, Bissell was employed as an engineer on several projects designed to improve roadways in the West.[49] He suggested to Frémont that to induce men to join the engineers, the men should be promised regular army soldiers' pay plus extra daily pay of forty cents for mechanics and twenty-five cents for laborers. The general agreed. This decision on Frémont's part was one of several that would get him in trouble with the War Department. A considerable amount of money passed through the Western Department's hands, and Washington called into question the department's expenses, including pay and contracts. In 1861, the three New York volunteer engineer regiments were paid as infantry, so naturally the army questioned Frémont's motives.

Frémont, like McClellan in the East, did have extraordinary respect for and appreciation of the work of engineers. As a member of the Corps of Topographical Engineers, he had spent seven years (1838–45) exploring and mapping the Rockies and California. It was during this time that he gained national prominence for his great adventures and was affectionately called "the Pathfinder." His chief-of-staff in 1861 was an engineer—Alexander Asboth.[50] General Frémont also understood the complex military situation he faced in the summer of 1861. St. Louis was threatened from the south and west by secessionists. Tennessee, Kentucky, Unorganized Territory (present-day Oklahoma), and Arkansas all bordered Missouri. St. Louis enjoyed railroad connections with Jefferson City and St. Joseph, located in the northwest corner of the state, but those lines dead-ended, and supplies coming from Kansas could arrive only by wagon trains. A direct line connected St. Louis to Cincinnati, but moving through lower Illinois might be problematic because many proslavery sympathizers occupied the southern region of Lincoln's home state. Men who could build roads, construct fortifications, and repair bridges were essential to maintain the flow of men and material into St. Louis and provide the Pathfinder with the resources needed to hold onto Missouri.

Bissell's Engineers, or the Engineer Regiment of the West, was composed of ten companies, with only three companies, A, D, and G, from St. Louis or Cape Girardeau County, Missouri. Three were from central Illinois, three from Iowa, and one from Michigan. By August, several companies made up of iron molders, railroad engineers, mechanics, and laborers were scattered principally in east St. Louis, working on fortifications there, or 115 miles south at Cape Girardeau along the Mississippi, building another fort to protect St. Louis from a land or water invasion by the Confederacy.[51] While the engineers were busy with their duties of cutting trees, building sawmills, leveling decrepit roads, and moving dirt, General Frémont's indecisiveness became a problem and, as a result, intensified the workload of Bissell's men.

After the Battle of Wilson's Creek on August 10, 1861, orders directed the Engineer Regiment of the West to operate at a frenetic pace. Company B recorded that it "had been engaged very nearly the whole time for the past two months [September and October] in working upon fortifications at this post [Cape Girardeau]."[52] The engineers mounted guns, built blockhouses, and welded iron to rim wheels to keep the railroads operating. As Bissell described, "The regiment had just completed a railroad bridge across La Mine River opening the Pacific Railroad to Sedalia . . . A detachment of two companies was at Jefferson City for ten days, building extra track and extra storehouses for facilitating military operations."[53] For the remainder of the year, the men of Bissell's Engineers helped prepare for an expected Confederate attack led by General Sterling Price, but none came. The forts were built and the railroads operated bringing supplies and manpower to Missouri. In December 1861, Missouri was under Union control, and the engineers had received hands-on training that prepared them well for the campaigns and demands to come.

Whether Kentucky would remain loyal to the Union in 1861 also caused President Lincoln considerable worry. On September 3, Confederate Brigadier General Gideon J. Pillow seized Columbus and violated the state's neutrality. Ulysses S. Grant responded by occupying Paducah. Both sides now began to recruit volunteers, and General Johnston established his extensive but thin defensive line in an attempt to prevent Northern incursions into Tennessee. The eastern portion of the Confederate line was anchored at Bowling Green, but Major General George B. Crittenden with his 4,000 men from the Military District of East Tennessee was ordered to guard the Cumberland Gap, the entrance into pro-Union east Tennessee.

Three major transportation hubs ran almost due east of Bowling Green: Glasgow, Columbia, and Somerset, Kentucky. A road network connected Somerset to Lexington to the north, London to the east, Columbia to the west, and the Cumberland Gap to the south. London also connected with Lexington to the northwest and the Cumberland Gap to the south. Lexington was tied to the state capital at Frankfort, to Louisville, and to Cincinnati. A supply line for any Union operation in eastern Tennessee depended on control of Somerset and London.

With rumors flying of impending doom, panic settled on Union-controlled Kentucky. Robert Anderson of Fort Sumter fame, now promoted to brigadier general and commander of the Department of the Cumberland, headquartered in Louisville, on September 19 wrote in desperation to an independent battalion of riflemen in Cincinnati, asking them to come quickly because "Kentucky has no armed men whose services I can command."[54] Anderson believed he was in great peril from Confederate forces prepared to strike Louisville, Frankfort, and Lexington, and to lay waste to the area around the Cumberland Gap. A calculated assessment of the situation was needed immediately to determine how a limited number of men should be deployed to do three things: (1) halt any Confederate attempt to capture three critical Union supply depots, (2) prevent the enemy from establishing a base of operations for a movement into Ohio, and (3) maintain some initiative by occupying areas closer to the Cumberland Gap that might support a movement into eastern Tennessee and pose a threat from the east to Confederate-occupied Bowling Green. It was a tall order, so the War Department sent an engineer.

On the day Anderson sent his message to Cincinnati, the War Department assigned Brigadier General Ormsby M. Mitchell to command the Department of the Ohio. Mitchell, an 1825 graduate of West Point, had already enjoyed a remarkable career. Besides his time as an infantry officer, he was assistant professor of mathematics at West Point and later became assistant professor of mathematics and philosophy and professor of astronomy at Cincinnati College. As a civil engineer before the war, he had worked on the Ohio & Mississippi Railroad and raised the money to build an observatory that held the second-largest refracting telescope in the world.[55]

Mitchell's keen-witted grasp of the situation was just the tonic needed to bring pro-Union forces to action. Within six days of his arrival, he wrote to the assistant adjutant-general of the army: "I deem the immediate occupancy of Kentucky as a matter of the greatest importance and the fall of Louisville as a disaster the consequences of which cannot be overestimated."[56] He immediately sent the few regiments he had to secure lines of communication by rail to Louisville and Cincinnati. Some were hurried to Lexington and Frankfort, and others were used to begin building fortifications around Cincinnati. Once Louisville was secure, Mitchell operated toward the Cumberland Gap to threaten Nashville and "commence active and immediate operations to drive Zollicoffer [Confederate Brigadier General Felix K. Zollicoffer] and Breckinridge [Confederate Brigadier General John C. Breckinridge] out of the State or to capture them."[57]

Among the Kentucky volunteers pouring into the camps, one outfit was Captain William F. Patterson's company of engineer soldiers. Known as Patterson's

Independent Company of Volunteer Kentucky Engineers, this unit might well have been folded into a volunteer infantry regiment, where men were needed, had Mitchell not intervened. The general understood the critical role that communication links would play when the army went on the offensive. Engineer troops would be necessary to repair roads and build defenses around potential advance supply depots. The army had to build these depots if it had any thoughts of moving through the Cumberland Gap and occupying eastern Tennessee. Patterson's engineers most likely trained at Camp Dick Robinson, seven miles north of Lancaster and astride the Wilderness Turnpike, sixty-five miles north of the Cumberland Gap.[58]

In October, Southern forces occupied Barbourville, a town sitting on the Wilderness Turnpike just ten miles from the gap. Lincoln was concerned that with the Virginia & Tennessee Railroad south of the gap open to Confederate operations as a vital supply line, Kentucky remained vulnerable to Rebel attacks from the southeastern part of Kentucky. Lincoln and Mitchell both wanted Zollicoffer driven from Barbourville and eventually from the state. In November, Union forces were successful in pushing the Confederates west, away from the turnpike and toward Somerset; Patterson's men were ordered to build defenses at Camp Hoskins and Somerset in preparation for another Union strike at Zollicoffer. Patterson's engineers would remain in Somerset for the remainder of the year.

If Kentucky's start to the Civil War in 1861 appeared confused, disorganized, and inchoate, and many of its leaders seemed nonplussed (Mitchell and Thomas were welcome exceptions), other states also experienced muddled situations that autumn. This was certainly true in Michigan, where men rushed to recruiting stations to sign on as volunteers to teach the secessionists a hard lesson. The chaos of Northern mobilization was especially evident in recruiting volunteer engineers. By January 1, 1862, Michigan's governor had sanctioned and the US War Department had mustered into service a unit that would claim great notoriety by the end of the war: the 1st Michigan Engineers and Mechanics. The five-month journey the engineers experienced from the first recruitment announcement to the first assignment was another example of how difficult it was to build the Union army and how fortunate the army was in choosing to form volunteer engineer companies and regiments.

In the summer of 1861, James W. Wilson of Chicago had begun recruiting a regiment of engineers, and because Michigan had organized more infantry companies than the War Department authorized, men of mechanical ability sought out Wilson's regiment. By September, companies for the Chicago regiment were being raised in Ionia, Marshall, Albion, and Grand Rapids, Michigan. In addition, Edwin P. Howland of Battle Creek had organized a company of engineers called the Battle Creek Corps, which proceeded to St. Louis, Missouri, and was mustered into the army on October 9, 1861.[59]

Two men recruiting the Grand Rapids company, one a surveyor and one a master carpenter, met with others in mid-September to decide whether it "would not be better to raise an entire engineer regiment within the state."[60] They agreed to ask a prominent civil engineer, William Power Innes, to organize and expand the efforts and to become the regiment's colonel. He agreed.

Innes sent telegrams to Secretary of War Cameron, asking permission to form the regiment, and to Michigan Governor Austin Blair, requesting that he send an endorsement to Cameron. Blair told Cameron that Innes was the general superintendent of the Grand Rapids Railroad, and this probably sealed the deal. Before the war, Cameron was involved in creating the Northern Central Railroad in Pennsylvania, and his assistant secretary, Thomas Scott, was vice president of the Pennsylvania Railroad—at that time the largest corporation in the world.[61]

In fact, Innes was not the superintendent of the Grand Rapids Railroad, but he was an ambitious engineer with a wealth of experience in the railroad business, not all of it worthy of celebration. His engineering skills were learned as a laborer on the Erie Railroad and as a civil engineer for the Oakland & Ottawa Railroad, and he was responsible for a line running from Ada to Lake Michigan, completed in 1858. During this time he had also served as chief engineer for the "prospective Grand Rapids and Northern Railroad Company." Finally, he had contracted the Amboy, Lansing, and Traverse Bay Railroad to build a line from Owossa to Lansing. As Innes surveyed the road, he discovered that a massive sinkhole blocked the route, and he convinced the owners of the company to invest considerable capital in an effort to span it. The project was suspended in 1860, and instead, Innes built a wagon road to carry passengers from the track for the remaining few miles to Lansing. The company was furious with Innes for the money wasted in trying to cross the sinkhole, and he was fired.[62]

That failure aside, Innes had considerable engineering skills, as did the officers and men under him. His officers included civil and railroad engineers, carpenters, surveyors, and a mechanic. In *"My Brave Mechanics,"* Mark Hoffman provides a superb statistical summary of the 1st Michigan. In 1861, fifty-eight percent of the recruits were originally from New England or New York, approximately twenty percent were from Michigan, and fifteen percent were foreign born. Fifty-two percent of men enrolled in the regiment were mechanics or artisans, and about thirty-five percent were either farmers or laborers.[63]

During the fall of 1861, the officers of the 1st Michigan continued to recruit soldiers to fill company rosters, and the men continued their training as engineer soldiers. By Christmas, General Halleck told the Battle Creek Engineers in St. Louis that they did not conform to Federal standards, and the men voted to disband, as did an undersized unit calling itself Chadwick's Engineers, from Marshall, Michigan. All that remained was the 1st Michigan, and the regiment finally received its first assignment. Kentucky was regarded as the state where the Union army in the West would launch its early 1862 offensive. So the Michigan men, eager to begin their work, were sent to the Blue Grass State and deployed to several locations, including Brigadier General George H. Thomas's division in Somerset.

As the Union army groped its way through the fall and early winter of 1861–62, it sought a plan to drive Confederate forces out of Kentucky and begin offensives against Tennessee and Virginia. The nascent Confederate army was also trying to build a defense perimeter along the border of northern Virginia and the state's coastline and west from Columbus, Kentucky, to the Cumberland Gap. Like commanders in the North, Southern commanders were in want of skilled engineers that would allow the government in Richmond along with the governor of Tennessee and General Johnston to begin constructing permanent fortifications. Unfortunately, given the prewar shortage of Southern-trained engineers and the few West Point engineers who had resigned their commissions to join the Confederacy, there were not enough engineer soldiers and officers to meet the need. For example, in November 1861, the secretary of war wrote to General Joseph E. Johnston requesting an engineer be sent to General Thomas J. (Stonewall) Jackson in western Virginia. "General Jackson is urging me to send him an engineer, and I have not one at my command. Have you one that you can possibly spare him?" Johnston responded the following day: "We have but one engineer officer, who is sick. We require more."[64]

The Confederate Congress had confirmed President Jefferson Davis's first appointments of engineer officers on March 16, 1861, and all were West Pointers. No orders or accommodations were made for engineer troops. Furthermore, these initial appointments included only two majors and five captains, leaving the role of chief engineer (a colonelcy) vacant. No engineer department was established. Major Josiah Gorgas, chief of the Confederate Ordnance Department, served as "Acting Chief Engineer."[65]

Gorgas found managing two major responsibilities very trying. Demands for engineers came from Tennessee, North Carolina, Georgia, and armies in the field. On April 19, 1861, North Carolina Governor J. W. Ellis telegraphed President Davis: "I am greatly in need of an engineer and artillery officers." With virtually no assistance from Montgomery, Alabama (the first Confederate capital), and later Richmond, governors turned to state engineers and local civil engineers, many of whom were incompetent. Gorgas complained to the Confederacy's first secretary of war, Leroy P. Walker, that he could not effectively perform both jobs. An engineer officer was needed to command the Engineer Bureau. Major Danville Leadbetter was finally ordered to Richmond in August to take control of the bureau as "acting chief."[66]

Immediately after Virginia seceded from the Union in May, the governor, John Letcher, took steps to build a defensive system using state engineers. Andrew Talcott staked out his chosen locations for batteries along the James River; Thomas H. Williamson marked defenses for Aquia Creek, the terminus of the Richmond, Fredericksburg & Potomac Railroad; and Alfred L. Rives and Richard K. Meade designed a defensive line on the peninsula, southeast of Richmond. Simultaneously, Talcott began to map out and construct the defenses of Richmond.[67] The city council's "Committee on Defense" was responsible for providing the workforce, and the provisional Confederate army was assigned the task of providing engineers to supervise the redoubts and entrenchments. Unfortunately, there were just not enough engineers to manage the construction. An Engineer Bureau memorandum dated October 28, 1861, listed just five officers supervising the defensive works in Richmond, while thirty-one others were assigned to other projects throughout the state.[68] There was also a shortage of laborers to build the fortifications.

On July 23, Governor Letcher asked an officer to discharge a Private George P. Hughes from his current duty station so "that he may be employed as overseer on the Richmond defenses."[69] The following day, engineers requested two balls and chains "for the benefit of runaway negroes," and the Engineer Department's "Slave Rolls" for July through October indicated that the vast majority of men who worked on the Richmond defense were slaves. Men named Buck Woods, Willie, Jack, Charles, and Bird worked twelve to fourteen hours a day for fifty cents.[70] Anywhere between forty and seventy slaves were working on the defenses at one time. Fewer than ten white laborers worked on the fortifications, earning $1.00 per day. There were usually an equal number of white and black carpenters, the former earning $2.00 and the latter earning $1.00 per day. Slaves who had no shoes were provided with a pair and had $2.50 taken from their pay. Finally, the Engineer Department listed six overseers on its rolls, each paid $40 per month.[71]

By December 1861, the Engineer Department appeared to be more concerned about reimbursing owners of slaves "at the rate of 70 cents per diem—50 cents only allowed when rations are furnished" than completing the defenses around the capital.[72] As the war began to heat up, Confederate organizers formed infantry regiments, artillery batteries, and cavalry battalions, but there were few skilled engineer officers to work on the fortifications along the coastline, to construct defenses in cities, and to serve in the field with the army. No pontoon trains were being built and no engineering companies formed. Moving dirt was to be the job of African American slaves. In the antebellum South, as we have seen, education was reserved for the elite, and those men with technical knowledge educated as engineers, especially at West Point or Virginia Military Institute, were considered even more highly educated than most. The paradox here was that most of society's wealthy made their fortune from the land and ensured that engineering and manufacturing did not interfere with the enormous wealth generated by plantation farming. Engineers were valued, yet there was limited incentive to become one. In *Staff Officers in Gray*, Robert E. L. Krick points out that of the 1,149 men who served as staff officers under Joseph E. Johnston and Robert E. Lee (both West Point–trained engineers), only 87 were engineers before the war. Of these, three were machinists, three were ironworkers, and four were carpenters.[73]

Out West in 1861, Confederate engineering operations were as problematic as they were in the East. General Albert Sidney Johnston's Army of Tennessee was not just establishing a defensive line to prevent Federal forces from invading the South west of the Appalachian Mountains; it was also protecting the Confederacy's largest producer of pig iron and bar, sheet, and railroad iron in the region.[74] There were more than seventy-five furnaces and forges within the "fifty-mile-wide belt" of the Cumberland and Tennessee rivers, where several hundred white laborers and thousands of slaves manned the operations.[75] Yet, interference from the Tennessee governor, poor engineering decisions, and a lack of interest in using white laborers to build defenses would cost the Confederacy the entire region and help make the first Northern hero of the war, Ulysses S. Grant. His reputation established, he would go on to defeat Confederate armies in the West and East and, in doing so, would put his own engineers and manpower to brilliant use.

The governor of Tennessee, Isham Harris, was a remarkable help and hindrance to the Confederacy in the summer and fall of 1861. The governor worked tirelessly and with considerable success to raise a fighting force, which consisted of twenty-four infantry regiments, ten artillery batteries, an Engineer Corps, Quartermaster and Ordnance departments, and an Ordnance Bureau.[76] He turned all of this over to the Confederacy and General Johnston. Like many politicians, especially in the South, Harris believed in his incontrovertible ability to understand and execute military strategy, and it was this characteristic that unlocked the monster within.[77] Harris believed in his ability to command, and he thought that his success in recruiting soldiers earned him that right. Placing great trust in General Gideon Pillow's assessment that because of Kentucky neutrality, Union forces would strike south from western Tennessee, Harris concentrated his forces at forts under construction along the Mississippi, leaving middle and eastern Tennessee vulnerable and defenses along the Cumberland and Tennessee rivers neglected. This was a serious blunder.[78]

In the spring, before Tennessee adopted its Ordinance of Secession, Harris had requested two civil engineers, Adna Anderson and Wilbur R. Foster, to identify sites for fortifications along the rivers. Anderson had served as an engineer for railroads in Connecticut and New Hampshire and was chief engineer of the Tennessee & Alabama Railroad and superintendent of the Edgefield & Kentucky. When the State of Tennessee voted to approve secession on June 8, 1861, Anderson offered his services to the Federal government and would soon become assistant engineer and chief of the Military Railroad Bureau's construction corps in Virginia. He would serve the US government with distinction throughout the war. Foster went on to serve in the Confederacy's corps of engineers.

So, when working for the State of Tennessee before secession, Anderson selected a site on a steep hillside near the town of Dover, overlooking the Cumberland River. The fort, protected by deep gullies, was started immediately with earthworks dug by slaves from the Cumberland Rolling Mills. After a "careful examination and study of all the topographical details," wrote Foster, "the first, or water battery at Fort Donelson was located by Mr. Anderson."[79] Anderson and his party then crossed the river and, moving west, paid studious attention to the flatlands between the Cumberland and Tennessee rivers. Accessing the high-water mark along the flood plain was crucial to identifying the most feasible location for the second fort. Anderson selected a site opposite the mouth of the Big Sandy River, a tributary of the Tennessee.[80]

The engineer lacked a labor force, so he returned to Nashville and presented his recommendations to Governor Harris. As it turned out, the site was an excellent location for a fort, but perhaps because the governor had started to question Anderson's loyalty, he wanted a second opinion. So he sent Brigadier General Daniel S. Donelson, a West Pointer (class of 1825, who resigned his commission in 1826), to investigate the situation and report back. Donelson's qualifications for this mission included serving several years in the state legislature, running a large plantation, and serving as brigadier general in the state militia. A comedy of errors ensued. Donelson disliked Anderson's choice and instead found an alternative site—Kirkman's Old Landing, twelve miles west of the fort being built at Dover. A third site also attracted Donelson's attention, across the river at Pine Bluff, Kentucky.[81]

The final position for what was to become Fort Henry was decided on in June 1861 when Colonel Bushrod Rust Johnson, a recently commissioned Confederate engineer, confirmed that Donelson's site at Kirkman's Old Landing was the best spot to construct the fort. Johnson was an 1840 graduate of West Point. He had served in the army during the Seminole War and as an infantry officer in the Mexican War, before being dismissed on charges that he was operating an illegal contraband business through the commissary department. He then taught as a chemistry professor at the Western Military Institute in Kentucky, and when the institute merged with the University of Nashville, he became superintendent of the school and a professor of civil engineering. There was no evidence that Johnson had any practical civil engineering experience; whether he did or did not, he selected a bad location to build a fort.

There were several problems. First, because of Kentucky's neutrality, the fort was built on the east bank of the Tennessee River. This meant that the fort was vulnerable to attack from the north and the west bank of the river. Furthermore, construction was in its early stages when Leonidas Polk, Episcopal bishop turned Confederate major general, moved into Columbus, Kentucky, violating the state's neutrality. If Confederate strategists had paid close attention and had understood the basic topography, the fort could have been shifted to the west bank or Kentucky side of the river, and the river would block land forces coming from the east. As it was, a range of hills on the west bank would allow enemy guns to command the fort's parapet. In addition, all the guns of the fort faced downstream.

Colonel Adolphus Heiman of the 10th Tennessee Infantry, who was to garrison the fort, sounded the loudest alarm about the problems at Fort Henry. Attached to his garrison troops was Captain Jesse Taylor's Company H, Fixed Artillery, from the state artillery corps. Taylor discovered the final and most disturbing fact about the fort: in the ordinary February rise of the Tennessee River, "the highest point in the fort would be under two feet of water and the lower river batteries [under] nine feet of water." "Arriving at the fort," Taylor continued, "I was convinced by a glance at its surroundings that extraordinarily bad judgment, or worse, had selected the site for its erection."[82] He expressed his concerns up the chain of command, which at the time was in confusion, and he concluded that his ideas "would receive but little consideration [compared] with those entertained by a West Pointer," Colonel Johnson.[83]

General A. S. Johnston, in overall command, asked his staff engineer, Lieutenant Joseph Dixon, to report on the progress of the two forts. Construction was moving at a snail's pace because there were not enough laborers to complete the defenses along the river. "None of the slave owners wished to lose harvests by renting slaves to the government . . . and an additional 2,000 were required for the twin river forts."[84] Dixon reported that neither fort was in an ideal location, but recommended that the work continue. Colonel Heimen, in the meantime, wrote to the state provisional army engineer and to Polk. Polk, the district commander, paid little attention to Heiman's complaints. The bishop was focused on his defenses at Columbus, so he wrote Heiman a dismissive note: "Your report of dispositions for defense of fort Donelson and Henry are satisfactory and I hope you will not relax your vigilance."[85]

On October 17. Johnston told Polk to "hasten the armament of the works at Fort Donelson and the obstructions below the place at which a post was intended."[86] Coinciding with the order was the arrival of Major Jeremy Gilmer, a former engineer in the United States army. Replacing Dixon as chief engineer, he was now responsible for the twin forts and for establishing a second line of defense on the Cumberland River near Clarksville. Gilmer had experience in map making, surveying, and constructing fortifications, including a fort in San Francisco, California.

Inspecting the twin forts with Dixon, Gilmer made five critical decisions. First, he agreed with Senator Gustavus Henry (the fort's namesake) that, with Heiman's garrison, Fort Henry was "in fine condition for defense."[87] Second, Gilmer agreed with Dixon that Fort Donelson would have been better located at Lineport, fifteen miles north, but "as the works at Fort Donelson were 'partly built' . . . he advised Dixon to complete the position."[88] Third, Gilmer believed the guns at Fort Donelson inadequate, and he ordered two naval guns, four additional 32-pounders, and "two 8-inch Colombiads or long range Parrott guns, all with garrison charges," to defend against Union gunboats. Fourth, he planned for river obstructions to be placed beneath the batteries at Fort Donelson. Finally, he moved south to Clarksville where he laid out plans to build a second line of defense as General Johnston had requested. Unfortunately, Gilmer did not personally supervise the construction at the forts, so he was unaware that a lack of slave labor and a reluctance of troop commanders to employ their soldiers on such work, along with illness among the men already on garrison duty, slowed the construction to a crawl.[89]

Manpower was a problem at the forts, so Johnston asked Polk if he could send 5,000 men from his command at Columbus. Polk responded that he could not because it would weaken his defenses. Johnston, who also had never inspected the twin forts and was operating in the dark, decided in mid-November to send Brigadier General Lloyd Tilghman to supervise the construction efforts on the two rivers. Graduating near the bottom of his West Point class of 1836, Tilghman had entered the dragoons and, after resigning from the army, worked as a civil engineer on several railroad developments. Johnston was delighted to have Tilghman on board, and the first thing Tilghman did was to stop construction of the timber obstructions that Gilmer had ordered to be built under the guns at Fort Donelson. Enraged at Tilghman's meddling, Gilmer told the local civil engineer in charge of building the obstructions to ignore Tilghman's order, which the civilian engineer did.

On November 20, Gideon Pillow, who had temporarily replaced the injured Polk at Columbus, ordered Lieutenant Dixon to move from Fort Donelson to Fort Henry and supervise the construction of a fort on the west bank of the Tennessee River. When Gilmer learned this and that Alabama slaves and whites were to be combined on construction gangs, he issued a formal complaint to Johnston, and Dixon was returned to Fort Donelson.[90] When the year finally ended, the forts remained vulnerable and incomplete. Confusion was ubiquitous. The chain of command was broken. Johnston, Polk, Pillow, Tilghman, Harris, and Gilmer all worked at cross-purposes. There were no laborers available to work on building the fortifications because, for the most part, slaves were the only Southerners who dug and moved dirt. White men joined the army to handle a rifle, not a shovel. Finally, the engineers had made several unfortunate decisions.

December 31, 1861, marked the final day of the war's first year. Since April, two major battles had been fought, one at Manassas Junction, Virginia, and the other at Wilson's Creek, Missouri. Considering what was about to come, the death toll at both battles was low. Nonetheless, wives had lost their husbands, children their fathers, parents their sons. Now, the new year would bring unmerciful suffering as the fighting, killing, and dying started in earnest. Both armies had spent the fall and early winter of 1861 preparing, mustering, training, and organizing the men who would fight the war.

The engineers, North and South, had played a supporting role in each side's mobilization efforts, their limited activities already demonstrating a difference in how Union and Confederate leaders anticipated the use of and need for engineers, and these early months revealed how the combatants would build an army that reflected different cultures and emphases. In the North, the army had formed an Engineer Battalion, four volunteer engineer regiments, and two volunteer engineer companies.[91] These engineer officers and troops had started building pontoon trains, worked at making repairs to railroads and bridges, and built fortifications, some as small as those in Somerset, Kentucky, and some as large as those that encircled Washington, DC. The officers of the volunteer regiments and companies were not West Pointers. These men were civil engineers, mechanics, railroad builders, and surveyors. The troops they led were made up of men from a variety of trades—mechanics, boat builders, carpenters, millwrights, masons, and farmers. Some were unskilled laborers, but none were averse to working with their hands, getting dirty, digging holes, and hauling equipment.

The Confederate army had also organized an Engineer Bureau and a Corps of Engineers, although President Davis had yet to assign a permanent head to the department. The president required that all officers be from West Point, but states enlisted the services of their own civilian engineers to help build coastal fortifications. There were no engineer troops. The labor used to construct the fortification along the Virginia coast, at Richmond, at Columbus, Kentucky, at forts Henry and Donelson, and along the Mississippi was predominantly African American slaves. Finally, because a West Point engineer was not available when it was time to consider laying out plans to construct forts on the Tennessee and Cumberland rivers, a civilian was sent. Adna Anderson was concerned about the flood plain in the region and reported this to the governor. The governor should have listened. Because Anderson was not a West Pointer, however, Major Bushrod Johnson was sent. He had little engineering experience to corroborate Anderson's work. He sited the fort in the flood plain.

Now battles loomed at forts Henry and Donelson, at Island Number 10, and on the southeastern Virginia peninsula; with railroad and pontoon bridges to build and repair and maps to make, the engineers prepared to march and equip troops for engaging the enemy.

CHAPTER FIVE

Early Successes and Failures

Fort Henry and Fort Donelson, Island No. 10, and Middle Tennessee

Our canal has been a gigantic work . . . Six miles, through a great forest of immense trees, which had to be sawed off 4 feet under water, and then through cypress swamp thickly studded with cypress knees, have furnished us with an amount of labor surpassing any one's belief who has not seen it. We have now a canal 50 feet wide, 4 feet deep, and 6 miles long, through which large steamers can pass and all our supplies be delivered to us.

Major General John Pope, commanding Union forces at New Madrid,
to Major General Henry Halleck, April 2, 1862

Beginning in 1862, the Union campaign to strike deep into the Confederate heartland took shape. Ingenuity and innovation led to the development of the United States Military Railroad and created a network for moving a massive volume of men, horses, equipment, ordnance, and supplies into and across challenging terrains. These logistical achievements enabled Union forces to strike hard at the Confederacy and weakened the South's ability to take advantage of its major strengths: internal lines of communication and a complex geography that should have made it extremely difficult for Union armies to operate inside the Confederacy. At the same time, Union engineer troops were available to support tactical field maneuvers in often critical ways, to an extent the Confederacy was unable to match.

Campaigns to capture Fort Henry, Fort Donelson, and Island No. 10 illustrate the scope of Union logistical creativity and the price of the Confederacy's inability to sustain strong engineering support during the war. Fort Henry, Fort Donelson, and Island No. 10 guarded the all-important Cumberland, Tennessee, and Mississippi rivers and denied the Union army control of these vital arteries into the South. The extraordinary story of the site selection and subsequent loss of Fort Henry in February 1862 serves as a foreshadowing of much that was to come for Confederate engineering operations and engineering-linked strategy and tactics. Moreover, the inability of the Confederate government to nationalize the railroads, or even to develop policies to manage them and keep them in repair, demonstrated a lack of understanding of the logistical effort it would take to win a protracted war.

The first year of significant fighting began with the Battle of Mill Springs, fought in southeastern Kentucky on January 19, 1862.[1] By Civil War standards, the battle itself was no more than a skirmish. Union and Confederate casualties combined were estimated at six hundred soldiers, including the death of Confederate General Felix Zollicoffer. The minor Union victory, however, had major implications for each side's strategic objectives, tactical maneuvers, and naval operations on the rivers. These objectives, maneuvers, and operations, in turn, would be successful only if Northern and Southern engineers could do their jobs with skill and imagination. Eighty-four years after Mill Springs, British Field Marshal Archibald Wavell wrote: "The more I see of war, the more I realize how it all depends on administration and transportation . . . It takes little skill or imagination to see where you would like your army to be and when; it takes much knowledge and hard work to know where you can place your forces and whether you can maintain them there."[2]

Mill Springs left the Union army in control of the Cumberland River to Carthage, Tennessee (about sixty miles from Nashville), and on the flank of Southern forces in Bowling Green. As a result of the Union army's position, Confederate General Albert Sidney Johnston shifted his western line of defense. Abandoning all of southeastern Kentucky, he anchored his right flank at Bowling Green and his left flank at Columbus, Kentucky, with the center of his line at forts Henry and Donelson. Johnston's ability to "know where he could place his forces and whether he could maintain them there" required four elements: solid intelligence of enemy movements, maintenance of the railroad so that he could move men and supplies quickly to the point of a Union attack, strong fortifications, and nerves of steel. His engineers were responsible for the forts and railroads. Unfortunately for the Confederacy, all four elements would fail Johnston, and the warning signs after Mill Springs turned into a current of defeat that swept Southern forces out of Kentucky and away from the vital manufacturing center of Nashville, Tennessee.

The lack of intelligence reports from spies and partisans in Paducah prevented Brigadier General Lloyd Tilghman and his Confederate defenders at Fort Henry from catching Grant's men unawares as they disembarked from their riverboats at Bailey's Ferry on the west bank of the Tennessee River, approximately three miles north of the fort. Three days before Grant's attack on Fort Henry, Tilghman and Major Gilmer of the engineers had left the fort to inspect the defenses of Fort Donelson, and they returned to Fort Henry only after receiving word from a courier that it was under attack. Tilghman wrote that, upon his arrival, "I soon became satisfied that the enemy were really in strong force at Bailey's Ferry, with every indication of re-enforcements arriving constantly."[3]

Tilghman was in trouble. The fort was poorly sited, poorly planned, and poorly made. And he knew it. With this in mind, did he stay and fight, buying time so that further preparations could be made to strengthen Fort Donelson? Or did he evacuate the fort immediately and march his small garrison to Donelson? Or did he move to Clarksville, eighteen miles east of Fort Donelson on the Cumberland River, and reinforce this critical supply depot that was also the final fort along the Cumberland before Nashville? With Federal gunboats within range of his position, he decided to stay and fight. It was an honorable choice—and a bad one.

In *Unconditional Surrender: The Capture of Forts Henry and Donelson*, historian Spencer C. Tucker describes Fort Henry as follows: "The fort covered three acres of ground with a five-sided earthwork parapet about eight feet high. Rifle pits extended to the river and along the water and from outside of Fort Henry's perimeter some two miles east toward Dover and Fort Donelson."[4] When Tilghman arrived in January 1862, one of his responsibilities was to finish building an additional fortification directly across the river, opposite Fort Henry, on the high ground overlooking the surrounding area. This was never accomplished. Yet, according to the chief engineer of the Western Department, Jeremy Gilmer, his boss was not to blame. When Gilmer arrived on January 31, "by the exertions of the commanding general, aided by Lieutenant Joseph Dixon, his engineer officer, the main fort . . . had been put in a good condition for defense, and seventeen guns mounted on substantial platforms, twelve of which were so placed as to bear well on the river." It was Dixon who had been assigned the task back in late November of building the additional fort on the west bank of the river, and he was promised that a large force of slaves from Alabama, with the troops to guard them, would soon arrive to do the work. Yet, "by some unforeseen cause the negroes were not sent until after the 1st of January last. Much valuable time was lost."[5] Gilmer finally reported that on February 1, the new fortification, Fort Heiman, on the west bank of the river was only a few days from completion.

When Grant and Flag Officer Andrew Hull Foote's attack came on February 6, Fort Henry, contrary to Gilmer's report, was not in good condition. The Federals opened fire at 11:45 a.m. and the battle was over just two hours later. The upper Tennessee River was lost to the Confederacy for the remainder of the war, and Tilghman, Gilmer, Colonel Heiman, and Lieutenant Colonel Milton A. Hayes, commander of Fort Henry's artillery, all agreed as to why. "The fault was in its location, not in its defenders."[6]

The predictions made by some engineers in the late autumn of 1861 that a rising river would doom the fort were prescient. In fact, as the Tennessee River crested thirty feet above normal, Union ironclads, with their limited cannon elevation, now found they were at eye level with the fort, making their fire very accurate. The ships moved in rapidly, floating over the mines that had been anchored to the bottom of the river when at its normal level. Confederate engineers had not considered adjusting the anchor cables as the river rose.[7]

Tilghman made it perfectly clear when he filed his official report of the bombardment of Fort Henry that "the wretched military position of Fort Henry and the small force at my disposal did not permit me to avail myself of the advantages to be derived from the system of outworks built with the hope of being re-enforced . . . The entire fort . . . is enfiladed from three or four points on the opposite shore, while three points on the eastern bank completely command them both, all at easy cannon range." The general made little note in his reports to Adjutant General Samuel Cooper of the rising river water that threatened to "eat away the earth and mud walls of the fort," although he did declare with hubris that "the history of military engineering records no parallel case" to match what he had endured in trying to defend such a poorly sited fort.[8] Colonel Heiman was less bashful. He wrote that additional torpedoes were sunk in the river but "were rendered utterly useless by the heavy rise of water." In addition, a "large force" tried to keep the water out of the fort. "The lower magazine had already 2 feet of water in it, and the ammunition had been removed to a temporary magazine above ground, which had but very little protection, but we had been at work day and night for the last week to cover it with sand bags and to protect it by a traverse."[9]

So, with Fort Henry in Federal hands, Grant and Foote moved in a combined operation to capture Fort Donelson on the Cumberland River. Donelson was constructed on the west bank of the river near the town of Dover, on high, formidable ground. This citadel was strong and had none of Fort Henry's flaws. It would take a two-day siege, ending on February 16, 1862, to capture the fort and most of its garrison, opening the Cumberland for a Union advance on undefended Nashville.

The fall of forts Henry and Donelson was a serious blow to the Confederacy. According to historian Richard D. Goff, "The consequences that followed the loss of Forts Henry and Donelson may well have been the greatest single disaster of the war."[10] A. S. Johnston had formed his western defensive corridor based on Jomini's principle of operating forces on internal lines of communication. The theory was simple: place your soldiers in geographical locations where men and supplies can be quickly moved from one point on the line to another so that the commanding general may at once concentrate his forces to exploit a weakness in the enemy's line or to withstand an enemy attack. Johnston's defensive position, from Bowling Green to Columbus, was on just such a line linked by three railroads.[11]

An effective use of the railroads might have allowed Johnston to shift men from one point along his defensive line to another, matching Grant's attacking army in manpower, maneuverability, and supply. Unfortunately, the railroad lines were severely taxed, and although some men recognized the management challenge in moving men and materials, no one took charge of the situation. This only exacerbated the problem. The superintendent of the Louisville & Nashville, George B. Fleece, told Johnston's quartermaster on January 2, 1862, that "at every station there is a large accumulation of freight, consisting of hogs, corn, flour, &c. The passenger travel [civilian] is also large. In addition to all, troops move in great numbers. In a word, the entire road is crowded with business to an extent unprecedented in the history of any branch of it." In addition, with only 10 engines, 120 boxcars, and 55 flat cars in operation, Fleece estimated that this was less than half the number needed to cover the entire 225 miles of rail lines. Under these circumstances, the maximum capacity of the road from Paris to Bowling Green each day was 12 freight cars.[12]

Fleece, hoping to do "justice to the army, the stockholders, or myself," proposed that he be granted permission to establish a new schedule "best adapted for the speedy, safe, and certain final accomplishment of all work" and be allowed to requisition engines and rolling stock from other roads, with the promise that the Confederate government would pay for the inconvenience and the use of the "machinery required."[13]

Any attempt to alter train schedules or procure additional engines would require voluntary cooperation from the civilian railroad superintendents, the general public, merchants, and army commanders. All these groups had different expectations and interests. Superintendents were responsible for operating the railroads efficiently and for making money. Money, in the winter of 1862, was a problem. All railroads were disinclined to take Confederate currency because it was difficult to determine its value. In January, the value of the currency fell as a result of discounting by Confederate purchasing agents, and consequently, the railroads were reluctant to substitute government business for private business. Next, the general public expected to move freely and in a timely fashion along the lines, just as merchants expected their freight to be delivered to the place the railroad promised to deliver it and not left rotting on a station platform because the army commandeered the trains. Finally, army commanders would argue over whose reinforcements and supplies should take precedence on what trains.[14]

This situation was a nightmare. All Johnston did, and perhaps all he could do, was to write to Adjutant General Cooper on January 8 requesting Cooper to send him "a full corps of competent Engineers and Machinists" from Captain John S. Butler's "Railroad Boys" with the 1st Tennessee Infantry, stationed in Winchester, Virginia.[15] Richmond did not answer Johnston's request, which left Johnston no choice but to ask for reinforcements from all quarters. Preparing for the Union offensive, Polk at Columbus, Tilghman at Fort Henry, and Buckner, Floyd, and Pillow at Fort Donelson sent no one.

Robert C. Black III, in *The Railroads of the Confederacy*, argues that Confederate military and government faced a major obstacle in fixing the dismal state of railroad transportation typified by General Johnston's dilemma during the winter of 1862. How was the government to manage the task of creating a network of transportation links essential to moving supplies and men efficiently to counter Federal incursions into the South? How could it support Confederate armies in the field and at the same time allow individual railroad companies their "rights" to operate freely, moving private freight and the general public along the same lines?[16]

The Provisional Congress offered a radical solution: nationalize the railroads that ran through Richmond, Nashville, Memphis, and Atlanta.[17] The committee assigned to investigate railroad transportation stated: "Great delay, inconvenience, and expense is caused by the numerous unconnected track, which, if joined by links, short in distance, would not only increase the facilities for transportation and the capacity of the roads, but would save much time, labor, and expense in transferring troops and freight."[18]

The committee's recommendation, on January 29, 1862, was both outside the comfort zone of the states' rights Confederacy and farsighted. Two questions remained. First, would the Confederate Congress and president embrace the committee's idea, one that represented the antithesis of the South's founding principles? Second, would the Confederacy have sufficient men with the management skills to execute such a plan? Few Southern lines exceeded 200 miles in length, which meant that superintendents managed a centralized operation including ticketing locations, repair facilities, refueling and water stations, and track maintenance. The chain of command was compact, with the man at the top able to watch over every facet of the business. Now the Provisional Congress (soon to give way to a permanent one) was asking for "proper management" to oversee an increase in the number of trains running each day and in the speed at which they traveled.[19]

Nationalization of the railroad required a management team with the ability to establish a decentralized chain of command because track and engine repairs, civilian and military schedules, and emergency operational adjustments would be spread over great distances and needed a structure similar to the one developed by the Pennsylvania Railroad before the war. The president, secretary of war, and quartermaster, ordnance, medical, and commissary departments, along with general officers in the field, would have to accept the central authority of the railroad head for the system to work. Finally, engineers and laborers would be needed to repair engines, build rolling stock, and properly maintain the track. Having few of these men before the war, and with the laborers among them being slaves, the South could hardly meet the rising demand that the war imposed.

The railroads did not serve General Johnston well as he attempted to defend, in early 1862, his extensive western front. More serious still were the disastrous results of the evacuation of Nashville. A manufacturing center and supply hub for Johnston's Confederate forces, the city began to move war material south to the interior of the country at the time of Fort Henry's capture. Johnston, whose direct contact with Polk's 17,000 men in Columbus, Kentucky, was now broken, had abandoned Bowling Green and ordered 15,000 men of Major General William Hardee's command to Fort Donelson, marching the remaining 7,000 men and supplies to Nashville.[20] Believing he would be unable to defend Nashville if Fort Donelson was captured, Johnston had been quietly shifting his major supply base south from Nashville.

The governor of Georgia, Joseph E. Brown, understood the importance of keeping precious Confederate supplies out of Union hands, and consequently, he loaned a number of locomotives and rolling stock from the Western & Atlantic Railroad to the Nashville & Chattanooga Railroad.[21] Unfortunately, the Nashville track "contained no less than 1,200 broken rails," so when Fort Donelson fell on February 16, Nashville citizens panicked. People pushed their neighbors off passenger cars, merchants demanded to load their goods onto freight cars, and quartermasters tried to move massive amounts of ammunition, clothing, medical supplies, and food from army warehouses to the trains. Wealthy citizens and high-ranking military officers demanded to board trains with their personal belongings and slaves. No one saw the president of the Nashville & Chattanooga and the person responsible for the military stores in the city, Major V. K. Stevenson, during the eight-day evacuation.[22] The lack of proper management and the failure to plan for the army's mass departure from Nashville cost the Confederacy vital war material that became increasingly difficult to procure or replace.

With a dysfunctional railroad system impairing Confederate western operations early in the war, Texas Congressman Peter W. Gray introduced a resolution that the House Committee on Military Affairs should "inquire whether further legislation is necessary to give increased efficiency to our interior lines of railroads."[23] Finally, on March 27, 1862, members of the committee presented a railroad bill to the House. The bill offered an effective organizational scheme to manage Southern railroads. The president would appoint a "military chief of railroad transportation." District superintendents reporting to the transportation chief were to have complete control of their own sections. In most ways, the House bill was modeled on the management systems developed by a number of Northern railroad companies before the war. The bill, however, stipulated that everyone involved in military transportation be given "military rank and military responsibilities."[24] This was designed to provide gravitas to railroad personnel, especially when wrangling with civilian operators. The bill was a positive first step toward nationalizing the Confederacy's railroads and addressing critical strategic, tactical, and logistic needs of the Southern high command. For some politicians, especially the fire-eaters, placing such control in the hands of the central government was an ideological abomination. Their goal was to prevent the bill from becoming law. They succeeded.

A storm of obstructions and amendments was forthcoming, ultimately drowning the bill, so that by April 21 when the Senate Committee on Military Affairs "reported it without amendment," the lack of interest in the bill's outcome resulted in its death.[25] The House struck out all the provisions of the original bill and then drafted a measure that required the secretary of war to consult with civilian railroad officials regarding military operations. Yet, even with a toothless railroad bill, Augustus R. Wright of Georgia and Thomas J. Foster of Alabama further challenged it: "We believe that this act . . . would be subversive of, and in direct contravention to, the great and fundamental principles of State sovereignty." The Confederate congressmen went on to claim that railroad superintendents were most cooperative with military officials in moving materials and that these same local managers were "more conversant with all the minute and complicated details of their roads . . . than the Executive or his military subordinates could possibly be."[26] That the army's experience in defending its western defensive line in the winter of 1862 suggested otherwise was ignored.

President Jefferson Davis was indecisive when it came to the railroads, vacillating between accepting the suggestion that Congress provide the capital to construct government facilities for re-rolling rails and building locomotives and leaving everything status quo. In a message printed in the *Wilmington Journal*, written just after the fall of Fort Donelson, Davis said that Congress could appropriate the money to create government railroad facilities, but it was "equally clear, that when the military necessity ceases, the right to make such appropriations no longer exists. To exercise this power when it exists, and to *confine it within the proper limits*, is a matter for the just discretion of Congress."[27]

The legacy of the Kentucky and Tennessee campaigns during the winter of 1862 presaged the problems to come for the Confederacy over the next three years. Few soldiers embraced the tedium of staff work that included the careful planning of logistics to support armies in the field. The Confederate Congress and the president believed that civilian railroad managers would place patriotic duty ahead of company profits and that the army would not have to waste valuable manpower resources on operating the railroads. But even if local railroad agents placed patriotism ahead of profits, most had managed only small companies and had no training or experience in operating in a larger system.

Finally, the Confederacy lacked competent engineers. Certainly some men in this capacity would serve the South well, but efforts such as the incompetent siting of Fort Henry and the lack of follow-through in fortifying Clarksville would persist throughout the war, and this would be costly. The lack of mechanics, machinists, carpenters, shipbuilders, and railroad workers in the ranks severely limited engineering operations. Southern engineers relied chiefly on slaves to make up their labor force, and white soldiers carried the Southern cultural attitude that it was beneath them to dig ditches.

Union engineering efforts and creation of efficient railroad operations in the western theater of operations stood in stark contrast to the Confederacy's difficulties. In the aftermath of forts Henry and Donelson and Federal control of the Cumberland and Tennessee rivers, Union General Henry Halleck's western army now set its sights on the Mississippi River and one of the Confederacy's major river barriers: Island No. 10.[28]

At the start of the war, Confederate authorities decided to send topographical engineer Asa Gray to New Madrid, Missouri, to ascertain whether the area had merit for establishing a line of defense below Columbus, Kentucky, along the Mississippi. Gray discovered that just ten miles south of New Madrid, at the base of a U bend in the river, sat Island No. 10, so named because it was the tenth island south of where the Ohio River met the Mississippi. As the river extended up the left arm of the U, it bent again and then continued south. New Madrid was located at the top of the upside down U. Gray reported that the island had "no superior, in my judgment, above Memphis."[29] The island was one mile long and 450 yards wide. It was positioned in the middle of the channel near the borders of Kentucky to the northeast, Missouri to the north, and Tennessee to the east.[30] The batteries on the island and on the eastern Tennessee shore presented a vexing problem for Union gunboats attempting to pass. Two forts at New Madrid, Fort Thompson west of the city and Fort Bankhead east of it, also commanded the river. The Mississippi would be at flood stage in February and March, inundating the mainland north of the island and consequently serving as a barrier to prevent Union soldiers from placing artillery near the island. The flooded area leaked into bayous and swamps. Four feet of water covered the base of tall trees and decayed stumps with jagged tops.

Island No. 10 was secure. General Polk's garrison in Columbus, Kentucky, at the left flank of Johnston's line was not. After the Confederate evacuation of upper Tennessee, General Halleck saw an opportunity to capture all of Polk's forces by cutting off the bishop's only escape route. Halleck's goal then was to occupy New Madrid. The man Halleck assigned to lead the operation was forty-year-old Brigadier General John Pope. Pope was a West Point graduate (1842) and had served as a topographical engineer in the regular army and fought in Mexico. Hiding his movements from Polk, Pope began his operation from Commerce, Missouri, northwest of Cairo, planning to move along the Sikeston Road south to New Madrid. Speed would win the campaign, which meant that a wagon train of two hundred teams filled with supplies had to keep pace with the advancing army.[31]

Colonel Josiah Bissell and his Missouri engineers were responsible for repairing the Sikeston road, and from the onset of the army's march, the weather conditions and geography forced the engineers to work at a frenetic pace. The banks of the Mississippi were overflowed, and the river's wash spread out on both sides of the roadway for miles. To make matters worse, the area the army moved through was known as the Great Mingo Swamp, described by Pope as "dismal and almost impassable."[32] In some places the water was ten feet deep, and it was never less than one foot deep. Walking through the swampy backwater was like attempting to pass through quicksand. A correspondent with the troops said the men "waded in mud, ate mud, slept in it, were surrounded by it."[33]

The weather was cold and wet, and drizzling rain and snow were the soldier's constant companion. "An old embankment upon which a corduroy road had been built, extended part of the way to New Madrid," Pope reported, "but the road had not been repaired for years, and was in a very bad condition, and in many places entirely impassable."[34] Bissell and his men gathered every piece of scrap wood they could find and every fence rail for almost eighteen miles.[35] The engineers built bridges out of fallen trees, ropes, and vines; in some cases, before they could rebuild roads and bridges they had to remove debris placed in their way by retreating Confederates. At one point along the march route, a reporter commented that "felled trees into the road and the bridges burned" forced the entire column to halt. Close to dusk, the men bivouacked for the night, and the engineers went to work by torchlight to open the road. Fifteen hours later the work was finished, and the march resumed with the engineers in tow.[36]

Bissell's Engineers were detached by companies and assigned other tasks throughout Pope's campaign on New Madrid. After removing "the obstructions thrown in the way by Rebel guerrilla chief [Meriwether] Jefferson Thompson," companies A and B, under Major Montague S. Hasie, proceeded to Point Pleasant, ten miles south of New Madrid, on the western bank of the Mississippi, and constructed earthworks on the riverbank.[37] Companies C, D, and K worked on repairing the Cairo & Fulton Railroad running east to west from Bird's Point, just south of Cairo, to Sikeston. Company G worked on the same railroad doing bridge repair. It was joined by Company H, which built a depot platform in Sikeston with lumber produced by Company F, which spent all of November and December 1861 operating a lumber mill.[38]

Pope's plan to cut off any escape route for 17,000 Confederate soldiers garrisoned in Columbus, Kentucky, did not come to fruition. General Polk sensed the trap and, starting on February 26, pulled his men out from the city's fortification and moved south along the Mississippi to link up with Brigadier General John P. McCown's 2,000 men in New Madrid. When Pope's men attacked New Madrid, McCown decided that his position was untenable and moved his troops across the river. This was a costly mistake, but not a mortal blow. Supplies for Island No. 10 would now have to be transported overland from Tiptonville, but the island remained unassailable from everywhere except by river south of Island No. 8. The Federal army at New Madrid had no means of crossing the river, and if they attempted to build small barges, an eight-ship Confederate flotilla under the command of George N. Hollins would make quick work of the effort. Island No. 10 commanded all river traffic to the south, and with a supply line open, the men operating the island's fifteen guns could hold out for a considerable length of time.

Pope had two options, and both required naval cooperation. First, he could order Flag Officer Andrew H. Foote's squadron of gunboats, steamboats, and mortars to bombard the island and shore batteries, making possible an amphibious assault. Second, Foote could run several ships past the Rebel stronghold and transport soldiers from New Madrid across the river and then attack the island from both sides.

There was a third option—or at least a suggestion—to cut a road south from New Madrid to the northern bank of the Mississippi, opposite Island No. 10, or to cut a rudimentary canal across the peninsula (from the right upright of the U to the left). Colonel Bissell thought it made perfect sense to build a canal. Pope weighed his options and decided the best plan was to run Foote's gunboats between the island and shore batteries. The general would send Colonel Bissell to assist Foote in preparing the gunboats with additional protection to run the Confederate batteries, and in addition, Bissell could reconnoiter the area to determine whether it was practicable to cut a road or dig a canal.

Pope was confidant that Foote's gunboats would succeed in passing the island.[39] Foote was not. For days his squadron had opened a ferocious cannonade that had little effect on the island, yet the Confederates' return fire damaged several of his ships and killed or wounded at least a dozen of his sailors. Foote was thus prepared for a long siege because he feared sending any of his gunboats to run the gauntlet between the island and the shore batteries. A gunboat might be sunk or, worse, fall into Confederate hands and be used against the Union flotilla.[40]

General Pope was furious to learn that Foote would not run his gunboats past Island No. 10. Bissell, for his part, was also frustrated because on March 19 he figured out that neither a road nor a canal was possible. Regarding the latter, he had explored the possibility that St. James Bayou, which entered the Mississippi River seven miles north of Island No. 8, connected to St. John's Bayou, which emptied into the Mississippi at New Madrid. It did not. Bissell returned to Pope's headquarters the following day with disappointing news.[41]

The following morning, while standing on the levee alongside the Mississippi waiting for the dug-out (canoe) and guide to arrive to take him back to Pope's headquarters, Bissell noticed that opposite him, directly across the submerged peninsula, stood an opening between large trees that appeared like a path through the woods. Bissell wrote: "This proved to be an old wagon road extending half a mile into the woods; beyond and around was a dense forest of heavy timber."[42] The guide determined that it was approximately two miles from the end of the old wagon road to the nearest bayou. The guide made a sketch of the area in Bissell's memorandum book, and the two men examined the wetlands until nightfall then returned to Pope's headquarters.

What happened next remains a mystery. Bissell claimed that General Pope and his staff had just finished supper when "someone said something about a canal." According to Bissell, Pope joked about the idea, stating that the entire countryside was under ten feet of water. Bissell then pulled out the map and announced that he "would have the boats through in fourteen days."[43] Pope was not impressed by Bissell's bravado, so the general took him aside and asked the engineer if he meant what he said. The answer was an unequivocal yes. After this briefing, Pope approved the project. General Schuyler Hamilton challenged Bissell's account of what happened, and there was considerable postwar bickering between the two men, each claiming responsibility for the idea of a canal.[44]

Regardless of who came up with the idea, the canal was an incredible feat of Civil War engineering and was recognized as such by General Pope. He wrote to Halleck on April 9: "Of Colonel Bissell, Engineer Regiment, I can hardly say too much. Full resources, untiring and determined, he labored night and day, and completed work which will be a monument of enterprise and skill."[45] The assembly point for the canal operation was Island No. 8, where six hundred men, four steamboats, six coal barges, four pieces of heavy artillery, axes, saws, rope, carpenter tools and tackle, and two million feet of lumber were gathered on March 23 to begin work.[46]

Bissell's men cut a break in the Mississippi River levee at Phillip's Plantation to access the cornfield. Although the barges could get through, the four stern-wheel steamboats drew thirty-six inches of water and the break proved too shallow for them to pass. So Bissell's men cut thirty more feet of levee and, jumping into the cold, waist-deep water, also removed stumps and logs while battling a swift current with a dangerous undertow. Within twenty-four hours, one steamboat, the *W. B. Terry*, and two barges made it through the cornfield and the half-mile-long submerged wagon road to the flooded forest, where the channel narrowed making further passage impossible. The task now assigned to Captain William Tweeddale of the engineers was to widen the channel and remove the stumps so the Union vessels could pass.

Tweeddale's process for carrying out his orders required men on rafts to

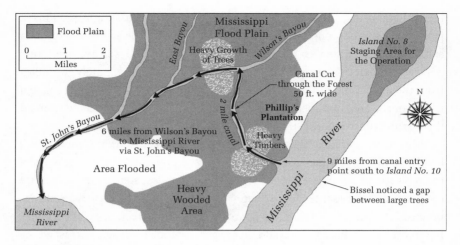

The Engineers Bypass Island No. 10

wedge springboards (on which to stand) into trees and then to cut the trees about eight feet above the water. Another set of soldiers on rafts tied a line around each downed tree and hauled it away with the use of a snatch block and steam capstan aboard the *W. B. Terry*. Next, a trailing raft was lashed to the stump, and an upright plank was also fastened to the stump. A frame shaped like an isosceles trapezoid was attached to the upright plank with a pivot pin so that the frame could swing like a pendulum. A saw blade was connected to the legs of the trapezoid, and with two men pulling on ropes, the blade could move back and forth to cut through the stump below the waterline, leaving a stump four and a half feet below the surface. Tree stumps less than two feet in diameter offered no problems, but larger trees such as elms "which spread out so much at the bottom" proved to be a challenge. In addition, since the water was muddy and murky, it was often difficult to determine what was interfering with the saw.[47]

The thicker trunks often pinched the saw, and then it required tackle to be attached to the top of the trunk and pulled so as to release the trapped saw. It could take as long as three hours to remove some trunks, and working twenty-four hours a day, it took the engineers eight days to clear a channel fifty feet wide and two miles long. This channel entered the first of three bayous, all of which were narrow and overgrown with trees. Now, as the engineers started through the bayou, Bissell spoke solemnly of the new danger awaiting them: "The river had begun to fall, and the water was running rapidly."[48] Men wore lifelines attached to their waists as they climbed out onto slippery logs to secure cables so that the downed obstructions could be pulled away. Some trees had to be sawed and axed into sections before they were moved, underbrush had to be cut and pulled to the embankments, and as the water dropped, stumps originally four feet under the water had to be recut.[49]

The exhausting work continued until April 6, when the six-mile-long channel opened to traffic. Once the engineers entered Wilson's Bayou, they eventually linked up with East Bayou and then St. John's Bayou, which emptied into the Mississippi at New Madrid. As the water continued to recede, only flat-bottom transports could navigate the canal, but some of Bissell's men had fitted these vessels with heavy artillery to create improvised gunboats.

General Pope had attempted to keep the canal a secret, but Confederate scouts detected the break in the levee opposite Island No. 8 as early as March 29. With the rapid drop in the water level, however, the Southern command did not believe the canal would be successfully completed. Some Confederate officers reported to a journalist that it "is believed utterly impracticable for the enemy to cut a canal from New Madrid across the bend to a point above the island." Another officer continued: "The trees in the Mississippi bottoms . . . are very large and grow together, and sent their roots deep into the soil. This growth of our swamps and bayous presents an impenetrable barrier to any such undertaking as that spoken of."[50] In Confederate commander Brigadier General William W. Mackall's final report of the operations around Island No. 10, he noted

This saw apparatus, designed to cut through tree trunks below the surface of the swamp, was used between March 23 and 31, 1862, during Island No. 10 operations. It demonstrated the adaptability and ingenuity of Colonel Joshua W. Bissell's Missouri engineers. *Battles and Leaders of the Civil War*, vol. 1 (New York: Century Company, 1887).

that he was told that the enemy "were endeavoring to cut a canal across the op-posite peninsula for the passage of transports, in order to land below the bend; that they would fail, and that the position was safe."[51]

The canal did not prove to be the decisive factor in the capture of Island No. 10. Yet, the canal enabled Pope to transport men and material to the eastern bank of the Mississippi, where he was able to block the only escape route the Confederates had—the road to Tiptonville. On April 4, the ironclad gunboat *Carondelet* ran past the Confederate batteries, and two nights later the *Pittsburgh* successfully did the same. Now, with gunboats attacking the island from front and rear, and the major road to and from the island blocked, Mackall's only recourse was to surrender. He did so on April 7, and the Mississippi was now open to Fort Pillow.

The canal and channel that Bissell and the Missouri engineers cut and dredged represented the ingenuity, confidence, and culture of Pope's army. When Bissell stood atop the Mississippi levee and spotted a flooded cornfield and old wagon road carved between the trees, he saw possibility, not impossi-bility. When trees sprang up from six feet below the waterline, a system was de-veloped, with the help of a makeshift saw, to cut them away. Many of the men in Bissell's Engineers were carpenters or builders, and in a former life had had to improvise to solve mechanical problems. Finally, Pope had an engineering background and could see the possibility of opening a canal. Although never discussed in histories of the New Madrid campaign, Pope could have said no to Bissell's plan. Although we can only speculate as to why Pope said yes, two factors had to have been his own background and his belief that Bissell's plan would work. Chances were that in his twenty years before the war as a surveyor in Florida, New Mexico, and Minnesota, and as the surveyor for possible south-ern routes for the first transcontinental railroad, Pope had considered creative possibilities to conquer challenging geography.

The canal and channel to bypass Island No. 10 would turn out to be one of the most demanding engineering feats of the war. There was no doubt that the canal scored psychological points for the North, as well. The Northern press celebrated the victory with headlines such as "How Colonel Bissell's Engineers Fores [Force] Their Way to General Pope" and "The Great Western Stump Cut-ter."[52] Even Southern newspapers marveled at the accomplishment. An article in the *Macon Telegraph* reported on the Yankee engineers' work: "The exploit is unparalleled, at least in this country, and is one of the most novel and marvel-ous which has been performed during the War."[53]

Bissell's regiment was not the only group of volunteer engineers engaged in a critical assignment in the spring of 1862. The 1st Michigan Engineers and Mechanics, encamped with Buell's army in Nashville, would be assigned the herculean task of repairing turnpike and railroad bridges destroyed by retreating Confederates south of the city as far as Murfreesboro. The Union army had to take the war into the deep South but could do so only if critical supply lines remained open and maintained. Every time several hundred partisans or Confederate cavalry burned a railroad bridge or destroyed rolling stock, 30,000 to 50,000 Union soldiers on the march would stumble to a halt. No supplies meant no movement. Moreover, Civil War armies, because of their size, needed to divide into smaller units (15,000 to 20,000 men, artillery, and wagons) and move on parallel roads to reduce congestion. A line of soldiers composed of 50,000 men, horses, artillery, and wagons, 124 miles long from head to tail, would take seventy-two hours to pass one spot. To adjust from columns of march to lines of battle (going on line) was a nightmare. For the engineers this meant keeping multiple roads open so the entire army arrived at the same place at the same time. If one group arrived and the others were stalled, the lone unit in advance of the others could be destroyed by a larger enemy force. Generals knew this to be a real danger, and it added additional pressure to the work of the engineers.[54]

In the early spring of 1862, Halleck's Department of the Mississippi was divided into three major components: Pope's army on the Mississippi, Buell's army in Nashville, and Grant's army in Savannah, Tennessee, northeast of Corinth, Mississippi, on the eastern bank of the Tennessee River. Halleck's slowly evolving plan was to have Pope's forces continue to slug their way south along the Mississippi toward Memphis, while Buell and Grant, under Grant's command, were to rendezvous at a place called Pittsburgh Landing and then strike Sydney Johnston's army at the important railroad junction at Corinth.

With his forces spread out over four states, Johnston had to decide where to concentrate his armies. He could not defend both middle Tennessee and the line along the Mississippi River. He chose the Mississippi defense and selected Corinth as the concentration point because of the railroads. The war was less than one year old and the fighting in the western theater three months old, yet both Union and Confederate armies understood the critical importance of the railroads. The quick concentration of troops, the movement of supplies and reinforcements over great distances, the ability to flank an enemy or block their retreat, and the ability to block an advance or, better yet, cut off an advancing army from its supply base made the locomotive a new factor in American warfare. The degree to which the railroads would influence strategic and tactical decision making was unprecedented and unanticipated. Both sides had to improvise, and ill-trained and ill-prepared generals and their staffs would have to manage and support operations over vast areas.

When Halleck ordered Buell to join forces with Grant in Savannah, Buell decided to move southwest from Nashville to Columbia along the Tennessee & Alabama Railroad and, from there, to move over a dismal road southwest to Savannah. On March 15, Buell sent multiple orders to the Michigan engineers. Colonel Innes was to take three companies north to Paducah, Kentucky, and then travel south on the Tennessee River to a spot called Lucas Landing, where they were to take possession of an abandoned railroad engine, six platform cars, and one boxcar. Innes and his men were to load these on barges and bring them back to Nashville.[55] Captain John B. Yates was ordered southeast from Nashville, along the Nashville & Chattanooga Railroad, to repair both the turnpike and the railroad bridge at Stones River just north of Murfreesboro.[56] Finally, Captain Perrin V. Fox, marching with Buell's army to the southwest, was called upon to repair a major railroad crossing at Franklin, Tennessee. Retreating Confederate cavalry had destroyed the bridge, which was a vital link in Buell's supply chain.

Buell and his staff were not ready to manage the multiple tasks assigned to the engineers, and the work was helter-skelter. Slow to repair the bridges, the engineers eventually completed the work, but their operations were not well coordinated. Buell had appointed John Byers Anderson, a civilian and former superintendent of transportation for the Louisville & Nashville Railroad, to manage operations north of Nashville. South of Nashville, Confederate forces had destroyed the two major lines from the city, both of which tied into the east-west Memphis & Charleston Railroad. For repair to these railroads, Buell tapped the colonel of the 13th Ohio Infantry, William Sooy Smith. Smith was an 1853 graduate of West Point, worked for the Illinois Central Railroad, and in 1857 had opened his own civil engineering company, Parkinson & Smith. In addition to repairing track and bridges, Smith had to acquire locomotives and cars.[57]

As Buell marched his men toward Pittsburg Landing, he left behind Brigadier General Ormsby M. Mitchell to guard middle Tennessee. For three months in the fall of 1861, Mitchell had command of the Department of the Ohio, and he was aggressive and ambitious. Without orders from Buell, Mitchell moved south to Murfreesboro where, with three companies of Michigan engineers and several infantry regiments, he supervised the rebuilding of two railroad bridges and one bridge on the Nashville Pike.[58] North of Murfreesboro, engineers were rebuilding a railroad bridge at Franklin, Tennessee, that had been washed out by high water. Michigan engineers, with the help of a civilian work crew from the McCallum Bridge Company of Cincinnati, the 38th Indiana Infantry, and an engineer company that operated a sawmill, built a temporary truss 110 feet long over the Big Harpeth River.[59]

Buell's major detachment heading for Pittsburg Landing also met up with a bridge problem. In Columbia, his columns came to an abrupt halt when it was discovered that the Duck River Bridge was destroyed and the river at flood stage. Their problem was exacerbated by the lack of engineer troops to work on the crossing. The engineers were scattered to the four winds, and all were functioning under a confusing command structure. Anderson, with no military rank, had no military authority. Colonel Smith, responsible for bridge repair and finding locomotives, had sent Colonel Innes, commander of the Michigan Engineers, north to bring engines and cars back to Nashville. Smith also assigned Captain Yates to rebuild the bridge at Mill Creek, but some of Yates's men were taking orders from General Mitchell at Murfreesboro.

Colonel Jacob Ammen, a brigade commander in Brigadier General William Nelson's division, was part of the column marching toward Pittsburg Landing, and his March 20 diary entry read: "Bridge over the Duck River at Columbia burned by rebels; river high; no boats. General McCook's division in advance, repairing bridge."[60] Alexander McCook's infantry worked on the bridge until March 27, when Nelson arrived and told Ammen that the river was falling and Ammens needed to find a "damn ford" before March 29. Nelson wanted his division to cross first and then have the honor of being the lead unit on the march to Pittsburg Landing. On March 29 Ammen wrote: "At 3am men wade stream. Cavalry in river to point out ford, break force of current, and protect infantry. Cold and disagreeable day. Bridge not completed."[61] Among all the confusion, Buell finally ordered some engineers who were working on the bridge at Big Harpeth forward to the Duck River. When they arrived on April 1, the bridge was near completion.

Buell's arrival at Pittsburg Landing came none too soon. After the first day of fighting at Shiloh (the name of a small Methodist church near Pittsburg Landing), Confederate forces had surged forward, almost driving Grant's army into the Tennessee River. Buell's arrival and Grant's audacity reversed Yankee fortunes on April 7 with a Northern victory. Now the Union army set its sights on the critical railroad junction at Corinth. The defending Southerners left the city on May 30 because they feared being trapped by a much larger Union force.

In his still highly regarded, classic study of railroads in the Civil War, George Edgar Turner describes the end of May 1862 as a significant time early in the war when Union forces held "a vital part of the Confederates' one line of railroad between Virginia and the Mississippi River."[62] They held the Nashville & Chat-

tanooga as far as Stevenson, Alabama, the three lines from Nashville to Decatur, the Mobile & Ohio from Columbus, Kentucky, to Corinth, and a direct line
from Louisville to Memphis. On May 31, Halleck instructed the 1st Michigan to
begin the all-important repair work along the Memphis & Charleston, between
Corinth and Decatur, Alabama. Turner no doubt would have agreed with historian Russell F. Weigley's later assessment of the railroads' importance in bringing about Union victory. Weigley argues that the Union's war-making capacity
at the start of the war existed only on paper. Converting those potential assets
into superior resources was one of the major roles of the railroads.[63] Historian
John E. Clark Jr. also emphasizes the value of the railroads: "The Union's successful use of the railroads neutralized a determined Confederacy's vast land
mass. They changed the nature of warfare by enabling the Union to shrink the
Confederacy to a manageable—and vulnerable—size." Clark continues: "The
modern management principles and procedures developed by antebellum
northern railroads produced outstanding managers whose ability and experience proved invaluable to the Union war effort."[64]

Turner, Weigley, and Clark are correct in their views about the virtues of
the railroad during the war. Yet, possession of the railroads was only part of
the challenge, as the managerial and logistical confusion facing Buell revealed.
Finding the right managers, establishing a proper chain of command, recognizing the value of the engineering workforce, creating an effective and authorita-
tive civilian organization, and curbing the massive egos of many a general—all
these were not givens in June 1862. The Civil War was to require a new type of
manager and new type of management system, one where both boss and em-
ployee had to overcome unforeseen, challenging, and changing circumstances.

In June 1862, in the western theater, there were at least two immediate problems. The first was a management issue. There was no system in place to control the resources of Buell's newly organized Army of the Ohio and no railroad/
engineer/management expert who had the authority to synchronize logistical
operations at both the strategic and tactical level. Buell, his corps, and his division commanders all had a chief engineer on staff. Most were West Pointers.
These engineers reported to both their commanding officers and the Engineer
Bureau in Washington, DC. Then there were civilian railroad men, such as John
Byers Anderson, and civilian workers. The 1st Michigan Engineers and Mechanics and Bissell's Engineers each had their own commanding officer. What
if a corps commander wanted to order a detail of engineers working on bridge
repair to work on a different project that suited the general but not the army's

overall efforts? Could the engineers disobey the original order? The answer was yes, but only if they were prepared to be called up on charges.

The second problem was a potential mutiny. At the time of their enlistments, the soldiers of the Michigan engineers had been promised the pay rate of regular army engineer troops, seventeen dollars a month, as opposed to the standard thirteen dollars a month for infantry. Engineers' pay was not forthcoming. Since January, Colonel Innes had petitioned the War Department, Missouri Congressman Francis P. Blair Jr., and Congressman Francis W. Kellogg of Grand Rapids, requesting back pay at the seventeen-dollar rate. In February 1862, Blair introduced legislation that would retroactively pay the volunteer engineers the regular army rate. The bill was finally voted down sixty-six to fifty-seven.[65] Regular army engineer officers also did not support the bill, believing that volunteer engineers were not as professionally trained, skillful, or disciplined as the regulars. The snob factor ran high.

Discontent among the soldiers of the 1st Michigan continued to swell. At the time of Buell's movement toward Pittsburg Landing, a number of men refused to march. Innes threatened arrest. Thirty noncommissioned officers were reduced in rank for their role in encouraging the men to stop working. The reduction was temporary for most.[66] Innes was finally able to quell the bad feelings and the men marched, but the pay they were promised was not forthcoming.

The pay issue would fester for six more months before it was resolved—an injustice in a war that generated many injustices. Notwithstanding the superiority complex displayed by regular engineers toward the volunteers and the volunteers' inadequate pay, the Michigan boys and their counterparts from Missouri and Kentucky did yeomen's work in the first six months of 1862. By June, the Union army had control of the Mississippi River to Memphis, thanks in part to Colonel Bissell's men at Island No. 10, and it controlled six major railroad lines, maintained by the engineers, which served both as points of access to the Confederacy's interior and as vital supply lines linking army depots to the front lines. The early months of 1862 had served as the first test for the North's engineering forces in the west. Now, about 600 miles to the east, General George McClellan was about to engage his volunteer engineers in their first test of the war in an area known for salt marshes and swamps called the Virginia peninsula.

CHAPTER SIX

McClellan Tests His Engineers

The Peninsula Campaign, 1862

When the current is rapid, the bridge is protected from the shock of
floating bodies by establishing a guard of observation above the bridge,
to arrest these bodies; by placing a stockade obliquely across the stream,
or by constructing the bridge by rafts, and withdrawing that part of the
bridge which is menaced, and allowing the body to float past.

The Union army's Manual for Engineer Troops *by Captain J. C. Duane*

In the spring of 1862, as Bissell's Engineers and Colonel Innes's 1st Michigan
cut canals through flooded lowlands and repaired and rebuilt damaged railroad
tracks and bridges in Tennessee, General George McClellan prepared his Army
of the Potomac engineers for the first great offensive of the war, targeting the
Confederate capital of Richmond. Making an amphibious landing in southeastern Virginia and using the York and James rivers as his supply lifeline, McClellan planned to move 120,000 soldiers northwest in an effort to enter Richmond
by the back door.

General McClellan spent considerable time trying to build an engineering
organization that could meet the challenges of the army's logistical demands.
These burdens required the coordination of troop movements and complex
supply functions. As daunting as this task was, McClellan faced several other
obstacles that made his efforts even more difficult. First, several West Point–
trained engineers opted, like McClellan himself, to accept field commissions in
the volunteer army. Prestige, promotion, and pay, all limited in the antebellum
army, were now waiting for the men who would lead the charge. Second, the
advent of the railroad and telegraph in warfare siphoned engineers away from
the field army and placed them in roles usually assigned to civilians. Third, regular army engineers balked at embracing volunteer engineers as equals, which

created tension between the two groups. Finally, the politicians in Washington saw war as a clash of arms and thus viewed with suspicion any attempt to raise troops for anything but infantry.

Under the circumstances, McClellan opted to continue to demand rigorous bridge-building training, and on February 26, regular army engineers waded into a frigid Potomac River across from Harpers Ferry and began construction on an 840-foot-long pontoon bridge. The water was 15 feet higher than normal summer levels and the day was raw and windy. The engineers used ship anchors and chain cables to lash together sixty pontoons, and the bridge was completed in eight hours. McClellan was delighted. "The bridge was splendidly thrown by Captain Duane . . . It was one of the most difficult operations of the kind ever performed."[1]

Captain Duane, who literally wrote the book on "Pontoon Drill" in his *Manual for Engineer Troops*, continued to express doubts about the usefulness of volunteer engineers and especially the officers who lacked a West Point education.[2] It seems probable that Duane was caught in an unenviable dilemma: recognizing that West Point–trained engineers were critical in the emerging modern war, but knowing there were not enough of them to respond to the growing needs. For Duane, doubts about the value of volunteer engineer officers were also personal. He had trained, studied, and earned his way into the most elite intellectual organization in the army and, perhaps, in the United States. The idea that men with no formal training could break the glass ceiling (promotions were hard to get in the prewar army) was both puzzling and threatening.

Even Brigadier General Joseph G. Totten, West Point class of 1805 and head of the Corps of Engineers since 1838, did not grasp the magnitude of change coming to the engineers, nor did he understand their expanding role in mobile warfare. In a letter to Stanton just one day before Duane's engineers built their bridge from Sandy Hook, Maryland, to Harpers Ferry, the head of engineers insisted that twelve of his officers stationed in places like New Bedford (Massachusetts), Oswego (New York), the Kennebec River (Maine), and Alcatraz Island (San Francisco harbor) should not be pulled from the forts to serve in the field with the armies. "Officers of some experience are required for duty of this kind which involves large disbursements and business practice, control of men and acquired knowledge of construction, as well as theoretical attainments as Engineers of fortifications."[3]

In mid-March 1862, Totten wrote to McClellan requesting two engineers from McClellan's Army of the Potomac to be assigned to coastal fortification in the North. McClellan politely declined the request. With preparation for the

Peninsula Campaign underway, McClellan reminded Totten that the peninsula operation "will require the services of a number of Engineer officers . . . If I can even keep them [engineers] until the question of [the siege of] Yorktown is disposed of I shall feel better satisfied."[4]

McClellan kept his engineers and restructured the department within the Army of the Potomac, adding to the Engineer Battalion the Volunteer Engineer Brigade, made up of the 15th and 50th New York Volunteer Engineers. Congress did not officially recognize these two regiments until July 17, 1862. The act included paying the New York volunteer engineers the same as engineers in the regular army.[5] Lieutenant Colonel Barton S. Alexander would continue to instruct the volunteers and serve on McClellan's staff. Colonel Charles B. Stuart was the regimental commander of the 50th New York, and Colonel John McLeod Murphy commanded the 15th New York. Their commanding officer was Brigadier General Daniel Phineas Woodbury, West Point class of 1836. Both Woodbury and Duane reported to McClellan's chief engineer, John G. Barnard.[6]

In April, as the engineers prepared to move to their staging area for the great campaign, McClellan also organized his topographical engineers. Not unlike the engineers, the mapmakers in the Army of the Potomac were made up of a cadre of both military professionals and civilians. The difference was that the topographical corps did not have a company of enlisted men, and civilians were already a part of fully formed mapmaking organizations, including the United States Coast Survey, the Smithsonian Institution, the Naval Hydrographic Office, and the Pacific Wagon Road Office, an agency within the Interior Department.[7]

The United States Coast Survey was the oldest scientific organization in the United States. Its superintendent at the time of the Civil War, Alexander Dallas Bache, was a West Point graduate (class of 1825) and had served as the first president of the National Academy of Sciences. Bache had established the first magnetic observatory in the United States, and by 1861 his agency had mapped the entire coast of the United States and posted numerous observing stations along the coastline. Throughout the war, Bache furnished the armies in the field with civilian mapmakers to augment the work done by topographical engineers on headquarter staffs.[8]

Throughout the Peninsula Campaign, McClellan's chief topographer, Brigadier General Andrew Atkinson Humphreys, along with five lieutenants, Coast Survey assistants, and civil engineers working under arduous conditions, skillfully mapped out the complex landscape over which McClellan was attempting to move 120,000 men. Humphreys's men made six maps, supplementing Coast

Survey maps of the James and York rivers and the state map of Henrico County, Virginia, to assist commanding generals in negotiating the territory.[9]

During the campaign, McClellan noticed considerable overlap between Humphreys's work and that of the army's chief engineer, Brigadier General Barnard. Reconnaissance of roads and positions of the enemy and construction of siege and defensive works were tasks "habitually performed by detail from either corps as the convenience of the service demanded."[10] Since both topographical engineers and engineers performed overlapping jobs, eventually, as an experiment, McClellan united the two corps under Captain Duane of the engineers. McClellan's decision, however, met with reproof from the War Department's Engineer Bureau chief, General Totten. As historian Earl J. Hess describes in *Field Armies and Fortifications in the Civil War*, "He [Totten] thought topographers were not true engineers and that adding them to his beloved corps would dilute its effectiveness."[11] Despite Totten's objections, McClellan believed the merger represented a more efficient operation, and eventually Congress made it official by folding the topographers into the Corps of Engineers in March 1863.

At the time of the Peninsula Campaign, the Union army's engineering and topographical engineering operations were a work in progress. Organizational structure apart, the skilled work of mapmakers proved decisive in the coming campaign. On the Confederate side, a lack of both adequate maps and skilled topographical engineers to make them prevented General Joseph Johnston from seizing an opportunity to severely damage the Army of the Potomac at the outset of the campaign.

Johnston saw the vulnerability of the Union IV Corps isolated south of the Chickahominy River, and he struck on May 31, 1862. But the Battle of Seven Pines (or Fair Oaks) became a pyrrhic victory for the South. The Northern soldiers left the field to the Confederates, but Johnston was not able to consolidate the victory. Of the 60,000 men that General Johnston had on hand to fight, the right wing of his army under General James Longstreet attacked the Union position with only six of the thirteen brigades available, and no more than four of the six brigades were engaged at any one time. The reason for this, according to historian Thomas B. Buell, was that the Confederate army on the peninsula lacked an understanding of the roads and terrain, which made it impossible to get troops into position in good time. A coordinated attack was impossible.[12]

When Stonewall Jackson broke off his Shenandoah Valley Campaign to reinforce the Southern army defending Richmond, Brigadier General Richard Taylor observed that "the Confederates knew no more about the topography of the country than they did about Central Africa. " "Here was a limited district, the whole of it within a day's march of the city of Richmond," Taylor continued, ". . . and yet we were profoundly ignorant of the country, were without maps, sketches or proper guides, and nearly as helpless as if we had been transferred to the banks of the [Congo]."[13] Taylor's comments pointed to a major flaw in Southern military operations throughout the war. The general understanding that fighting on their home territory gave the mapping advantage to the Confederates was inaccurate. Southern commanders could rely on local citizens to provide information about roads, but this knowledge was inadequate when it came to campaign planning or moving thousands of men and material. What was seen as passable by a local farmer was not necessarily adequate for an army corps of 25,000 men.

General Johnston and General Robert E. Lee, who replaced Johnston after he suffered a wound on June 1 at Seven Pines, did not have a blatant disregard for maps. What they did have was a lack of mapmakers and formal military bureau to establish, catalogue, and distribute maps. Before the war, the United States Army had an established and functioning organization prepared to adapt to wartime conditions, namely the Corps of Topographical Engineers. The Confederate War Department was never able to match the Union's topographical bureau.

When the war began there were forty-five topographical engineering officers in the United States Army, and seven decided to join the Confederacy. Of those, only four—Charles Read Collins, William Holding Echols, Joseph Dixon, and Joseph Christmas Ives—served as Southern topographical engineers during the war.[14] Jed Hotchkiss would become the most famous Confederate topographer of the Civil War. A schoolteacher and self-taught mapmaker and surveyor during the antebellum period, he joined General Jackson's staff and provided the eccentric and aggressive general with a map of the Shenandoah Valley in March 1862, and after the Valley Campaign he continued to work on the map for at least another eighteen months. Hotchkiss continued his service with the Army of Northern Virginia after Jackson's death, but he never received a military rank from the Confederate government.

After the fiasco at Seven Pines, the new commanding general, Robert E. Lee, an engineer himself, initiated a plan to make sure all engineer officers in the field army, including topographers, reported to one man: the army's chief engineer at Lee's headquarters.[15] This organization prevented engineers assigned to army corps and divisions from acting on different sets of orders. Lee appointed Lieutenant Colonel Walter Stevens of the engineers as head of the Department of Richmond, responsible for the city's defenses. Then on June 6, 1862, Lee commissioned Albert H. Campbell as captain of engineers and asked him to report directly to Stevens.[16] By December, Campbell's office in Richmond became the map bureau of the engineering department.[17] Hotchkiss, for his part, placed little faith in Campbell, remembering him for the "large amount of bad work he had done and for the long time he took to do it in."[18] Yet, even the few well-trained mapmakers of the Confederacy, such as William Willis Blackford and James Keith Boswell, experienced great difficulty in procuring the proper equipment. Besides pencils, tracing paper, and field notebooks, topographical engineers needed telescopes, prismatic compasses, odometers, barometers, T-squares, and boxes of colors.[19] Other priorities and the Union blockade made these items hard to get.

Maps of various areas of the country were available in Richmond, but these county maps, made in peacetime, were worthless for wartime armies. Most of the maps identified towns, were small in scale, and displayed no terrain features except mountain ranges. Military maps not only needed to include accurate distances of roads and intricate terrain features such as woodlands and swamps, they also needed to identify springs, cultivated fields, pastures, and orchards. The difference between a road and a farmer's cow path was substantial, and if marked incorrectly on a map could mean the difference between a successful flanking movement and a disaster.[20]

Poor maps had proved problematic for the Southern army at Fair Oaks and Seven Pines. Yet, the lesson learned on the Peninsula for the Army of the Potomac was that volunteer engineers had the skills necessary to function as well as the regular army engineers. Union officers also learned that even some infantry regiments could be called upon to build a bridge. These lessons, however, took time to sink in.

The regular army engineers' skepticism remained keen as the Army of the Potomac set sail for Fort Monroe, on the southern tip of the Virginia peninsula, to begin McClellan's Peninsula Campaign. Almost immediately upon their arrival the rains began turning roads into slop, which made the march miserable. By April 5, the Union army made contact with Confederate lines protecting Yorktown. McClellan, using limited intelligence information and his cautious nature, determined that he faced 100,000 Southerners positioned behind well-constructed trenches and fortifications. Consequently, he ordered his engineers to construct siege lines from which his artillery could bombard Yorktown and drive out the Confederate defenders.

During the ensuing siege of Yorktown, the volunteer engineers began to earn the respect, albeit tempered, of the regular engineers. There were still signs of ambivalence. In Brigadier General Barnard's report on engineering operations during the siege, he wrote: "Captain Duane, with his command, and Lieutenants Comstock and McAlester, have superintended the siege works. All of these officers have exhibited great energy, industry, and courage." Conversely, he wrote this about the volunteers: "During the siege operations, General Woodbury, with his brigade, has been mainly engaged on the construction of roads and bridges, making gabions and fascines, and constructing Battery No. 4 (13-inch mortar)."[21] This was the proverbial understatement. Barnard's journal of the siege was somewhat more specific and pointed to the magnitude of the volunteer brigade's achievement. For April 24 he noted: "The northern approach to the upper pontoon bridge, 1,200 feet in length, is nearly finished, and will be completed probably tomorrow. Crib bridge, floating bridge, and middle pontoon bridge are all in working order . . . The roads in the two branch ravines above dam, with the secondary roads leading up to the plateau, will . . . be completed to-day."[22] The brigade corduroyed 5,000 yards of road and built two engineering depots while at Yorktown.[23] Duane's battalion deserved credit from Barnard for the monumental work it did, and would continue to do, throughout the campaign. Yet, so did Woodbury's men.

The Yorktown siege ended on May 3 when Confederate General Joseph Johnston's army slipped out quietly and moved northwest toward Richmond. McClellan had spent the entire month of April preparing the siege and accomplishing little. Johnston, on the other hand, had bought more time to prepare his army and the defenses around the Confederate capital for the coming fight, and he placed burgeoning demands on McClellan's expanding supply lines.

With Yorktown abandoned, McClellan reneged on his promise to Secretary of War Stanton to keep General Irvin McDowell's corps in Washington and instead ordered McDowell's men to move south toward Richmond. This forced President Lincoln to intervene. Demanding that a sizable force be left to defend the city, the president compromised with McClellan and sent Major General William B. Franklin's division of 11,000 soldiers to West Point, Virginia, a place where the Mattapony and Pamunkey rivers joined to form the York River at its northwestern end. From there McClellan would move his giant army west along the Pamunkey to White House. The Richmond & York River Railroad also connected West Point at the head of the York River to White House. From this new base Little Mac would orchestrate his operation against Richmond, approximately seventeen miles away.

A detachment of the 15th New York was assigned to assist in Franklin's river landing operation. The challenge was to prevent the artillery pieces from bogging down in the mud and sand of the beach. The engineers placed the guns on two canal barges lashed together and floated them to shore, where the barges were connected to boardwalks and wheeled ashore. Pontoons were used to bring the men to the beach. Some barges were grounded into the beach while others were strung together to extend 220 feet into the river.[24] The 50th New York was divided into four detachments to assist with the barges, prepare a pontoon train, construct trestle bridges, and repair a railroad bridge and a portion of track. Woodbury summarized the work done from May 19 to May 29: "One bridge, single span, 26 feet, at Black Creek; one bridge, two spans, 18 and 20 feet, respectively, roadway 12 feet, at Mill Creek . . . three bridges . . . beyond the White Church over streams 8 feet in width and 15 inches in depth, built with stringers laid on crib abutments; two trestle bridges 120 feet in length across the Chickahominy at Bottom's Bridge."[25]

These challenging engineering feats anticipated the engineers' almost miraculous contribution to the preservation of the Union IV Corps at Seven Pines (Fair Oaks) on May 31 in the first major battle of the Peninsula Campaign. Since landing on April 4 at Fort Monroe, the engineers had worked ceaselessly to accomplish two tasks. First, they opened avenues of transportation for thousands of men to move through a scarcely populated southeast Virginia, with many culverts and swamps traversed by few roads. Second, they maintained roads and bridges to meet logistical demands. Depots were built to establish an operational supply base. The engineers built roads, bridges, harbors, and warehouses to sustain over 100,000 soldiers and 30,000 horses and mules.[26] Most roads had to be widened. Until this happened, two wagons traveling in opposite directions could not pass each other. Furthermore, many of the roads were so muddy that engineers were required to construct three or four layers of corduroying before logs would stay above the surface of the mud.[27] The quartermasters complained that the log surfaces of the roads were "exceedingly rough, and the consequence is that the wear and tear of our transportation has been very great," but with so many roads to attend to, little else could be done.[28]

The traffic on the roads, the length of the supply line, and the rain and mud with which they had to contend provided the engineers with remarkable on-the-job training. On May 28, Brigadier General Silas Casey's division of the Union IV Corps alone reported 194 wagons in use, 84 of which were used by the fourteen regiments of the division, 30 for infantry ammunition, 13 for forage for artillery animals, and 10 for artillery ammunition.[29] Supplies originated in New York, Philadelphia, and Baltimore, traveled by steamers to West Point, "then about another 45 miles up the tortuous Pamunkey. Once at White House, the supply line stretched inland 15 miles to the Chicahominy and, eventually, across that stream to within six miles of Richmond."[30] The month of April experienced eighteen days of precipitation in the Tidewater Region, and in May, fifteen days, compounding the tactical and engineering problems.[31]

When the fighting began at Seven Pines, it became apparent at once that the Union IV Corps was in danger of being destroyed. A key reason that approximately 30,000 Yankees escaped and McClellan's army avoided disaster proved to be a group of men from the 1st Minnesota Infantry and 5th New Hampshire Infantry who knew how to build a bridge. Five bridges were constructed over the Chickahominy River, west of McClellan's base of operation at White House, one by the 1st Minnesota and 5th New Hampshire.

The Chickahominy flowed through an area of heavily timbered swamp, generally 300 to 400 yards wide.[32] When the battle began on May 31, Union forces were deployed on a northwestern line along the river, with the exception of Brigadier General Erasmus Darwin Keyes's IV Corps, which had crossed to the Richmond side of the river and advanced as far as Fair Oaks Station on the Confederates' right flank and Seven Pines on their left flank. Keyes's men were isolated, and Confederate General Johnston decided to take advantage of the situation and attempt to encircle Keyes and destroy his entire corps. Realizing Keyes's vulnerability, McClellan was frantic and ordered Brigadier General Edwin Vose Sumner's II Corps to cross the river immediately and reinforce the retreating Keyes; but Sumner faced a disaster waiting to happen. Three days of heavy rains had turned the river into swollen rapids, and the flooring of the bridge built by the Minnesotans and New Hampshirites was rising, soon to tear apart under the pressure of the water.

Longstreet's corps was to envelop the Union right by taking the Nine Mile Road. D. H. Hill was to make a secondary attack along the Williamsburg Road, and Benjamin Huger was to protect the Confederate right flank along Charles City Road. Without good maps, Longstreet's corps managed to get on the wrong roads, the ones assigned to Hill and Huger. As a result, approximately 30,000 men were congested along a five-mile stretch of road and were unable to move. Johnston ordered the attack for 6:00 a.m., but it took Longstreet seven hours to sort out the confusion and bring six of his brigades on line. The critical time lost by Longstreet's men, along with the rickety bridge built by two companies from the 1st Minnesota under the command of Captain Mark W. Downie and Second Lieutenant Christopher Heffelfinger and the 5th New Hampshire under the command of Colonel Edward E. Cross, saved Keyes's corps from destruction and the Army of the Potomac from humiliation.

When Sumner reached the bridge, as Downie predicted, high water was affecting the already questionable stability of the structure. Colonel Alexander of the engineers implored Sumner not to cross.

> "General Sumner, you cannot cross this bridge."
>
> "Can't cross the bridge? I can, sir. I will, sir!"
>
> "Don't you see the approaches are breaking up and the logs displaced? It is impossible," Alexander implored.
>
> "Impossible? Sir, I tell you I can cross. I am ordered!"[33]

The approaches at both ends of the bridge were quagmires, forcing soldiers to push wagons onto the swaying, slippery bridge. The rain pelted down, and the menacing sound of the water below reminded the men of the deadly consequence of falling. Yet, led by Sumner, Brigadier General John Sedgwick's division and Brigadier General Israel Richardson's division crossed throughout the night, and just as the last soldier walked off the bridge, the structure collapsed with a deafening sound into the river. Sumner's moxie sufficed, however, as his men arrived in time to save Keyes's left wing from collapse.[34]

After Lee took command of the army, he ordered a general withdrawal from the Fair Oaks area. This gave both armies time to reorganize and allowed McClellan to repair and build more bridges. One of the most remarkable bridges constructed by men from the 15th and 50th New York Engineers was Woodbury Bridge. In six days of rain, the engineers constructed an approach road to the bridge, fifteen feet wide and through a swamp. The roadway and bridge were approximately one mile long, and embankments at both ends of the bridge elevated the structure to protect it against rising water.[35]

The Engineer Brigade, with the assistance of the 3rd Vermont Regiment, also built a bridge opposite Dr. Peterfield Trent's home, McClellan's headquarters; this bridge was 1,080 feet long and constructed with forty cribs and six trestles.[36] Another officer from the 15th New York, Captain William A. Ketchum, almost lost a bridge to rising water, but after removing a damaged trestle and skillfully replacing it, finished his project in nineteen hours. Yet, he issued a complaint in his report to Major General Woodbury, commander of the Engineer Brigade, which would reveal early problems with the Union army's command structure: "I would also beg leave respectfully to report that I was very much annoyed by the constant interference of officers higher in rank than my-

During the Peninsula Campaign in June 1862, men from the 15th and 50th New York Volunteer Engineers built the impromptu Woodbury Bridge across the Chickahominy and a log roadway, approximately half a mile long, from each end of the bridge. The innovative log work underneath each end of the bridge served to support raised embankments, which protected the bridge against rising water. Library of Congress, Prints & Photographs Division, Civil War Photographs, LC-DIG-ppmsca-33363.

self, who came to me ordering me to hurry up the work, and representing that they had the authority of the general commanding."[37]

The restoration and refitting of both armies finally came to an end on June 25 at Oak Grove. In an attempt to drive Confederate units from his front and begin a progressive assault on Richmond, McClellan ordered an attack that was repulsed by a larger Rebel force. There were 626 Northern and 441 Southern casualties. The next day, Lee seized the initiative in what later would be called the Seven Days Battles. The ferocity of the Confederate attacks at Mechanicsville on June 26 and Gaines's Mill on June 27 demonstrated two things: McClellan did not have the stomach to fight, and Lee and his men did. On the same day that the fighting began at Oak Grove, McClellan considered moving his base of operation from White House to somewhere along the James River near the navy's gunboats. Barnard sent Woodbury to White Oak Swamp Creek to build three bridges and to corduroy a road through the swamp. The latter was approximately five miles southeast of the Grapevine Bridge. The creek was narrow and not difficult to bridge, but it was bound on each side by approximately 200 yards of swamp. Cutting trees, building a raised corduroy road, and doing it all in twenty-four hours was the challenge. On June 27, McClellan ordered the army to "change the base of operation" to the James River, and Lee's intention was to cut them off before they got there.

Fighting continued at Savage's Station on June 29, but the dénouement was to be at White Oak Swamp. If Stonewall Jackson could cross over to the left or south bank of the Chickahominy River and then strike the retreating Federals from the flank or rear, he would destroy McClellan's army. Jackson, however, had one major obstacle in his path. To carry out Lee's orders he had to rebuild a bridge across the river. Union engineers had destroyed all the structures they had worked so hard to construct and maintain. From New Bridge near Gaines's

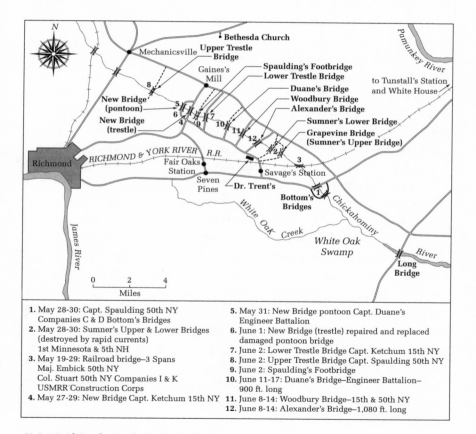

1. May 28-30: Capt. Spaulding 50th NY Companies C & D Bottom's Bridges
2. May 28-30: Sumner's Upper & Lower Bridges (destroyed by rapid currents) 1st Minnesota & 5th NH
3. May 19-29: Railroad bridge–3 Spans Maj. Embick 50th NY Col. Stuart 50th NY Companies I & K USMRR Construction Corps
4. May 27-29: New Bridge Capt. Ketchum 15th NY
5. May 31: New Bridge pontoon Capt. Duane's Engineer Battalion
6. June 1: New Bridge (trestle) repaired and replaced damaged pontoon bridge
7. June 2: Lower Trestle Bridge Capt. Ketchum 15th NY
8. June 2: Upper Trestle Bridge Capt. Spaulding 50th NY
9. June 2: Spaulding's Footbridge
10. June 11-17: Duane's Bridge–Engineer Battalion– 900 ft. long
11. June 8-14: Woodbury Bridge–15th & 50th NY
12. June 8-14: Alexander's Bridge–1,080 ft. long

Union Bridging during the Peninsula Campaign, 1862

Mill to Bottom's Bridge east of Savage's Station, everything was destroyed. On June 29, Jackson ordered Major Robert Lewis Dabney, former schoolteacher and pastor, to rebuild the Grapevine Bridge.[38]

Dabney served as Jackson's assistant adjutant general, and in that role he was the general's advisor, confidante, and spiritual counselor. But as an engineer, Dabney was incompetent. The only logical reason that Jackson selected the major was that no engineers were available. To compound Jackson's problems, Dabney chose as laborers men who knew nothing of carpentry and construction, and these men were not pleased to be used as work gangs. This became a common theme throughout the rest of the war—Southern soldiers did not sign up to dig ditches and labor at menial tasks; they signed up to kill Yankees.[39]

Help arrived around mid-morning. Captain Claiborne Rice Mason, a road builder before the war and a self-taught engineer, arrived with his slaves and took over the work on the bridge. A martinet, Mason was confident that his slaves would do exactly as told and the bridge would be finished before noon. Unfortunately for Mason, the Chickahominy River remained high and rebuilding the bridge required some improvisation by workers who were not allowed to think for themselves. The problem was with the center section.

Union engineers had also had problems with the center section. After the first Grapevine Bridge tore apart due to the high and swift current, McClellan immediately ordered it rebuilt, and the assignment went to Captain Ira Spaulding of the 50th New York Volunteer Engineers. After his men had been in the cold water for at least nine hours, frequently diving beneath the surface to place the legs of the trestles firmly in the river bottom, Spaulding realized the rapid current was shifting the legs of the center portion of the bridge. It was only a matter of time before the trestle section in question would be carried down the river. Spaulding's solution was to dismantle the center section, about forty feet of bridge, and then anchor and tie off pontoons in the middle of the bridge. The floating pontoons could now ride the current, provided they were lashed to the anchored trestle sections. So the new center was connected to the side rails of the trestle sections, one coming from the south bank and the other from the north bank. After twelve hours the bridge was completed. Spaulding's regimental commander, Colonel Stuart, marveled at what was accomplished and commended the officers and men for their skill and endurance.[40]

Now, Confederate Captain Mason had to rebuild the Grapevine Bridge, and it is not clear how he did it. Twenty-four hours after Dabney started construction, Mason finished the rebuilding on the morning of June 30. Meanwhile, the main body of the Confederate army was attacking McClellan's forces at Frayser's Farm. The Union line formed a right angle along White Oak Swamp Creek and was waiting when Jackson's men struck around noon. Jackson's tardiness eliminated any hope of a pincer movement and of cutting off the retiring army from its supply base. Historians have faulted Jackson for arriving late and making a poor showing during the Seven Days Battles. Some historians blame it on his fatigue after fighting a brilliant Valley Campaign. Others argue that personality conflicts with subordinates slowed him down. Yet others describe it as a mystery, citing the man's reflective mind and observable eccentricities. Perhaps a lack of army engineers and soldiers willing to labor all day in waist-high cold water should be added to the list of reasons why Jackson arrived too late to cut off McClellan's army from its supply base and perhaps altering the course of the war.

The final battle of the Peninsula Campaign was fought on July 1 at Malvern Hill. Lee continued to pound away at McClellan's army, but at a high price— 5,355 killed or wounded. A combined total of 8,569 casualties at Malvern Hill forced both armies to break off the fighting; Lee moved back toward his Richmond defenses and McClellan toward his new base at Harrison's Landing on the James River. The Union army would not move back to the Washington, DC, area until August, but the Peninsula Campaign was over.

As generals evaluated what went right and what went wrong with the campaign, it was not clear in July 1862 whether the Union's significant advantage in engineering, technology, and logistics would really matter. For the Confederacy, there were just enough trained army engineers to mark out and supervise the construction of major defenses around key geographical points in the South, making it difficult for Union armies to invade the Southern heartland and making it possible to prolong the war and to force the war-weary Northern public to demand peace and recognition of the Confederacy. Yet, a lack of engineers and skilled laborers within the army had already hampered the South. The disaster at Fort Henry, the inability to establish a management protocol for the railroads, the lack of attention to establishment of an engineering bureau in Richmond, poor maps, few soldiers willing to labor on engineering projects, and no engineer battalion or regiment in any of the armies in the field—all foretold future logistical and tactical problems and raised concerns about how creative and aggressive Southern commanders could be with their forces.

From the standpoint of the Union engineers, the Peninsula Campaign demonstrated their mettle. Their last bridge, at Barrett's Ferry, was built under the direction of Captain Spaulding of the 50th New York and lieutenants Cross and Comstock of the Engineer Battalion. Made up of five spans of trestle and ninety-six pontoons, it was 1,980 feet long and, despite the deep water and strong tidal currents, was completed in less than twenty-four hours. Quartermaster General Montgomery Meigs reported that 5,899 horses and 8,708 mules drawing 2,578 wagons and 415 ambulances, as wells as 12,378 artillery and cavalry horses, moved over it.[41] General Barnard's report made it clear that he had finally come around to appreciating the skill and commitment of the volunteers: "On the Chickahominy and on retreat to the James the duties of the brigade were arduous, as have been described, and I found in its chief throughout the campaign an officer prompt and fertile in expedients, daring and assiduous in execution, and always exhibiting a wise foresight."[42]

Barnard's report was also unique among the thousands filed by commanding officers throughout the war. It was a polemic as much as it was a report. Not only did it recap the operations of his men, but it also reflected on the challenges ahead. The report asked two major questions: what went wrong during the campaign, and what needed to change to maximize the contribution the engineers would surely make to bringing about the defeat of the Confederacy?

Barnard was candid. The campaign failed because the army lost the initiative from the moment it landed at Fort Monroe. Morale and power rested with the army, but the siege of Yorktown and the fact that the "troops toiled a month in the trenches or lay in the swamps . . . took a fearful hold of the army, and toil and hardship, unredeemed by the excitement of combat, impaired *morale*."[43] The disasters of the campaign were self-inflicted, deprived the army of élan, and diminished manpower and materials.

To improve the engineering function within the army, Barnard made four recommendations: (1) promote engineers to adequate rank, (2) provide proper recognition for engineers' distinguished service, (3) recruit more engineer forces and organize them properly, and (4) provide the engineers with the authority to deliver and maintain the proper tools. The general argued that engineers carried great responsibility. The efforts to get ordnance, ammunition, food, and medical assistance to the front lines most often depended on temporary bridges and roadways built under dangerous conditions and over inhospitable ground. Engineers fixed defensive positions and indicated the "points of attack of fortified positions." Accurate maps had to be prepared to prevent commanders from taking the wrong road and getting lost at a crucial time during a battle. "Adequate rank," Barnard wrote, "is almost as necessary to an officer for the efficient discharge of his duties as professional knowledge . . . To give him the proper weight with those with whom he is associated he should have, as they have, adequate rank."[44]

This lack of rank Barnard equated with a lack of recognition, distinction, and respect. Field commanders seldom wrote about engineering operations in their reports, and George McClellan, once an army engineer himself, took the work of the engineers for granted. Barnard argued that this lack of recognition was unfair. He also believed that by limiting promotions for engineers, the corps had lost good men like George Meade, George Thomas, James McPherson, and William Rosecrans to field commands.[45]

As Barnard suggested, more engineer soldiers were needed and a better chain of command had to be implemented. Each army corps had to have its own engineer troops, a pontoon train, and tools. Under the current system, the commanding general detailed engineering soldiers, yet corps and division commanders also demanded work from the same soldiers. For example, the Grapevine Bridge was being rebuilt when General Sumner was ordered to cross it. The engineers did not know he was coming and had not determined the safety of the bridge. Sumner did not care. He planned to cross the river. An engineering party from Sumner's corps could have arrived at the bridge earlier and focused on its construction. As it was, the engineers working on the bridge were undermanned, and as a result, infantry soldiers from the 1st Minnesota and 5th New Hampshire were helping to build the bridge. In addition, the tools used on the bridges were scattered everywhere. Because tools were distributed by the Quartermaster Department, it was difficult to maintain any system of responsibility for them. In some cases, engineers had problems constructing causeways and bridges because they had no tools or very few of them.

The ordinary soldier, Barnard continued, with the 1st Minnesota, the 9th and 22nd Massachusetts, the 5th New Hampshire, or the 3rd Vermont infantry regiments that built bridges during the campaign found an axe or a shovel "a very convenient thing to have at his camp, and carried one off with him. When the army moved he found it inconvenient to carry it and threw it away."[46]

More engineer soldiers were needed, as well. Barnard in his report lamented the shortage of engineer troops and tools, yet astute infantry commanders soon recognized that there were men within the volunteer infantry regiments who had the skills to build bridges. The antebellum labor force in the North had produced a great number of men with mechanical or construction skills. They had worked with their hands in shipyards and locomotive plants, and they had learned how to fix things sometimes just by improvising. These men would prove to be the Union army's greatest advantage.

Finally, Barnard evaluated the pontoon equipage used during the campaign to determine what the army should adopt in future offensives. In the fall of

1861, Duane had recommended the use of French pontoons, and now the general finally agreed. "The Birago trestle, of which I had formed so high an opinion," he wrote, "proved itself dangerous and unreliable—useful for an advance guard or detachment, unfit in general for a military bridge." He also ruled out American India-rubber and Russian canvas pontoons.[47] The advantage to the Birago trestle was its light weight. The problem was its strength. On the other hand, although the engineer soldiers had trouble maneuvering the French pontoons into position, these pontoons had what Barnard described as "floatation power." The pontoon was designed to carry a significant amount of traffic. "No make-shift expedient, no ingenious inventions not tested by severe experiment, nor light affair, of which the chief merit alleged is that it is light, will be likely to do what is required, and what the French pontoon has so often done."[48]

General Barnard was correct when he wrote to the chief engineer of the United States Army, Brigadier General Joseph Totten, in January 1863 that the "movements of the whole army were determined by the engineers."[49] So it was apparent to some in July 1862, although perhaps taken for granted by field commanders, that engineers were essential to Northern logistical operations and the army's movements. Bridge building, road making, mapmaking, and technological improvisations had all been attempted with a fair amount of success.

It was also clear in July 1862 that some Union war planners were relying on engineers to figure out how to utilize the taproot of all technological developments of the mid-nineteenth century: the railroad. The creation of the United States Military Railroad began in an inconspicuous fashion in the spring of 1861 when Secretary of War Simon Cameron seized all the commercial telegraph lines around Washington. He then asked his assistant, Thomas Scott, to help him manage the telegraph and railroad lines around the capital.[50] The story of what happened next, creating an organization to manage and maintain military railroads, is remarkable.

CHAPTER SEVEN

The Birth of the United States Military Railroad

Thomas Scott, Daniel McCallum, and Herman Haupt

That man Haupt has built a bridge across Potomac Creek, about 400 feet long and nearly 100 feet high, over which loaded trains are running every hour, and, upon my word, gentlemen, there is nothing in it but bean-poles and corn stalks.

Abraham Lincoln, as told to Herman Haupt, May 1862

The Union army's Engineer Brigade and Battalion were not alone in facing considerable and sometimes overwhelming challenges in the spring of 1862. With supply lines critical to McClellan's potential success, the railroads would be pressed into service. To operate efficiently, they would require sound management drawing on the resources available. Northern railroad operations would not be the problem, but the Union army's ability to use railroads abandoned by Virginia and to cope with forces that would militate against that use was a challenge. In the prewar years, rapid expansion of the railroad produced men with the skill sets necessary to build locomotives, construct bridges, manage men and material, and adapt to circumstances and problems. The decision was made by the Federal government early in the war to tap these manpower resources and to use the railroad as a war tool. Fortunately for McClellan's army on the Peninsula and for the Union's war effort in general, the idea of nationalizing the railroads and establishing a bureau to coordinate civilian and military use developed just two weeks after the war started.

With the fall of Fort Sumter, panic struck Washington. The Lincoln administration feared that with proslavery Maryland and Virginia on Washington's borders, the city was vulnerable to invasion by Confederate forces. Furthermore, any attempt to send Northern troops to the capital by railroad meant they had to pass through Baltimore and thus could be blocked by fire-eating Marylanders. This worry soon appeared prescient. As Union Brigadier General Benjamin Butler's soldiers arrived in Baltimore by train from Massachusetts and New York, mobs greeted them with rocks, garbage, and musket balls. The soldiers returned fire, and at least a dozen locals were killed. The soldiers managed to work their way to the west end of the city where they boarded a train for Washington. Lincoln had some troops in Baltimore, but the pro-Southern faction held the city.[1]

Maryland Governor Thomas Hicks wanted no trouble and suggested to Lincoln that troops from the North should bypass Baltimore and travel by boat to the capital. That option was too slow for Lincoln. Governor Hicks, therefore, in an effort to avoid a likely confrontation between Federal soldiers and Southern sympathizers, blocked all the railroad bridges into the city. For Union troops there were two choices: force their way into the city (and lose Maryland to the Confederacy) or take regiments by steamer to Annapolis, then take the train to Elkridge, Maryland, just south of Baltimore, and reconnect with the Baltimore & Ohio Railroad. The problem was that pro-Southern Marylanders had torn up the track and destroyed several locomotives.

Congressman David Wilmot of Pennsylvania and Major General Robert Patterson, commanding Pennsylvania volunteer troops, suggested to the War Department that the railroad between Baltimore and Havre de Grace be taken over by the Federal government. Instead of waiting for an answer from Secretary of War Simon Cameron, General Butler took the initiative and took possession of the Annapolis & Elkridge Railroad.[2]

Just a short time later, Cameron seized all the commercial telegraph lines around Washington, and he then asked Assistant Secretary of War Thomas Scott to help him manage the telegraph and railroad lines around the capital.[3] One of the first things Scott did was to take charge of the Annapolis & Elkridge Railroad and create the United States Military Railroad Service, in practice if not in name. Next, Scott asked Andrew Carnegie, superintendent of the Pittsburg division of the Pennsylvania Railroad, to assist him, and under Carnegie's guidance the first government telegraph line was built connecting the War Office with the Navy Yard.[4] Scott then called on four telegraph operators from the Pennsylvania Railroad to report to Washington to form the nucleus of the Signal Telegraph Corps. The corps functioned independently except for supervisors who received a nominal military rank because they had to draw funds from the Quartermaster Department. All the operators received orders directly from the secretary of war. The corps eventually employed more than 1,500 civilian operators and construction workers.[5] After the Peninsula Campaign, the Telegraph

Corps started to build more permanent lines hung from poles, which were protected by Union cavalry patrols. Eventually, copper wire insulated by vulcanized rubber was used to protect the lines from the elements, and by the end of the war, the Construction Corps had strung 15,000 miles of wire.[6]

As the former first vice president of all operations of the Pennsylvania Railroad, Scott was a valuable asset: he had engineering skills, business acumen, and management capability. He saw the potential in rail transportation for the army, and he anticipated the problems that the military would face if the government did not nationalize the railroads. In early January 1862, he prepared a report on the military's transportation needs—anticipating an act of Congress passed on January 31, 1862, authorizing the president to take possession "of any and all the railroad lines in the United States" and all the telegraph lines in the country until "the suppression of this rebellion."[7]

Scott argued that although technically subject to military control, Northern railroads needed to remain under civilian management. Nonetheless, he made it clear to railroad operators that they were to act as "direct adjuncts" of the War Department or he would force them into doing so. He promised them that the latter was not in their best interests.[8] Scott, like his new boss, Secretary of War Edwin M. Stanton, made thoughtful and decisive decisions. This decisiveness would become the expectation among those in the military railroad service. For example, while visiting a Pittsburgh foundry in 1862, Scott learned that a city ordinance required the foundry to haul finished work to the Allegheny River by horse-drawn wagons. To rectify this waste of valuable time, Scott ordered a railroad siding to be built from the foundry to the river. This minor expense was recuperated tenfold as iron efficiently reached the river and was carried to waiting quartermaster depots, shipyards, and railroad track and engine facilities.[9]

The next step for the War Department's newborn authority to take over railroads, as necessary, was to appoint a central administrator. Stanton wasted no time. On February 11 he appointed Daniel C. McCallum as military director and superintendent of railroads in the United States. McCallum would be responsible to the War Department and report directly to Stanton.

McCallum was an excellent choice for the position. As superintendent of the Erie Railroad before the war, he had designed a new management system to adjust to the increasing length of company railroads. It was virtually impossible for one person to travel and inspect the entire 340 miles of track and the equipment, repairs, and ticket operations of the railroad. Recognizing the limitations, McCallum had developed a decentralized management structure, which included three radical changes to the way a company did business.

First, the delegation of authority was essential for the railroad to successfully operate all along its line. Local managers had to solve problems as they arose, making important decisions on complicated and critical issues. Senior managers were expected to allow these local managers to manage and not interfere.[10] Second, McCallum's new system demanded *"personal accountability through every grade of service."*[11] Third, frequent written reports were required to keep upper management informed on each section's work. For military railroads, besides reporting on fuel supplies or schedules, there were additional burdens: documenting the replacement of destroyed locomotives, bridges, and track; resolving conflict between civilian passenger and freight demands and the War Department's demands; weighing generals' individual needs against the greater needs of the army; and sorting out army priorities to determine what should be shipped first: men, artillery, supplies, or the wounded.

It took time for McCallum to organize his new department, and an unanticipated crisis at the outset of the Peninsula Campaign disrupted his early management planning. In all wars, but certainly in the Civil War, where the number of soldiers and volume of supplies exceeded anything Americans had experienced in past wars, unanticipated problems are the norm. So when General McClellan asked Irvin McDowell's corps to join him and threaten Richmond from the north, the problem McCallum faced became readily apparent. McDowell assured the War Department that he needed to establish a reliable supply line before he would move against Richmond. This meant that the Richmond, Fredericksburg & Potomac Railroad needed serious repairs, from Aquia Creek to Fredericksburg. Not only was the track of the railroad torn up, but for a three-mile stretch it was impossible to determine whether there had been a track at all.[12] The need to rapidly rebuild or, in some cases, build the railroad and wharfs at Aquia Creek left Stanton and McDowell nonplussed. The secretary of war sought advice, so on either McCallum's or Scott's recommendation, he called to Washington first Daniel Harris, and then Herman Haupt.

Daniel Harris was the former mayor of Springfield, Massachusetts, and president of the Connecticut River Railroad. A highly skilled engineer, he was known for his supervision of the construction of twenty-seven bridges along the Hartford, Providence & Fishkill Railroad, his cautious business style, and his strong opposition to one of Herman Haupt's major investments, the Hoosac Tunnel project.[13] Stanton explained the work required, and Harris asked for time to consider a decision. Stanton needed an immediate answer, so he called Haupt to Washington and offered him the same assignment, and Haupt accepted.[14]

The workforce was initially made up of soldiers detailed from infantry regiments, many of whom were disinclined to spend time as laborers. Some of these men had skills that Haupt put to good use, but others needed simple and careful instructions about what they were supposed to do. For example, Haupt devised a simple instrument made of sticks to level the rail bed. In three days, working around the clock, his work crews relaid three miles of rails and then prepared to rebuild two bridges—one over Accokeek Creek and the other over Potomac Creek.[15]

The Accokeek Creek trestle bridge was 150 feet long and was built over a 30-foot-deep chasm. Fifteen hours after construction began, General McDowell rode over the finished bridge in a locomotive. It was a remarkable achievement, and congratulations came in from the War Department. The most incredible work, however, was yet to come as Haupt moved immediately to Potomac Creek and started plans for a bridge "that would span a gorge almost four hundred feet wide with a maximum height of ninety feet over the water, 'a frightful looking chasm.'"[16]

Taking soldiers from the 6th and 7th Wisconsin and the 19th Indiana Infantry regiments, Haupt asked the officers to make a list of each soldier's prewar occupation. From the list Haupt organized the men "into teamsters, choppers, carpenters, mechanics, and laborers, and formed them into squads, each under a non-commissioned officer."[17] As the crews went about their business, Haupt began to formulate the idea of a civilian construction corps. Having operated with untrained and in some cases noncompliant soldiers, he had already asked General Halleck whether the general would be open to the idea. Halleck, a West Point–trained engineer and scholar, believed experienced engineering soldiers like those from Captain Duane's Engineer Battalion should eventually be assigned the work. For the time being, Haupt deferred to the general even as he continued to ruminate on his plan. Now, as he organized his work gangs, a nascent structure for a construction corps was formed.

To speed up construction on the Potomac Creek Bridge, Haupt "hastened transportation of the logs from the woods to the bridge site by building a wooden tramway, for which he procured several sets of rollers." Instead of building the lower section of the bridge with trestlework, he had the soldiers lay crib work: because "many of the men were accustomed to building log houses and were not carpenters," he later wrote, "I put them at work [with] which I supposed they were familiar."[18] It rained for nine days straight, and the men handled approximately two million board-feet of lumber, but the bridge was completed and the first locomotive crossed. Lincoln and various cabinet members came out to see this wonderful accomplishment on May 28. The president

In May 1862, Herman Haupt's Construction Corps rebuilt the Potomac Creek Railroad Bridge in nine days. When President Lincoln saw the bridge, he is alleged to have said: "That man Haupt has built a bridge across Potomac Creek, about 400 feet long and nearly 100 feet high, over which loaded trains are running every hour, and, upon my word, gentlemen, there is nothing in it but beanpoles and cornstalks." Library of Congress, Prints & Photographs Division, Civil War Photographs, LC-DIG-ppmsca-11749.

said: "I have seen the most remarkable structure that human eyes ever rested upon. That man Haupt has built a bridge across the Potomac Creek 400 feet long, nearly 100 feet high and running trains over it. There is nothing in it but bean poles and cornstalks."[19] McDowell said of the bridge: "It is a structure which ignores all the rules and precedents of military science as laid down in books."[20]

A third bridge, over the Rappahannock River, was completed at the same time that Haupt finished the Potomac Bridge, its construction supervised by Daniel Stone, a respected railroad contractor. Stone's was also an impressive achievement. The War Department had succeeded in finding a handful of brilliant railroad men, especially Haupt, who worked miracles during the month of May 1862. Yet the organizational component of military railroad operations was in its infancy. The question of how to develop and where to find skilled work crews was unresolved. Furthermore, Stanton's management of this fledgling railroad bureau was perplexing and impulsive. As the Peninsula Campaign entered June, Confederate General Stonewall Jackson began his advance up the Shenandoah Valley, forcing Lincoln to withhold McDowell from moving on Richmond and instead move him to Front Royal on the Manassas Gap Railroad.[21] To supply McDowell's army, the Orange & Alexandria Railroad had to be repaired from Alexandria to Manassas, and the Manassas Gap Railroad had to be rebuilt west to Front Royal.[22] The problem was that Haupt, Stone, and McCallum had overlapping authority.

The order to convey McDowell's army to the Shenandoah Valley exposed a major flaw in the young military railroad department's command structure. Stanton had given McCallum the authority to "enter upon, take possession of, hold and use all railroads . . . that may be required for the transport of troops, arms, ammunition and military supplies."[23] These same instructions had been given to Haupt and Daniel Stone, and now Haupt needed to operate in Stone's territory. Stone vehemently objected to Haupt's assumed authority. Haupt, puzzled by the management confusion and conflicting authority, took his concerns to McDowell, who informed Stanton about the predicament. Without hesitation, Stanton wired Haupt giving him supreme authority over the railroads "within the geographical limits of the Department of the Rappahannock."[24] This management decision still left a problem because, theoretically, Stanton gave Haupt authority over Stone and over McCallum who was the head of military railroads.

Fortunately, serendipity brought together McCallum and Haupt. They understood each other's gifts and talents. McCallum was a skilled engineer with considerable administrative skills, and he understood the magnitude of the work ahead and the need for overall control of his department. Northern Virginia was his immediate focus, but soon the army's western theater of operation would require his attention as well. Conversely, Haupt wanted to be working in the field, engaged in the construction of railroads and the transportation problems that frequently arose. The men tacitly accepted each other's role, and this turned into a boon for the North.

Haupt, now as chief of construction and transportation for the Department of the Rappahannock, started an organizational process that revealed his true genius. He analyzed the number of deficiencies in railroad operations up to that point in the war and discovered that there was significant military interference in running the trains, that equipment was insufficient, and that supply depots refused to unload cars in a timely fashion and never returned them. Operators were not always at their posts, and there were no timetables. Trains used for military personnel and equipment did not run on a schedule. Consequently, Haupt issued his first set of general orders on June 2, outlining his expectations of personnel and establishing operating standards for army commanders. No military officers except McDowell were to interfere in the running of the trains. The trains would run on a schedule, one to which everyone was to adhere. Before this, military trains had operated without timetables and all movements were communicated by telegraph, so if the telegraph lines were cut the trains did not run. All trains would now depart on time, fully loaded or not. Conductors and agents were to report to Haupt daily, noting exact arrival and departure times of the trains.

Of course, Haupt's general orders did not eliminate his headaches. Just two days after he published his regulations, he received an order from Quartermaster General Meigs directing him to report to a quartermaster officer at a depot to expedite a shipment of supplies to Union cavalry units that had been without for several days. Haupt turned the telegram over to McDowell. McDowell then made a decision that no doubt altered the course of the war. He telegraphed Stanton the following: "I beg that the Quartermaster-General's telegram, directing Colonel Haupt to report to Colonel Rucker, may be revoked . . . With the broken-down road, and weak, worn-out locomotives, bridges going down with the freshet, and insufficient assistance, he has difficulty enough without adding to them by placing him under an officer who has had no experience in the business of railroad management, of which Haupt is the head." Understanding that Haupt would not tolerate serving under someone with fewer skills than himself, McDowell added this final sentence: "I shall lose [him] to all intents and purposes if he is placed under an officer who is not under my command, and who knows comparatively nothing of the business he is to superintend."[25] Stanton rescinded Meigs's order. This was, without exaggeration, one of Stanton's most important decisions of the war.

With the support of McDowell and Stanton, Haupt continued to provide structure to railroad operations. As he supervised various construction crews in mending track, building bridges, and repairing locomotives, the idea of a permanent construction corps continued to form in his business-like mind. Just one week after McDowell's telegram to Stanton, Haupt issued his blueprint for a construction corps. Laborers recruited from civilian life and from regiments within his department would be paid extra for their services in exchange for hard work. The expectations were not ambiguous: "Men who are not willing to work, even for sixteen hours continuously, when required, are not wanted in the Construction Corps of the Rappahannock and are requested to leave it and return to their regiments at once."[26] Men were formed into squads of ten men commanded by a noncommissioned officer, and either a civilian foreman or an army lieutenant commanded two squads. Multiple squads were led by a super-intendent or army captain.

Haupt kept up a furious pace and did not care whom he offended. He was confident in his ability to deliver a superb railroad system to the army and took whatever actions necessary to accomplish the task. New regulations came out from his office at the end of June. Quartermaster and commissary officers were forbidden to load cars without proper authority. The common practice of ship-ping materials used by officers for their own private use was no longer permit-ted. Haupt made many an officer angry and unhappy, but he remained impervi-ous to their demands.

McCallum and Haupt continued to implement the military railroad system as Major General John Pope was given command of the newly reorganized De-partment of the Rappahannock, known as the Army of Virginia. Fresh from his western theater victories, Pope arrived in Virginia ready to go on the offensive. Although aggressive, Pope possessed a McClellan complex. The hero of Island No. 10 believed he was called east to regain the initiative lost by the Army of the Potomac's efforts on the Peninsula. He believed the president and War Depart-ment selected him because he had proved himself to be the top general in the army. Pope anticipated crushing Lee's army and ending the Civil War.

Pope made McDowell corps commander, but he refused to recognize Haupt as part of the army organization and made clear that "a separate and indepen-dent department for the construction and operation of the railroads was unnec-essary."[27] Since the railroads were used to transport supplies, Pope placed his quartermaster department in charge of them. Haupt went to Washington, ex-plained his displeasure with the arrangement, and asked to be relieved. Stanton granted his request, and the railroad man went back to his home in Massachu-setts just as the Army of Virginia started its summer offensive.

If Pope hoped to destroy Lee's army somewhere in Virginia, he would need control of the railroads to accomplish it. His major supply line from Washing-ton extended 25 miles to Manassas Junction along the Orange & Alexandria Railroad, where warehouses and railroad cars stored a wealth of food, clothing, shoes, and ammunition. His forward base of operation was another 30 miles south at Rappahannock Station. The Army of Virginia continued south until August 9, when it clashed with 24,000 men under Stonewall Jackson at Cedar Mountain, a Confederate victory that prompted Union Assistant Secretary of War P. H. Watson to telegraph Haupt: "Come back immediately; cannot get along without you; not a wheel moving on any of the roads."[28]

In Haupt's absence, confusion reigned. Generals interfered with timetables, quartermasters neglected to unload and return cars, and large supply stores sat untouched. During this period of chaos and confusion, Haupt decided to return. General Pope and the War Department realized their grave mistake in forcing his resignation, but Haupt's return, while restoring a degree of order, did not eliminate the meddling conduct of generals and staff officers. For example, on August 23, Brigadier General Samuel Davis Sturgis announced to Haupt that he, Sturgis, was to assume military control of the railroads for the benefit of his division. Haupt immediately contacted the War Department and sent a telegram to Sturgis: "You have now detained 5 trains 2½ hours. You kept the sick from coming to Alexandria where the Medical Director has long been waiting with ambulances. Some of the sick have died in the cars. The engines may be ruined by standing so long fired without the ability to move." Enclosing a copy of Pope's general order dated August 18, 1862, giving Haupt complete control over the railroads, he told Sturgis that unless ordered by Stanton, Halleck, or Pope, the general's men were not to board the train that evening.[29] Halleck then telegraphed Haupt and Sturgis to say that no one was to give orders to the railroad superintendent's subordinates except through Haupt, and no one was to interfere in the running of the trains.[30]

When Sturgis received Haupt's message countermanding his order to take control of the train, his anger was heard and felt by everyone. "God damned son of a bitch," he bellowed and then threatened to shoot Colonel Haupt if he continued to cause problems. He would have the colonel arrested and disgraced. How dare he embarrass a superior officer. "I want to see how the damn son of a bitch looks in the face."[31] Sturgis sent provost marshal guards to bring Haupt to the general's headquarters. By the time Haupt arrived, under guard, Sturgis was drunk. At the same time, Halleck's dispatch arrived giving complete authority over the railroads to Haupt. Haupt then had a difficult time convincing the intoxicated general that the note was from Halleck, not Pope. Throughout the next few hours Sturgis just kept repeating: "I don't care for John Pope a pinch of owl dung."[32] Sturgis finally wore himself out and Colonel Haupt went back to work, only to deal with the next crisis.

Five days after the Sturgis incident, the Army of Virginia met disaster at Second Bull Run. Under Haupt's direction, however, the railroads continued to achieve some noteworthy accomplishments. On the single-tracked Orange & Alexandria, operators managed to maintain a steady flow of supplies and men to the battlefield and to evacuate the wounded and support the army's withdrawal. The Construction Corps made repairs to the lines, corrected derailments, and built bridges. In a twenty-four-hour period, the effective management of the rails allowed the army to transport 15,000 soldiers, ammunition, food, forage, and the wounded. Haupt worked for days without sleep, and Stanton, Lincoln, and the cabinet were most grateful for his efforts. On September 5 he was promoted to brigadier general. Haupt, however, did not sign the commission because he never wanted to be under binding military orders.[33] Stanton would tolerate this arrangement for only a short while longer.

Meanwhile, Haupt continued to suggest to Stanton ways in which the military railroad bureau could be improved, especially in the western theater. Haupt did not see McCallum as the head of all military railroads and instead saw him as just the chief administrator. Now, Haupt suggested that McCallum be promoted to brigadier general and placed in charge, while Haupt, who saw himself as the current director of military railroads and head of the construction bureau, would remain chief of construction and transportation and would report to McCallum "all matters appertaining to the office details."[34] Stanton, who was upset with Haupt for not signing his commission, filed the latter's recommendations and did nothing to revamp the railroad bureau. It would continue to operate as it already did throughout the summer of 1862, with men assigned as repair engineers and superintendents to specific railroads and all reporting to Haupt. By December 1862, the average number of men per month working for the Construction Corps was between 750 and 1,700, used as the situation dictated.[35]

Construction squads were augmented by numbers of runaway slaves that first trickled and then started to flood into Union camps. McClellan insisted the runaways be returned to their owners, and many field commanders complied with the general's order. Evidence suggests that railroad superintendents were less willing to dismiss the impoverished blacks and instead put them to work on railroad construction. In early October, an engineer in charge of railroad repairs, Erasmus L. Wentz, wrote to McCallum imploring him to send shoes so the contraband could continue to work. "The negro force that I have on the Norfolk and Petersburg Railroad in Government employ, are so poorly shod that I find it impossible to work them any longer without furnishing them with shoes . . . I find it impossible to procure the necessary permission from Military Authority here."[36]

In the summer and fall of 1862, under Haupt's leadership, the United States Military Railroad remained in its adolescent phase but accomplished two things that would guide the service throughout the rest of the war. First, military personnel would not interfere with railroad operations. Civilian experts, some now in uniform, recruited by McCallum and Haupt, would control the United

States Military Railroad. Officers were prevented from commandeering cars for their own use, and soldiers were stopped from making campfires out of wood intended for locomotive fuel.[37] Second, strict schedules and protocols were maintained for moving men and material and loading and unloading cars, and shipping priorities were established: subsistence, forage, ammunition, hospital supplies, veteran troops, and then new recruits.[38]

Unlike the Federal army, the Confederate Army of Northern Virginia had neither skilled manpower nor support from its government for control of the railroads, even though both Robert E. Lee and Stonewall Jackson understood the railroad's value. After disengaging from actions around Richmond in the summer of 1862, Lee had utilized the railroad to operate against Pope's army. Furthermore, during the Second Bull Run campaign, Jackson managed to capture seven locomotives of the United States Military Railroad.[39] What happened to the locomotives, however, was a sad commentary on Confederate railroad management, and this would beleaguer the South for the remainder of the war.

A brief background is in order here. Shortly before First Bull Run, President Davis appointed William Shepperd Ashe, major and assistant quartermaster, to be in charge of Confederate railroad transportation in Virginia. Ashe was president of the Wilmington & Weldon Railroad, and just before the outbreak of hostilities, he had discussed plans for creation of a Southern rail system. His problems in the fall of 1861 and winter of 1862 were monumental. Many of the major railways were, like the Orange & Alexandria, only single track, rolling stock and engines were limited, freight was backed up at stations in Tennessee with cargo headed for Richmond, railroad presidents focused on making a profit, and governors protected their sovereign status under the principles of states' rights.[40] Under these conditions, any attempt by military officers to interfere in the operation of the railroad or in the repair of locomotives and track often was met with a scathing letter from the president of the line.

Encouraged by Secretary of War George W. Randolph, just a week before the Second Battle of Bull Run, President Davis recommended the appointment of someone to coordinate railway operations throughout the South, but Quartermaster General Abraham C. Meyers opposed the idea. Government control, he argued, would anger railroad personnel, cost too much money, and confuse public and private accounts. Without further discussion on the matter, on September 1, 1862, generals Jackson and Lee presented the captured locomotives to the quartermaster's department in Richmond.

It was decided that the seven engines would be divided by lots, and the results were that the Orange & Alexandria secured three and the Virginia Central and Richmond & Danville got two each. This infuriated the president of the O&A, but the new railroad manager, Captain Mason Morfit, countered that the O&A had less need of additional locomotives to rent to the government.[41] The president of the O&A, John S. Barbour Jr., and a number of his employees had taken Jackson's captured equipment across the Rappahannock to safety, so Barbour felt entitled to his choice of engines. The unfolding dispute over how they were distributed was a microcosm of the Confederacy's transportation squabbles.

Theoretically, the captured locomotives belonged to the army because Lee and Jackson confiscated them. Yet, there was no transportation sage within the War Department to dictate the terms of their use because private companies operated all the railroad lines. The management problem was even more cumbersome when supplies came from outside the state. For example, Morfit's sphere of influence did not extend south beyond Roanoke. North Carolina had its own coordinator of military railroad traffic—the president of the North Carolina & Atlantic Railroad, John Dalton Whitford—as did the other states in the Confederacy.[42]

Also, it was torturously difficult to construct new lines. During the 1850s, the South had witnessed an expansion of lines primarily designed to move agricultural goods, not freight or passengers, to coastal centers for export overseas. This meant that when the war broke out, serious gaps existed between important supply areas, especially connecting the heartland with the outer rim of the Confederacy, which would see the heaviest fighting and largest concentration of troops. One such gap was between Danville, Virginia, and Greensboro, North Carolina. Greensboro was connected to Wilmington, Charleston, and Atlanta by rail, and although a line 130 miles east of the city connected central and eastern North Carolina with Petersburg and Richmond, there was no western link running along the edge of the Appalachian Mountains.

It made sense to build the western line, but powerful business interests in North Carolina had initial doubts because it would divert western commerce to Virginia rather than to the North Carolina coast. All parties finally agreed to build the line, after months of haggling between the Confederate government and Richmond and North Carolina businessmen and politicians. The Piedmont Railroad Company, subsidized by the government, selected Captain Edmund T. D. Myers, son of the quartermaster general, to supervise the construction. Young Myers was a capable engineer and a good choice. The 28-mile line was successfully completed by late 1863, but not without major frustrations.[43]

The first problem was labor. Funds from the Richmond & Danville were used to purchase slave labor. When that effort failed to produce enough workers, the War Department cited a Virginia state law allowing it to impress slaves from owners to work on the line, but the governor of North Carolina, Zebulon Vance, refused to assist in contributing to the labor pool. His aide-de-camp wrote to the War Department in November 1862: "His Excellency must decline authorizing or recommending the legislature to authorize the drafting [of] slaves for this purpose." It was suggested that government contractors should ask slave owners directly for bondsmen. Of course owners would consider hiring out their slaves, but at exorbitant prices.[44]

Second, Myers could not get supplies and iron fast enough because private companies, including the long-disused Roanoke Valley Railroad, blocked his attempts to secure the tracks. On November 5, 1862, John Jones, a clerk in the Confederate War Department, wrote: "It is alleged that certain favorites of the government have a monopoly of transportation over the railroads, for purposes of speculation and extortion!"[45] As a result of these difficulties, Myer wrote to Jeremy Gilmer, chief of the Engineer Bureau, in July 1863, advocating a complete takeover of the project by the Engineer Bureau, and Gilmer agreed. Gilmer endorsed Myers's suggestion and presented it to Secretary of War Seddon. The secretary, who understood the political climate in the capital and among the governors, acknowledged receipt of the idea and then ignored it.[46]

In the summer of 1862, the South did enjoy one bright moment of success in its railroad operations. After the Union captured Corinth, Mississippi, and began moving slowly east through northern Alabama, Confederate General Kirby Smith realized that with considerable reinforcements, he could mount a surprise offensive from Knoxville and Chattanooga into northern Kentucky, striking at the Union army's critical western supply area. For General Braxton Bragg to join Smith on this potentially decisive campaign, he would have to travel a circuitous route to avoid Don Carlos Buell's eastbound forces. Bragg would move 30,000 men over 770 miles using six different railroads to Chattanooga and then launch a fall offensive with Smith. This offensive culminated, in early October, in a tactical Confederate defeat at Perryville, Kentucky, where Bragg sustained such heavy losses that he was forced to retreat to Tennessee.[47]

The cooperation and coordination required to move Bragg's men over six different railroads was an anomaly for Confederate war planners. More typical was the strain placed on the network of railroads used by plantation owners attempting to move goods, by citizens evacuating areas in a war zone, and by the military moving men and material to and the wounded from the front. These competing interest groups continued to squabble over who should have transportation priorities as the railroad infrastructure continued to crumble. The Confederate military needed to establish a strong railroad management system, and to accomplish that, President Davis needed to nationalize the railroads. Neither happened. The Confederacy's military struggle for independence was always in conflict with each state's struggle for political and economic independence. This conflict was never more evident than when trying to move vital supplies and men over the railroads. Furthermore, the cultural construct of slavery, established more than five decades before the war, was so ingrained in Southern citizens that it was almost impossible to challenge even as the war demanded change. Slavery in the South had created a culture dedicated to the proposition that manual labor was beneath most whites. The wealthy were in charge and expected to get their way. When the war started, there were soon too many chiefs. Generals, plantation owners, businessmen, governors, cabinet members, politicians, bureau chiefs, and the president—all crossed paths and all practiced a management style that wavered between obstinate and obstructive on the one hand and unquestioningly deferential on the other.

Conversely, by the late summer of 1862, the Union army was still developing a more refined management system for the railroads, but the foundation was laid. Haupt was on the verge of creating the railroad Construction Corps, and McCallum was building the United States Military Railroad Bureau. In the western theater, the army's use of captured Confederate rails and engines was more helter-skelter than in the Virginia theater. Yet, as capable minds worked to find solutions to the North's transportation and supply problems, a more immediate crisis was the initiative seized by Confederate forces. Braxton Bragg's Army of Mississippi was moving toward Kentucky, and Robert E. Lee's Army of Northern Virginia was marching into Maryland.

CHAPTER EIGHT

Summer–Fall 1862

Maryland, Kentucky, and Tennessee

> When I reached the Potomac the army was crossing the ford at Shepherdstown. Artillery, Infantry, ambulances, wagons, all mixed up in what appeared to be inextricable confusion in the water; and the ford, too, was full of large bowlders.
>
> *Lieutenant William Miller Owen, CSA, September 18, 1862*

In the eastern theater in the late summer and fall of 1862, engineering failures cost both Confederate General Robert E. Lee and the new Union commander, Major General Ambrose E. Burnside, who replaced McClellan in October, the chance for great battlefield victories. In Lee's case, these failures serendipitously provided President Abraham Lincoln with the opportunity to issue a preliminary proclamation that radically altered the meaning of the war. Regarding the use of engineers, Lee committed a sin of omission and Burnside a sin of commission. In the west, too, in the fall of 1862, engineering played a vital role. A pontoon bridge over the Ohio River allowed Union troops from Ohio and Indiana to reinforce northern Kentucky against a Confederate invasion of that crucial border state. Furthermore, the fighting in Kentucky led the Union high command to the conclusion that a change in leadership was necessary. The new officer selected, Major General William S. Rosecrans, had by year's end introduced such significant and innovative changes to the operation and organization of his engineers that other Union army commanders would follow suit. Adopting Rosecrans's principles, commanders deliberately built a sophisticated and efficient engineer service, which would positively affect the remaining campaigns of the war.

The Second Battle of Bull Run (August 29–30) and the Battle of Chantilly (September 1) were major victories for the Southern army, in what were a series of successes for the Confederates under General Lee. When Lee took over from the wounded Joseph Johnston in June, Union forces were ensconced on the southeastern Virginia peninsula and lumbering toward Richmond. By September 1, the Confederates had not only driven McClellan off the peninsula but had taken the fight to Union Major General John Pope's Army of Virginia as it attempted to move on Richmond from the north. After turning back Pope, Lee concluded, first, that the fighting had ravaged the farms of northern Virginia that summer, and second, that his army now held the military initiative and had an opportunity to carry the war into Maryland. This, he hoped, would force the Union army to cross north of the Potomac River, relieve pressure on the Virginia countryside, enlist new recruits from the ranks of the many Marylanders sympathetic to the Confederate cause, and allow famished Confederate soldiers to feed off the abundance of farms in the north.[1]

The Army of Northern Virginia was in high spirits, but it was in ghastly physical condition. Lee wrote to President Davis before the campaign began: "The present seems to be the most propitious time since the commencement of the war for the Confederate Army to enter Maryland." Yet, two paragraphs later, he stated: "The army is not properly equipped for an invasion of an enemy's territory. It lacks much of the material of war, is feeble in transportation, the animals being much reduced, and the men are poorly provided with clothes, and in thousands of instances are destitute of shoes."[2]

The enigmatic Lee was vague in describing his strategic intentions to President Davis, but Confederate Major General John G. Walker claimed that in a conversation with Lee on September 7, "Marse Robert" was unambiguous: "You remember, no doubt, the long bridge of the Pennsylvania railroad over the Susquehanna, a few miles west of Harrisburg. Well, I wish effectually to destroy that bridge, which will disable the Pennsylvania railroad for a long time."[3] If General Lee truly intended to move on a south-north axis from Harpers Ferry to Hagerstown and points farther north, using the mountains as a screen, his logistical planning for such a move was sloppy and costly.

Lee was well aware that he did not have the logistical support network to carry out such an audacious plan. He would need the railroads, and he would need engineers. He had neither. On September 5, Lee telegraphed President Davis announcing his supply route for the campaign. The general expected his army's foraging parties to collect enough food for the men, but he required the quartermaster and ordnance bureaus in Richmond to forward additional clothing, food, and the necessary ammunition. Winchester, Virginia, would serve as the forward supply depot. Wagons would carry the necessary provisions from

Richmond to Culpeper Court House, then travel west through the Blue Ridge Mountains to Luray, and then proceed north to Front Royal and finally Winchester.[4] The one-way trip from Richmond to Winchester was 162 miles. Trains could have transported supplies from Richmond to Front Royal, with wagons carrying the material the last 20 miles to Winchester, but during the Maryland Campaign the Confederate army did not use a single railroad—for two reasons.

First, the Manassas Gap Railroad had been torn up for miles, and the Orange & Alexandria had been damaged, including the railroad's bridges over the Rappahannock and Rapidan rivers. Second, there were no workers to repair bridges. Lee requested that President Davis take the necessary steps to repair the bridges, and the general asked "the president of the road to have timber prepared for that purpose."[5] It is unclear whether the president of the O&A prepared the timber, but the bridges remained in disrepair. Railroad presidents were reluctant to assign their own white workforce to such projects because executives feared that they would assume the cost of the repairs. The lack of work on the tracks, rolling stock, and bridges meant that Confederate authorities could not ship the rich harvest from the lower Shenandoah Valley along the Virginia Central Railroad to the O&A and on to Lee's ravenous soldiers.

Confederate railroad management in September 1862 remained a decentralized operation. Individual railway presidents determined the cargo their trains would carry and absorbed the entire cost of the railroad's operation. The government could not offer to repair engines damaged by the army's use because the government did not operate any repair facilities. Consequently, moving military supplies instead of civilians and local plantation owners' goods was not a top priority.

So, as the Army of Northern Virginia, dirty and starving, crossed into Maryland, on a steady diet of green corn and apples, large quantities of provisions often sat rotting on railroad station platforms. North Carolina, for example, had food enough to meet Lee's needs, but no one in the army had the authority to order it to Virginia, and no one had the authority to stop civilian traffic, supersede a local general's orders, and requisition a train to take essential war material to Richmond.

In addition to Lee's supply difficulties, as he marched into Maryland with the possible objective of reaching the Susquehanna and destroying the bridge, he had only one engineer on his staff, no engineer troops, and no pontoon train. The Confederate army's engineer organization, like railroad management, was decentralized. Individual army, corps, and some division commanders decided whether to keep an engineer officer on their staff. Major Thomas Mann Randolph Talcott, a former engineer with the Ohio & Mississippi Railroad, served as General Lee's aide-de-camp and engineer officer, along with Major Jeremy F. Gilmer.[6] The latter served as both chief engineer for the Department of Northern Virginia and "acting head" of the Engineer Bureau of the War Department, a position he would hold until October 1862 when he was appointed as the Confederacy's first chief of the bureau. At the time of the Maryland Campaign, Gilmer was in Richmond attending to the defenses of the city and trying to devise a more professionally functioning administrative bureau and, at the same time, support armies in the field. In the fall of 1862, there was no central bureau that coordinated the building and maintaining of bridging equipment and tools, nor did engineers enjoy any status within the Confederate War Department.

One of the first things Gilmer tried to do as "acting head" was persuade General Lee to form bridge-building companies within each field army. The companies would be made up of men with mechanical skills recruited from regiments, and their organization would be permanent.[7] Thus far, when roadways and bridges needed repair, field commanders had taken men from the ranks and formed them into pioneer details. The army defined pioneers as units that cut roads through forests or built bridges, much like engineer troops. In 1862, the term *pioneers* usually referred to temporary groups of soldiers who would return to their infantry regiments once the assigned task was completed. Gilmer discovered that many men assigned pioneer duty balked at doing backbreaking work normally associated with slave labor. As a result, most of the work was not done well.

Lee thought Gilmer's idea of bridge-building companies an intriguing one, but many of Lee's field commanders resisted because, they argued, a permanent organization would remove too many men from the infantry or artillery. The commanding general would let the idea rest for six months before he revisited it in the spring of 1863. Meanwhile, Gilmer continued to work at supporting the Army of Northern Virginia and improving the general quality of engineering operations. In September, as Lee began his Maryland Campaign with no pontoon train, the Engineer Bureau attempted to provide the army with captured equipment.[8] In addition, Gilmer began to supervise the manufacturing of a pontoon train in Virginia, and he sent engineer officer Major James Nocquet to Chattanooga to start constructing pontoons for the Army of Tennessee. Gilmer managed to build about six hundred feet of bridging by the winter of 1863 and stored it at Gordonsville, Virginia, 67 miles northwest of Richmond.[9]

The lack of engineers and proper engineering equipment during the Maryland Campaign revealed what Lee chose to ignore: his strategic goals in crossing the Potomac River were untenable from the start. Without pontoon bridges the army was forced to ford the Potomac, and this included moving wagons weighted down with supplies, especially ordnance. Other ordnance wagons remained empty throughout the campaign. In Richmond, ordnance officers could not procure the different types of munitions needed, load them on trains, and see that they reached the forward supply depot at Winchester. The Engineer Bureau had no manpower to repair the railroad bridges over the Rapidan and Rappahannock, and no army officer was able to coordinate train schedules. During the Battle of Antietam on September 17, Captain James Reilly's North Carolina Battery could not refill its ammunition chests, and one of his sections remained out of action the entire day because it had no ordnance. William Owen of the Washington Artillery observed that the ammunition issued to the artillery "was not enough for a long engagement," and he deemed the situation dangerous.[10]

Those wagons that did contain ordnance and ammunition were difficult to maneuver at a ford because no approach roads were dug at the riverbank and the rocky and muddy river bottoms made the crossing time-consuming, labor intensive, and tedious. For example, on the evening of September 18, one day after the battle, orders were issued to pull the army back across the Potomac from Sharpsburg to Shepherdstown. Boteler's Ford was the crossing point.

General Jackson's chief quartermaster, Major John A. Harman, labored, cursed, and yelled most of the night as he and his men struggled with the heavy wagons to negotiate what was "a very high and almost perpendicular bank, and except for the still greater danger from behind [Union cavalry], was a descent that no prudent wagoner would ever have attempted to make." Infantrymen were in the water, dragging emaciated horses and cumbersome wagons through the river. Lieutenant William Owen observed, in utter disbelief, the total confusion: soldiers, horses, artillery, and wagons were bottlenecked in the Potomac River among the rocks and boulders.[11] Jackson praised Harman, writing: "The promptitude and success with which this movement was effected reflects the highest credit upon the skill and energy of Major Harman."[12] Pontoon bridges, however, would have made each of the army's crossings more efficient and swift—although they would also have made the general's rare approbation showered upon Harman unnecessary.

Another example of the cost to Lee's army of the lack of proper engineering support came at the Chesapeake & Ohio Canal Aqueduct No. 2, which crossed the Monocacy River before the canal emptied into the Potomac. The canal served as an important transportation link for the Union army and Northern commerce, so Confederate Major General John G. Walker was ordered to destroy it. Confederate General D. H. Hill had attempted to destroy the aqueduct days before Walker arrived on the scene but failed to do so because he lacked the proper equipment and because the lock keeper, Thomas Walter, suggested that by breaching the levees, Hill's men could drain the water from the canal.

When Walker's division arrived on September 9, they were determined to destroy the aqueduct. They quickly discovered that the bridge was a "solid mass of granite." "Not a seam or crevice could be discovered in which to insert the point of a crow-bar," Walker wrote, "and the only resource was in blasting. But the drills furnished to my engineer were too dull and the granite too hard, and after several hours of zealous but ineffectual effort the attempt had to be abandoned."[13] Although Walker mentioned an engineer, there was no one on his staff with engineering background. The most likely candidate for the job was Captain William Augustine Smith, who attended the Virginia Military Institute and, as a result, probably studied some civil engineering.[14] It was also possible that Walker called upon an infantry or artillery officer with engineering experience. Either way, the incident at the Monocacy Aqueduct points to a pattern of poor planning and support.

With insufficient army engineers, no engineer regiment, no pontoons, and a tentative supply line, the Confederate army had a difficult time sustaining a demanding campaign. The army's control over the railroads was nonexistent. An article entitled "Gross Mismanagement" in the *Memphis Daily Appeal* read: "Some day since we advertised to the fact that the government officials did not do their duty in the important matter of forwarding to Richmond and distributing donated clothing to the soldiers in the army of the Potomac [Lee's army]." The *Columbia (South Carolina) Guardian* asked rhetorically: "Whose fault is it that boxes of clothing are lying in railway depots or lost on the way, or not delivered at their destination? That of government officials. Many of these men use railways for their own purposes of speculation, and many more are both insolent and careless."[15]

As a result, men starved and suffered. The consequence of this was that some men decided to walk away from the army or lie down by the side of the road and die. On September 21, Lee wrote to Davis that the army's "present efficiency is greatly paralyzed by the loss to its ranks of the numerous stragglers . . . A great many men belonging to this army never entered Maryland at all; many returned after getting there, while others who crossed the river kept aloof."[16]

Lee believed that when his army crossed into northern Virginia after the Battle of Antietam, it would take only several days of rest and recovery before he could cross back into Maryland. He soon recognized that his wishes were unrealistic and that his top priority was to move his army toward Staunton where food and clothing were available, although in limited amounts. On November 6, War Department clerk John Jones observed that "I believe the commissaries and quartermasters are cheating the government . . . The Commissary-General to-day says there is not wheat enough in Virginia (when a good crop was raised) for Gen. Lee's army, and unless he has millions in money and cotton, the army must disband for want of food. I don't believe it."[17]

The Army of Northern Virginia, however, did not have a monopoly on problems. After the Maryland Campaign, President Lincoln attempted to persuade General George McClellan to take the fight to the battered Southern army. He refused. Lincoln finally relieved McClellan of command on November 7 and replaced him with Rhode Island native Major General Ambrose Burnside. Known for his loyalty, honesty, and sideburns, Burnside understood that Washington was looking for a decisive battle, and he immediately set to designing a plan to bring about just such a battle.

In November, Lee's forces were stationed around Culpeper and Gordonsville, Virginia, both on the O&A Railroad. At Gordonsville, approximately 30 miles south of Culpeper, the O&A connected with the Virginia Central Railroad, which was the Southern army's critical supply link with its depots in Richmond on the eastern end of the line and in Staunton on the western end. Burnside proposed to feint an attack on Lee's lead elements in Culpeper and slip west to Fredericksburg, cross the Rappahannock River, and then, moving east again along the turnpike through Chancellorsville, strike at Lee and cut off his supply line. On paper it was an interesting plan, but the execution required shifting 120,000 men without tipping off the enemy. Crossing the Rappahannock quickly required the engineers to arrive at the river with the pontoon trains at the same time the army arrived. A delay would give Lee's cavalry time to report the whereabouts of Burnside's forces so that Lee could prepare a formidable defensive position.

By the end of 1862, the Union army had worked out some command and control issues that plagued it earlier in the year. But the Battle of Fredericksburg was about to reveal that the engineers still functioned under an inchoate command system.

The *dramatis personae* during the Army of the Potomac's offensive in December 1862 helps explain the blunders. At the head of the army was General Burnside, and he was coordinating his efforts with General-in-Chief Henry Halleck. Halleck was not the commanding general of all the Union armies in the field; rather, he served Lincoln and Stanton as a chief of staff. He could make suggestions to Burnside, and he would convey the president's concerns and wishes to army commanders in the field, but he had no command responsibility. It was at best a confusing arrangement. General Meigs, head of the Quartermaster Bureau, and General Totten of the Engineering Bureau reported to Stanton, but quartermasters and engineers attached to various field armies did not necessarily report to the bureau chiefs but rather to their commanding generals. Haupt operated on his own, informing General McCallum, superintendent of the military railroads, of his supply needs, and both men reported to Stanton.

Here was where it got complicated. After taking command, Burnside appointed Lieutenant Cyrus Ballou Comstock to his staff as chief engineer, Army of the Potomac. Comstock graduated first in his West Point class of 1855 and was seven years younger than Captain Duane who commanded the Engineer Battalion. General Daniel Woodbury commanded the Engineer Brigade, and both Duane and Woodbury had reported to General Barnard—but now Barnard was chief engineer in the Department of Washington, DC. So, both General Woodbury and Captain Duane now reported to Lieutenant Comstock.

During the Peninsula and Maryland campaigns, Comstock served as chief engineer for General Edwin Sumner's II Corps. Now, as chief engineer for the entire Army of the Potomac and anticipating a late fall offensive, he ordered the Engineer Battalion to move to Falmouth, Virginia, just northwest of Fredericksburg, to await the arrival of the pontoons. On November 12, Burnside met with Halleck, Meigs, and Haupt to discuss the best possible supply line for the upcoming campaign, and Haupt suggested to the generals that the Potomac River landings at Aquia Creek and Belle Plain afforded the army a more secure route than the turnpike from Alexandria to Falmouth. Wharfs could be rebuilt, and the Richmond, Fredericksburg & Potomac Railroad that ran from Aquia Creek to Falmouth was in good condition.[18]

Lincoln approved the change of base from Falmouth to Aquia Creek and Belle Plain on November 14. Haupt reported to Burnside at 11:00 a.m. on November 17 that in addition to 800 feet of wharves to be built, small railroad cars with tools would be pushed to a damaged bridge and civilian carpenters would make repairs. When the larger cars and engines arrived, they would be unloaded and placed on the track to start delivering supplies to Falmouth. Smith shops and machine shops complete with lathes, planers, portable engines, and small tools were also built.[19] In eleven days the rail line was opened, and Haupt started to think ahead about repairing the track and bridges beyond Fredericksburg. He wrote to Burnside that he had tried to procure more civilians to continue work on the wharves and bridges but that Halleck did not favor "my idea of forming a

Herman Haupt, the lynchpin of the Union army's transportation efforts, designed this ingenious barge, or ark. Railroad cars loaded with material for bridge construction were rolled onto the barge at Alexandria then floated to the Union supply base at Aquia Creek Landing. At Aquia Creek, the cars were unloaded directly onto railroad tracks, a locomotive was coupled to the cars, and they were hauled to army depots near the front. Library of Congress, Prints & Photographs Division, Civil War Photographs, LC-DIG-ppmsca-10301.

construction and transportation corps . . . for our work." Haupt continued: "He [Halleck] thinks that the engineer troops, who have been enlisted, and receive double pay for this particular duty, should attend to it."[20] Haupt then suggested that several engineering companies be turned over to his control for a labor force that he would train. Burnside did not respond to this request.

Ignoring the axiom "the devil is in the details," Burnside, after his meeting with Halleck on November 12, made two costly assumptions. First, when Halleck told him that he would instruct General Woodbury to move the pontoons to Falmouth, Burnside assumed Halleck would give specific instructions regarding the urgency of the movement. He did not. Second, Burnside assumed that the pontoon train with Captain Duane had moved from McClellan's former headquarters to Washington. It had not.[21]

Finally, on November 14, Comstock contacted Woodbury and asked about the location of the pontoon train. Surprised that the pontoons were needed so quickly, Woodbury recommended to Halleck that the entire operation be delayed at least five days. When Halleck refused to listen to anything about a delay, Woodbury told Halleck that he would leave immediately, provided the quartermasters furnished him with the 270 horses or mules he needed to move the train to Falmouth. The horses arrived five days later, and the first set of pontoons rolled into Falmouth on November 25. As originally scheduled, Burnside's army began its march and arrived in Falmouth on November 20, where it waited five days for the pontoons.

The second pontoon train began its movement south under the command of Major Ira Spaulding. Woodbury did not provide the major with a sense of urgency, so as his train became bogged down in molasses-like mud along the turnpike road from Alexandria to Falmouth, he moved his pontoon boats to the Potomac and floated them to Belle Plain. His wagons carrying the chess, balk, and tools continued to slog through the mud toward Falmouth. Comstock was not aware that Spaulding's pontoons were destined for Belle Plain, so no wagons were there to greet them and move them to Falmouth. Burnside could not understand why it took so long to deliver the pontoons, and he was furious with Woodbury. The delay in Burnside's entire movement gave Lee's army ample opportunity to prepare defensive positions on the southern bank of the Rappahannock at Fredericksburg and then to wait and see what the Union general would do next.

As more pontoons arrived at Falmouth, Burnside met with Lincoln on November 27 to discuss his plan, which Burnside had revised since the Army of Northern Virginia had occupied the city of Fredericksburg and the hills beyond it. Although the president remained skeptical, he admired Burnside's unquestioned determination. Lincoln wrote to Halleck stating that the commanding general believed he could cross "the river in the face of the enemy and drive him away, but that, to use his own expression, it is somewhat risky."[22] With chronic pressure from the Northern press about the army's inactivity, and living with the torturous idleness of the army under McClellan, Lincoln was emotionally ready for a fight. Intellectually, he was not so sure of Burnside's initiative, but it was probably refreshing to work with a general who wanted to engage the enemy rather than endlessly maneuver.

Burnside's revised plan called for the engineers to throw six bridges across the Rappahannock. Two pontoon trains had arrived with approximately seventy-six bateaux, yet each bridge would require eighteen to twenty pontoons.[23] Once Lincoln gave tacit approval to Burnside's plan, the general's assistant adjutant general, H. W. Bowers, telegraphed Captain O. E. Hine of the 50th New York ordering him to send an additional forty-three pontoons, thirty-eight wagons, and sixteen sets of trestles to Belle Plain. Hine was in charge of the engineers' workshop near the Washington navy yard, and after he sent the extra pontoons on December 3, Bowers requisitioned up to eighty more bateaux, some with chess and balk.[24]

Then, on December 5, winter came to northern Virginia like a lion, with rain, sleet, and three inches of snow. For the next three days bitter cold settled, with temperatures dropping to as low as sixteen degrees as men huddled around camp fires to keep warm. The fear now was that the river would freeze.[25] Yet, by December 10, temperatures turned somewhat milder and Burnside decided to launch his attack. The next day at 3:00 a.m., screened by the fog floating atop the Rappahannock, the Engineer Battalion, ordered to build a bridge for General Franklin's Left Grand Division, started moving east, away from the center of the Union line. A steep embankment prevented the engineers from bringing the wagons to the river's edge, and instead, they had to haul the 1,600-pound pontoons 200 yards to the water.[26] Construction of the bridge, along with the two next to it built by the 15th New York Volunteer Engineers, led by Lieutenant H. V. Slosson, took the entire day. Ice on the river had to be broken, men suffering from exposure had to be pulled from the icy water, and two wounded men, shot by Confederate skirmishers on the opposite shore, had to be carried to safety.

The engineers building the center and western bridges were less fortunate than Lieutenant Slosson's men.[27] The 50th New York began construction on the center bridge, but under constant enemy fire the entire day, they could not complete the work. By late afternoon, men from the 15th New York were called upon to finish the bridge. Using several of the pontoons as boats, the engineers rowed men from the 89th New York Infantry to the western bank of the river, and these soldiers were able to drive off Confederate sharpshooters; the bridge was finished by dusk.[28]

The two western bridges now remained to be built, directly across from the town of Fredericksburg, and this meant that Confederate soldiers could use buildings and cellars to hide from the Union artillery and infantry trying to prevent them from shooting at the engineers. At 4:00 a.m. on December 12, pandemonium broke out at the bridge. Captain Wesley Brainerd wrote of the chaos: "At the signal we started on a double-quick, ten men besides myself . . . When I reached the end of the bridge but five of my men were with me, the other five had either been killed or were wounded and were crawling off." Within minutes

Brainerd was the only man left on the bridge, and then his left arm jerked over his head and he thought he "had been hit with a bar of hot iron," and collapsed onto the bridge. Somehow he managed to get to the river's edge and, with the help of his comrades, made it to an aid station before bleeding to death.[29]

Finally, men of the 7th Michigan, 89th New York, and 19th and 20th Massachusetts regiments crossed the river in pontoons "and carried handsomely the houses and shelters occupied by the Confederates," and this gave the engineers the freedom to complete the bridges. The next day, Burnside's army crossed all six pontoon bridges and attacked the entrenched enemy. The wanton killing that ensued at Marye's Heights was devastating to the Army of the Potomac, and Burnside lost his command. General Woodbury was exonerated for his role in transportation of the pontoons in the weeks leading up to the battle, but he was dismissed as commander of the Engineer Brigade and sent to Fort Tortugas in the Florida Keys, where he contracted yellow fever and died the following year. The Engineer Brigade's losses at Fredericksburg were nine killed, fifty wounded, and two captured.[30]

The situation in the western theater in the late summer and fall of 1862 demanded the same attention to organizational detail as did the Union's eastern armies, and the task was more challenging given the size and scope of operations. Whereas McClellan and Burnside had the Engineer Battalion and Brigade to build bridges and roads, supplemented by Wrigley's Independent Company of Engineers formed in Philadelphia and assigned to the defense of Washington, General Buell's Army of the Ohio had the 1st Michigan Engineers and Mechanics, the Missouri Engineers, and a company of engineer troops from Kentucky, covering five states and more than twenty-five different railroads. It was 345 miles from Memphis to Chattanooga and 300 miles from Chattanooga to Louisville. All of this territory had to be covered, and tracks and bridges had to be repaired and built without the assistance of a General Haupt or civilian Construction Corps. Improvisation and the skilled management of resources were required to develop an organizational model that would support further strategic initiatives necessary for tightening the grip on Confederate resources. These strategic initiatives had to be accomplished without the Federal army suffering serious setbacks, reversals, or a major defeat.

Setbacks occurred regularly, however, as Confederate cavalry led by Nathan Bedford Forrest and John Hunt Morgan raided isolated Union outposts, destroyed bridges, mangled track, cut telegraph wires, and sliced through vital supply lines, preventing General Buell from controlling eastern Tennessee. By September 1862, Buell's Army of the Ohio was moving on a west-to-east axis along the Memphis & Charleston Railroad. The 1st Michigan Engineers worked all summer on maintaining and repairing this railroad as well as the Nashville & Chattanooga Railroad as far as Stevenson, Alabama. Before Buell's army could enter Chattanooga, a 2,000-foot pontoon bridge had to be thrown across the Tennessee River, but his engineers did not have a pontoon train. Instead, operating two local saw mills, Colonel Innes and his engineer soldiers had to build it.

Building the pontoons and the train to transport them was difficult because Innes's men had no nails and the oakum and pitch used as caulking had yet to arrive from Louisville and Cincinnati.[31] There was more bad news. Confederate raiders had captured several engineers and runaway slaves who were working with the Michigan men. North of Nashville, Morgan's men had destroyed a railroad tunnel, bringing supplies from Kentucky to a halt, and Bragg, who had skillfully used the railroad to move his force of 40,000 around Buell, arrived in Chattanooga ahead of the Army of the Ohio. Now Bragg and Kirby Smith, using the Cumberland Mountains as a screen, raced north toward Louisville and Lexington. Buell had no choice but to follow, abandoning northern Alabama and middle Tennessee as far as Nashville. Forced to destroy machinery in the railroad machine shop and the partially built pontoons, the Michigan engineers, who were scattered from Nashville, Tennessee, to Huntsville and Stevenson, Alabama, had to march north on the double-quick to the rendezvous point at Bowling Green.[32]

Before Buell finally entered central Kentucky in force, Smith's Rebel army of 11,000 men closed in on Lexington, Kentucky, and Cincinnati, Ohio. It was a real threat, which only intensified when Smith's forces nearly destroyed a Union army under the command of Major General William "Bull" Nelson, a former lieutenant commander in the navy, at the Battle of Richmond, Kentucky. Two days later, on September 2, Smith's army marched victoriously into Lexington. Rumors now spread among Unionists that Smith was successfully recruiting thousands of Kentuckians for the Confederate cause. The reality was just as frightening. Union Major General Horatio G. Wright, department commander, did not know Bragg's whereabouts, and because Union soldiers from the Battle of Richmond retreated to Louisville, Cincinnati went undefended.[33]

Cincinnati was a central supply depot for all Union armies operating in the western theater. Railroads carried grain from Wisconsin and Illinois and iron from Pittsburg to Cincinnati. Second to New Orleans, it was the largest city west of the Appalachian Mountains. The city needed to be defended at all costs because if it fell into Confederate hands, millions of dollars of valuable Union army supplies would be lost. President Lincoln, furthermore, desperately searching for a military victory to accompany the announcement of his Preliminary Emancipation Proclamation, would view the capture of Cincinnati as a humiliating defeat. Wright needed someone he trusted to recruit militia men from all over Ohio, Indiana, and Illinois and to strengthen a series of batteries opposite Cincinnati, in the hills of Covington and Newport, Kentucky. He selected Major General Lewis Wallace.

Wallace worked with Colonel Charles Whittlesey of the Corps of Engineers, who had originally designed the defenses in northern Kentucky. These batteries proved to be skillfully employed. The Licking River ran perpendicular to the Ohio, and with three gun emplacements to the west of the Licking and five to the east, fifteen batteries covered all the approaches to Newport and Covington and hence to Cincinnati. For example, batteries Shaler and Kearny could lay down a deadly crossfire on the Alexander Turnpike (now US 27).

Wallace's real problem, however, was how to get the necessary manpower from Cincinnati over the Ohio River and into the defenses. He had been ferrying men and supplies across the river on barges, but the process was too slow. He had asked the builders and carpenters in the city if they could build a pontoon bridge, but no one knew how. Finally, he asked a city architect named Wesley Cameron if he could build a bridge using the barges that were ferrying troops. Cameron was alleged to have said: "You get me the material and manpower, and I'll get the job done."[34] Using the labor of slaves and free blacks, the barges were held in place by anchors, lashed together, and covered with planking. Within thirty hours the bridge was completed. By September 4, Union soldiers began to cross the Ohio and take their positions in the carefully dug defenses.

On September 10, Kirby Smith sent 8,000 Confederates north to probe the Union position in Covington. Sometime between September 11 and 12, they left without firing a shot. Cincinnati released its collective breath. Eventually, Bragg's army would be discovered near Louisville. Buell, now knowing the location of the invading Confederates, finally caught up with Bragg on October 8 in the vicinity of Perryville, Kentucky. At the Battle of Perryville, Buell's army stopped the Confederate attempt to capture the wealth of army stores in Kentucky and Ohio and thus score a major strategic victory. Buell did not pursue the retreating Southerners, and this cost him his job.

His replacement was William S. Rosecrans, recent victor of a fierce fight at Corinth, Mississippi, where he drove off 23,000 Confederates under the command of General Earl Van Dorn. Now Rosecrans, a West Point graduate and army engineer before the war, sharp and ambitious, prepared to reorganize the army, which included a stronger and more efficient engineer service. Rosecrans would enjoy some success and one major failure before being laid in the tomb of the mostly forgotten generals. Foremost, he brought about changes to the structure and operation of the engineers that would reverberate throughout the entire Union army for the remainder of the war.

As head of Buell's former command, renamed the Army of the Cumberland, Rosecrans evaluated his engineering needs. To Rosecrans, engineer soldiers were essential, and he had too few of them. The 1st Michigan Engineers and Mechanics had done well but during the summer and early fall of 1862 had been spread too thin. They had no pontoon train and no experience throwing one across a river, and the regiment stubbornly refused to drill. Furthermore, the men were still not paid for engineer duty, and rumors spread that a mutiny was possible. Members of Rosecrans's staff did not trust the Michigan regiment, including Innes and the other officers. The 70th Indiana, known as the "Railroad Regiment," had worked at repairing tracks and bridges, but Rosecrans wanted a more permanent unit.[35]

The Army of the Cumberland had only one officer from the Corps of Engineers, Captain James St. Clair Morton. Morton was an iconoclast of the first order. As historian Philip L. Shiman points out in his essay "Engineering and Command: The Case of General William S. Rosecrans 1862–1863," Morton was not afraid to challenge conventional wisdom. Before the war he publicly criticized the use of masonry fortifications along the coast and advocated instead the use of earthwork fortifications, "a heresy to a corps whose prime mission was the construction of great forts of stone and brick."[36] Rosecrans, who was raised a Methodist and while a cadet at West Point, a bastion of Episcopalism, converted to Catholicism, was an iconoclast himself, so there is no doubt he sensed a kindred spirit in Morton.

The Army of the Cumberland also had just one topographical engineer, Captain Nathaniel Michler, who was responsible for mapping middle Tennessee and central Kentucky. Morton, Michler, and John B. Anderson, a civilian railroad engineer that Rosecrans inherited from Buell to serve as superintendent of railroads, were not enough to satisfy Rosecrans's needs over such a large geographical area. All three men were highly skilled professionals. Anderson had hired a civilian construction crew made up of carpenters and laborers who worked from a special construction train to repair railroad bridges, but the overwhelming amount of work required hindered the army's movement.[37] Rosecrans asked the War Department for more engineer officers, and he told the new general-in-chief, Henry Halleck, that he would not launch an offensive to regain control of the Nashville-to-Chattanooga corridor until he was satisfied that the army's supply link from Nashville to central Kentucky was repaired and secured. "The Army of the Potomac cannot possibly be as much in need of engineers as I am," he wrote to the War Department, and within a month two more engineers arrived.[38]

Rosecrans energetically set about addressing his engineering problems. First, he ordered every brigade and division commander to assign an officer to topographical duty. It was a stroke of brilliance. Rosecrans was aware that these officers would not have the technical ability to produce detailed maps, but they could be taught what to look for when in the field. Thus, these officers recorded information about roads, bridges, and geographical features such as culverts, streams, wetlands, and open fields. This information was forwarded to Michler who, with several assistants following his careful instructions, made and reproduced maps. Michler did not like the idea of untrained officers collecting data for the mapmakers. Topographical engineers required a keen sense of observation and the ability to measure distance. Officers assigned to this task from infantry brigades and divisions would make costly mistakes, Michler argued. Rosecrans did not like or appreciate his objections, and Michler was forced to work within this unorthodox system.

Next, the commanding general assigned Morton the task of training and commanding a Pioneer Brigade. Engineer soldiers were paid more than infantrymen, but infantry occasionally detailed to serve as road builders or on bridge repair were not, so an army commander did not need permission from the War Department to organize pioneer units. Rosecrans ordered that twenty men from each regiment, half mechanics and half laborers, be detailed as pioneers. "The most intelligent and energetic lieutenants in the regiment, with the best knowledge of civil engineering will be detailed to command, assisted by 2 non-commissioned officers."[39] Most regimental colonels were pleased to comply with Rosecrans's order, believing that their best men detached to the pioneers would serve only temporarily and for the most part would remain with their infantry regiments. This proved to be a wrong assumption.[40] Approximately 2,000 men were organized into three pioneer battalions and became permanent units. Unlike the 1st Michigan whose men refused to drill, the pioneer battalions would train not only as engineer troops but also as infantry and were brigaded together with an attached battery, the Chicago Board of Trade Battery.[41]

To Rosecrans's satisfaction, Morton created a command structure for the Pioneer Brigade, even though it was designated as a temporary organization. He assigned someone to the role of quartermaster and another officer to adjutant,

and he made acting majors and captains out of men who were still officially lieutenants.[42] Soldiers in the brigade, such as William Wesley Perkins III, found the
work hard, yet better than having "the cannon boll flying over a fellow head."
Private Perkins described the duties in a letter to his brother Rice: "We are ordered to hull all the wood we can in camp[;] some goes every day from this company. There is someone from every regiment for pioneers and macanic."[43]

Morton also continued the practice of forcing free African Americans and
runaway slaves to work on the fortifications around Nashville. Union conduct
toward blacks in the South often was as cruel and callous as that of the plantation owners from whom they had fled. President Lincoln's Preliminary Emancipation Proclamation, issued after the Battle of Antietam, had no effect on
racial attitudes, especially among army generals, most of whom, like Rosecrans,
were Democrats. Despite this malicious treatment and the deplorable conduct
of Northern soldiers toward blacks, many African Americans believed that in
laboring for the Union army they were actively contributing to their emancipation and freedom.[44]

Pioneers could not be paid as engineers, but they did receive extra pay authorized for fatigue duty: twenty-five cents extra per day for soldiers working with
common tools such as an axe or shovel and forty cents extra per day for those
who worked with carpenter tools such as augers.[45] Members of the 1st Michigan
Engineers and Mechanics had also been promised extra pay, but not a cent had
been forthcoming. Since the passage of legislation on July 17, 1862, retroactively
recognizing volunteer engineer troops and placing them on the same pay scale
as regular army engineer troops, the men of the 1st were anxiously awaiting
their back pay. The money arrived in camp in September, but Paymaster Major
Charles T. Larned refused to pay the men without specific orders from the War
Department because he was not certain when the regiment's status officially
changed. Fractious soldiers began to talk of mutiny. Ezra Stearns recorded in
his diary: "Excitement is increasing . . . the men talk of making a strike," and on
November 10 they did. Many of the men blamed Colonel Innes for not working
hard enough on their behalf, and others were just tired of serving for over a year
without pay. Given the circumstances, it is surprising that only one-quarter of
the regiment refused to work. Many feared arrest and others were talked out of
striking by their officers. Finally, a telegram arrived from the War Department
in late November ordering Rosecrans to pay the regiment engineers' wages
under General Orders No. 177, and one week later the men were paid. Most of
the mutineers served thirty days of hard labor without pay and then returned to
the regiment.[46]

As the problems of pay, manpower, and mapping were sorted out, Rosecrans continued to gather supplies in Nashville for an eventual forward movement south toward Chattanooga. Nonetheless, he still had one major problem: there was no pontoon train. On November 22 he wrote to the United States Army's chief engineer, Brigadier General Totten, requesting 700 yards of iron pontoons. Perhaps displaying his knowledge of military equipment or perhaps being a contrarian, Rosecrans insisted on the iron instead of the wood-framed pontoons. The wooden ones leaked and the iron ones were better, Rosecrans said, and he wanted them at once.[47] Iron pontoons, developed by the Prussian and Austrian armies, were nonexistent in the United States.[48]

Totten responded by saying it would take the Engineer Department six weeks to build wooden bateaux and at least ten weeks to build the iron pontoons, and the Engineer Department had no model for iron bateaux, and no one had experience in building them. Rosecrans was not pleased with Totten's response.[49] Fifteen days after the initial request was submitted, Totten's chief assistant, Brigadier General George W. Cullum, finally convinced Rosecrans that wooden bateaux, built in Cincinnati under the supervision of the Department of the Ohio's chief engineer, Lieutenant Miles Daniel McAlester, were the best alternative.[50] Unlike the pontoons Rosecrans described as leaky, which had been built and then destroyed by the 1st Michigan because of the army's sudden retreat north, the new wooden bateaux would be built with seasoned wood. Unfortunately, Rosecrans did not provide a sufficient number of wagons to move the entire train over roads. Instead, the railroads transported a number of the pontoons when needed, especially during the Murfreesboro Campaign.[51]

The year ended, and the Army of the Potomac was no closer to capturing Richmond or defeating Robert E. Lee's Army of Northern Virginia than it was when the year began. Out west, the Army of the Cumberland gained back some territory along the Nashville-to-Chattanooga corridor, and now attention would turn to the Mississippi River and the city of Vicksburg. Nonetheless, the Confederacy still held the critical coastal ports of Charleston, Wilmington, and Mobile and the entire lower South. Confederate and Union soldiers were fighting with skill and determination, and both sides had some officers who led well and others who led less well. The most significant difference between the two sides was in their ability to manage an emerging modern war. The size of the armies dictated new, untried methods of moving men and material over complex terrain. There was no book on how to do this, but as historian Bruce Catton pointed out, "The volunteer army [Union] was teeming with men quite capable of playing the part of military engineers if some capable officer directed them."[52] Some officers had recognized mechanical skills in their men and had started to utilize these assets. McClellan had formed two volunteer engineering regiments, Rosecrans established a Pioneer Brigade, and the War Department turned over the operation of the military railroads to Herman Haupt. Now, as the war entered its third year, Union officers started to recognize that by tapping into their soldiers' mechanical skills and ingenuity, anything was possible and a great deal could be accomplished.

Part Three

Applied Engineering

CHAPTER NINE

Vicksburg

While work [was] underway General Rawlins, Dana, and I spent time together passing from bridge site to bridge site encouraging officers and men in their novel and necessary work and with admiration for the volunteer soldier and his unequaled capacity for practical bridge building.
Lieutenant Colonel James H. Wilson, Corps of Engineers, USA

I learned that General McPherson was using mortars made of trunks of trees (gum trees being the best) to throw 6 and 12-pound shells.
First Lieutenant Peter C. Hains, Acting Chief Engineer, XIII Corps, USA

The strategic significance of Vicksburg has been a topic of debate among historians for several decades. In the epic narratives of the war by James McPherson, Bruce Catton, and Shelby Foote, the capture of Vicksburg tore the Confederacy in two, making it impossible for Southerners living in the eastern half of the country to reap the benefits of important supplies that, with Vicksburg still in Confederate hands, would have flowed from the west and sustained the war effort. This thesis held sway for many years until the work of historians such as Herman Hattaway, Archer Jones, and, most recently, Albert Castel proved that the amount of material goods coming from the western half of the Confederacy did not contribute measurably to supplies in the east. Of course, this revelation raised another challenging question. Was the capture of Vicksburg as important as we had come to believe? The answer is yes, it was. The capture of Vicksburg accomplished four critical strategic objectives.

First, it gained control of the Mississippi River for the Union navy. This was important because both the Union Army of the Gulf and the navy's Mississippi River Squadron needed to work against Confederate forces, including guerrillas operating in Louisiana. In addition, after Vicksburg, the navy was able to move large quantities of supplies along the river to points where these supplies could be unloaded and transported overland to Union armies operating deep within the heartland of the Confederacy. Second, the capture of Vicksburg was a great boost for Union morale and, conversely, a significant setback for Southern soldiers and civilians. Third, the victory eliminated an entire Confederate army from the war. Finally, it elevated the status of General Grant in the eyes of the Northern public and especially in the eyes of President Lincoln. Lincoln would select Grant to rescue the beleaguered Union forces inside Chattanooga, which the general would accomplish; after that, Grant would be promoted to lieutenant general and placed in charge of all Union armies. That decision also turned out quite well.

Vicksburg was the most important campaign of the war, and Grant's determination to capture the "Gibraltar of the Confederacy" and the risks he took to do so also made the campaign the most remarkable of the war, and perhaps even the most remarkable in American military history. Vicksburg's capture, however, would not have been accomplished without an extraordinary engineering effort, made even more astonishing by the fact that when the campaign began, Grant had at his disposal only three army engineers and a company and regiment of volunteer engineers. The story of what happened to bring about the Union army's success at Vicksburg captures the essence of how the North engineered victory during the war.

Operations around Vicksburg did not begin well for the Union army and navy. Yet, a willingness to try different approaches to get at Vicksburg spoke to the army's increasing confidence and faith in volunteer soldiers' ingenuity, proficiency, and versatility. If an idea could be imagined then it could be attempted, even though there was no guarantee of success. This was true in the early summer of 1862 when Admiral David G. Farragut, Commodore David D. Porter, and General Benjamin Butler hatched a plan to conquer Vicksburg.

Porter's mortar schooners and 3,000 soldiers sent by Butler under the command of Brigadier General Thomas Williams would ascend the river, and once below the town, the mortars would lob their 200-pound shells in a bombardment that would destroy the city and knock out all the Confederate batteries. Williams's men would then occupy Vicksburg. If this attack failed, Williams would attempt to dig a canal across De Soto Peninsula, a narrow neck of land about 5,000 feet wide on a sharp bend in the Mississippi River. Viewed from the heights of Vicksburg, the Mississippi looked like an inverted C. This provided an opportunity for Union engineers to dig a cutoff and permanently alter the course of the river. In his carefully researched *Grant's Canal*, civil engineer

David F. Bastian wrote of the canal project: "The slope of the projected water-way would be much greater than the natural course and, once completed, gravity would propel the water across the peninsula rather than around it."[1] The goal was to change the course of the Mississippi and to turn Vicksburg into an inland town.

The idea of changing the course of the Mississippi was not new. In the 1850s, the US government hired civil engineer Charles Ellet Jr. to study and recommend how flood prevention and navigational improvements could be achieved along the Ohio and Mississippi rivers. One of Ellet's suggestions was a cutoff over the De Soto Peninsula, opposite Vicksburg.[2] Such a cutoff would have done irreparable damage to Vicksburg's economy, however, and bowing to extensive lobbying by the city fathers, the state legislature passed a bill in 1858 outlawing efforts at a cutoff.[3]

Farragut and Porter's force managed to slip eight ships past the relentless cannonade from the city's guns, but sixteen sailors were killed and another twelve were wounded. Running warships past Vicksburg was possible, but nothing was accomplished by doing so. Confederate guns would continue to rain down terror on any Union gunboat attempting to pass the heights, and supplies to the city would continue to arrive overland from as far away as Mobile, Alabama. Confederate quartermasters could ship goods from Mobile north along the Mobile & Ohio Railroad as far as Meridian, Mississippi. Supplies were then shipped west on the Southern Railroad of Mississippi to Jackson and on to Vicksburg. Confederate forces under General Sterling Price also controlled the Mobile & Ohio from Meridian as far north as Tupelo. Under these circumstances, moving the mighty Mississippi seemed like the best option.

Williams set about surveying the site and selecting the path the canal was to take. With the help of his 3,000-man force and slaves from surrounding plantations, the excavating began on June 28 with initial success. The work was brutal. Men moved dirt in excessive heat and humidity, had very little in the way of clean drinking water, and fought constantly with mosquitoes. The ditch was fifteen feet wide and approximately three feet deep. An embankment or levee on either end blocked the river from entering the ditch, and the idea was that when ready, the levees would be broken and millions of gallons of water from the river would rush into the canal, flooding the entire area and forcing the river to form a new channel. Unfortunately, as the men were doing the thankless and back-breaking work of cutting trees, removing stumps, and shoveling dirt, the river began to fall at a rate of a foot per day.[4] A correspondent for the *Chicago Tribune* wrote that "when the levee at each end of it [canal] was cut through, it was found to be above the level of the water."[5] To rectify the problem, the center of the ditch was dug five feet deeper and the excavated earth was thrown onto three-foot-deep sections, raising them closer to the surface.

By late July the river had fallen twenty-five feet, the summer heat remained oppressive, half of Williams's workforce was sick and disabled, and Farragut was eager to leave the area. The project resulted in failure, and the men were evacuated from the peninsula. Vicksburg continued to stand as the South's Rock of Gibraltar. The Union army had no further plans to solve the military problem of how to capture Vicksburg until October, when General Halleck turned the Department of Tennessee over to General Grant. Within one week, Grant was ready to move on Vicksburg in the most direct way possible.

Any map would have revealed that Grant's operational directives made perfect sense and, if successful, would allow him to besiege the city. He would move south to Holly Springs, Mississippi, approximately 150 miles from his primary supply base in Columbus, Kentucky. To place additional pressure on Confederate commander General John C. Pemberton, Grant would send 32,000 men under General William T. Sherman from Memphis up the Yazoo River to the heights north of Vicksburg known as Chickasaw Bluffs. This plan would force Pemberton to divide his force of 32,000 men, hence giving the advantage to the Federal attackers. The major flaw in this strategy, and why it failed, was that Grant failed to protect his own extended supply lines.

The overland advance came to an abrupt halt in the last week of December 1862. Confederate cavalry under Earl Van Dorn and Nathan Bedford Forest raised havoc with Grant's plan. Van Dorn's horse soldiers captured Grant's critical advance base at Holly Springs, and Forest destroyed 60 miles of railroad track along the 180 miles of roadway between Holly Springs and Jackson. This disaster forced Grant to return to Grand Junction, Tennessee, and reconsider his options. Furthermore, the Confederate army, in entrenched positions at Chickasaw Bluffs, stopped Sherman's diversionary force in its tracks. Sherman gathered his wounded, buried his dead, and retreated back to Memphis.

In *War and Peace*, Leo Tolstoy wrote: "If in the accounts given by historians . . . we find that wars and battles appear to follow a definite plan laid down beforehand, the only deduction we can make . . . is that these accounts are not true."[6] During the Vicksburg Campaign, anyone who believed that Grant was following a distinct plan was only kidding himself. Grant had once remarked to a staff officer: "In war anything is better than indecision. We must decide. If I am wrong, we shall soon find out, and can do the other thing. But not to decide wastes both time and money, and may ruin everything."[7] Grant took to heart his own advice. After his initial plan to capture Vicksburg died at Holly Springs and Chickasaw Bluffs, he would operate under the notion that trial and error was the only feasible method of capturing the city. Improvisation was the highlight of the next six months of campaigning.

Vicksburg's location and the geography of the surrounding area were the obstacle preventing Grant from having any opportunity to capture the South's most important bastion along the Mississippi. Without cutting the central supply artery from Jackson to Vicksburg, the latter could be defended indefinitely. Getting to Jackson and then moving his army west to besiege Vicksburg was the all-consuming problem. The bluffs along the eastern riverbank ran north and south for approximately 100 miles. To the north, the Yazoo River, its delta, and the Mississippi floodplain stretched for 175 miles north and south and 60 miles east and west. To the south of Vicksburg, batteries sat atop the bluffs as far as Grand Gulf, 40 miles below the city, at the junction of the Big Black River and the Mississippi.[8] Finally, on the Louisiana side of the river, a maze of bayous, swamps, and lakes, all part of the Mississippi flood plain, made passage all but impossible, and Confederate cavalry and saboteurs were prepared to build obstructions in an area already deemed impassable.

Grant, however, had confidence in his engineers' ability to improvise and move his army and their supplies where he imagined moving them, or at least making the attempt. This was truly remarkable, first, because the terrain was treacherous; second, because he had only three engineer officers and two engineer units when he began operating against Vicksburg in January 1863; and third, because he would need to rely on infantry as engineers and pioneers.[9]

One group of engineers was Captain William F. Patterson's Kentucky Company of Mechanics and Engineers, and the other was Colonel Josiah Bissell's Engineer Regiment of the West. Both were undersized units that alone were not capable of building the causeways and bridges required to move the army where Grant wanted to move it. Nonetheless, Grant and his corps commanders soon found the additional skills they sought among the men serving in the infantry. To augment the engineer troops, he had the army form pioneer companies, draft infantry regiments, and use runaway slaves under the supervision and guidance of officers and noncommissioned officers serving as engineers.

For two months in the late winter and early spring of 1863, Grant ordered four simultaneous attempts to get his men and supplies on terra firma east of Vicksburg. With Milliken's Bend, 20 miles northwest of Vicksburg on the Mississippi, as the advance supply base, four expeditions were launched. First, there was an effort to reopen Williams's original canal and then to dig one at

Duckport, Louisiana. The second effort was to connect Lake Providence, 60 miles north of Vicksburg, with the Tensas, Black, and Red rivers, bypassing a series of forts and batteries guarding approaches to the city from the Mississippi River. Third, an attempt would be made to blast a hole in the levee to open the Yazoo Pass, 200 miles north of Vicksburg, to access the Tallahatchie River and then approach the city from the northeast. Finally, the engineers and navy would endeavor to enter Steele's Bayou, 40 miles north of Lake Providence, and strike at the city from the north. All four of these expeditions would keep the army's supply lines connected and protected. All four would require ingenuity and innovation, and all four would fail.

Grant placed little faith in the Yazoo Pass option. The failed Chickasaw Bayou venture taught several important lessons. Confederates had mined the river, which resulted in the sinking of the ironclad *Cairo*, and had chained together floating logs to block the navigable channel leading to Yazoo City, 40 miles north of Vicksburg.[10] Grant was also aware that Confederate guns along the bluffs commanded any potential landing spot for a move on Vicksburg. Further, the winter rains had elevated the river and the runoff flooded small streams and swamps.

Yet, the flooding could work to the Union's advantage. The Yazoo Pass was a former ship passage providing transports from the Mississippi with access into Moon Lake on the east bank of the river, 305 miles north of Vicksburg. From Moon Lake, ships could access Coldwater River, which ran into the Tallahatchee River and eventually dumped into the Yazoo. By using this route, Grant's army might get to dry land just north of Haynes' Bluff and attack the stronghold from its flank. It was a long shot. The pass had become a mosquito-infested wetland overgrown with vegetation. The levee was eighteen feet high, and there was a difference in height of eight and a half feet between the water in Moon Lake and the pass. Lieutenant Colonel Wilson of Grant's staff reported that if a small crevasse could be cut, the force of water would enlarge the gap and, within days, ships would be able to pass through.

On February 3, Wilson placed a mine in what would become the mouth of the cut, and two crevasses were dug about twenty feet apart. The mine was detonated, sending debris everywhere and widening the gap in the levee as torrents of water powered through. The next day Wilson wrote: "The Pass is open, and a river 75 or 80 yards wide is running through it with the greatest velocity ... By 11pm the opening was 40 yards wide, and the water pouring through like nothing else I ever saw except Niagara Falls. Logs, trees, and great masses of earth were torn away with the greatest of ease." Within a few days the power-

Vicksburg Operational Area, Spring 1863

ful flow of water would subside, and the navy would send shallow-draft vessels through the new channel to explore the region and determine the next move. Wilson wrote to Grant's Chief of Staff John A. Rawlins: "The work is a perfect success."[11]

For ten grueling days, steamboats slugged their way through the pass as the men on board cut cottonwoods and sycamores that reached completely over the stream. Some trees were four feet wide at the base and weighed thirty-five tons. Because of the flooding, there was no more than a three-foot-wide strip of dry land along the bank on which to work. Wilson reported that "our greatest difficulty so far has been to obtain tackle strong enough to resist strain . . . to lift the heaviest logs."[12] It was 6 miles from the Mississippi to the levee cut by Wilson on the west side of Moon Lake, and from there through the pass to Coldwater River another 14 miles. Grant wrote to Halleck on March 7 that Commander Watson Smith had traveled the 30 miles of the Coldwater and entered the Tallahatchee with "2 ironclads, (*Chillicothe* and *Baron De Kalb*), 2 rams, and 6 light draught gunboats, . . . and . . . 14 transports with 6,000 soldiers."[13]

Commander Smith, Wilson, and Grant remained optimistic as steady movement continued, yet from the moment the levee was cut in early February there was no disguising the Federals' intentions, and Confederate General Pemberton had time to respond. Partisans felled sycamore, oak, elm, and pecan trees into the streams, and these, with driftwood, formed taxing obstacles to remove. Using slaves from surrounding plantations, Confederates also hastily built a fortification at the confluence of the Tallahatchee, Greenwood, and Yazoo rivers at Fort Pemberton. High water resulting from flooding of the pass and fourteen days of rain since February 1 eliminated any hope of landing troops on dry ground to flank the batteries. The Confederates quietly waited for Smith's small fleet to arrive.[14]

The naval assault began on March 11 and continued until the thirteenth, with no appreciable damage to the fort. The gunboats approached to within 800 yards and no further. Wilson blamed Smith and the navy: "I can see a disposition on the part of the Navy to keep from a close and desperate engagement. I tried to give them backbone but they are not confident. Smith not the equal of Lord Nelson."[15] In fairness to Smith, the tight channel, the obstacles in the water, the lack of repair facilities, and no pressure on the fortification's flank by Union infantry all allowed the Confederate guns to focus on the small river fleet. Wilson devoted considerable attention to the Yazoo Pass operation, but it was at a dead end. The flotilla reversed course and started back to the Mississippi. Grant, though, was not ready to give up on the idea of attacking Vicksburg from the north. Reconnaissance confirmed the efficacy of a route through Steele's Bayou to Black Bayou and Deer Creek. Entering the Yazoo River there, an attempt could be made to land troops between Yazoo City and Haynes' Bluff, where the army could attack the Confederate army's right flank at Vicksburg.[16]

From the outset, vessels under the command of Acting Rear Admiral David

Dixon Porter had problems in navigating Steele's Bayou. Shallow water, smoke-stacks ripped off by overhanging trees, overgrown channels just wide enough for gunboats to pass, and eerie sounds that frightened superstitious sailors made navigation almost impossible. When Confederates felled trees, mice, rats, and snakes hit the deck of the gunboats, adding to the deplorable and frightening conditions in the bayou. Porter forged ahead, traveling half a mile an hour until he finally reached Deer Creek. Behind Porter's boats was General Sherman's detached force made up of his army's Second Division pioneer corps, the 8th Missouri Infantry, and two companies of the Missouri Engineers. "Deer Creek is a narrow, sluggish stream, full of willow bushes and overhanging tree limbs inhabited by animals," Sherman wrote. ". . . Porter's ironclads move like snails."[17]

At Rolling Fork, Deer Creek turned 180 degrees and emptied into Sunflower River and eventually the Yazoo River. Porter and Sherman had successfully by-passed Haynes' Bluff and were approaching Yazoo City. Then a disaster struck that almost cost Porter his entire operational fleet. As the admiral moved about one mile from Rolling Fork, his boats became trapped. With Confederate obstacles in the bayou and sharpshooters along the banks, Sherman reported that "an avalanche of water from Rolling Rock" created a logjam for Porter's ships. Sherman pointed out that the floodwater "actually came from Colonel Wilson's act to cut the levee on the Mississippi."[18]

When Sherman learned that Porter's fleet was in trouble, he sent infantry through the bayous to rescue the navy. Once the soldiers secured the area and the pioneers, engineers, and infantry cleared the obstacles from the narrow channel, sailors disconnected the tiller ropes and tackles to the boats' rudders and backed the boats down the bayous to the safety of the Mississippi.[19] The Steele's Bayou affair was over, and Grant learned that he could not get at the enemy from the north. The confusion of the expedition was reflected in one commanding officer's attempt to write his after-action report. Under normal campaign conditions, officers wrote their own reports without assistance, but in this case one of Sherman's staff officers was asked to assist. The commander said to the staffer in a vexatious tone: "I want you to tell me where I have been, how I went there, what I did, and if I came back the same way I went, or if not, how did I get back."[20] The break in the Yazoo Pass levee did serve the purpose of flooding the entire area north of Vicksburg and limiting the defenders' supply lines to those along the roads and railway due east to Jackson. Otherwise, Wilson's venture and Steele's Bayou had been confusing and failed operations. Yet, while Grant was trying to get his army north of Vicksburg, he was also scheming

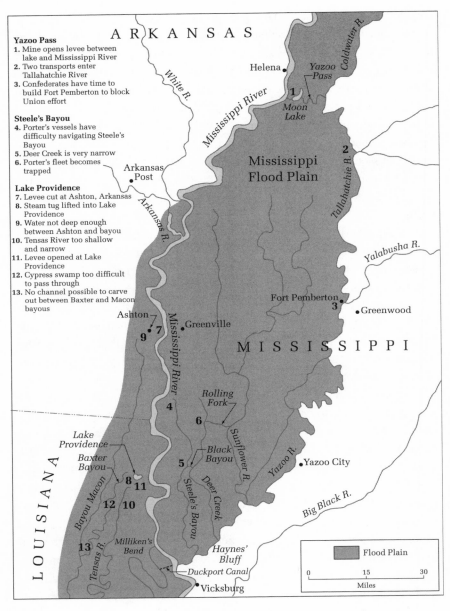

Yazoo Pass
1. Mine opens levee between lake and Mississippi River
2. Two transports enter Tallahatchie River
3. Confederates have time to build Fort Pemberton to block Union effort

Steele's Bayou
4. Porter's vessels have difficulty navigating Steele's Bayou
5. Deer Creek is very narrow
6. Porter's fleet becomes trapped

Lake Providence
7. Levee cut at Ashton, Arkansas
8. Steam tug lifted into Lake Providence
9. Water not deep enough between Ashton and bayou
10. Tensas River too shallow and narrow
11. Levee opened at Lake Providence
12. Cypress swamp too difficult to pass through
13. No channel possible to carve out between Baxter and Macon bayous

ARKANSAS

White R.

Mississippi River

Helena

Yazoo Pass

Coldwater R.

Moon Lake

Arkansas Post

Arkansas R.

Mississippi Flood Plain

Tallahatchie R.

Yalabusha R.

Fort Pemberton

Greenwood

Ashton

Greenville

MISSISSIPPI

Mississippi River

Rolling Fork

Sunflower R.

Lake Providence

Black Bayou

Baxter Bayou

Deer Creek

Steele's Bayou

Yazoo R.

Yazoo City

Bayou Macon

Milliken's Bend

Tensas R.

LOUISIANA

Haynes' Bluff

Duckport Canal

Big Black R.

Flood Plain

0   15   30
Miles

Vicksburg

Yazoo Pass, Steele's Bayou, and Lake Providence Operations

to move his army south and to attack from dry land south and east of the city. Three attempts were made, and the first one was to rebuild the original De Soto Peninsula cutoff.

In January, Grant ordered Colonel Josiah W. Bissell, the intrepid officer responsible for constructing the Island No. 10 canal, to survey Williams's cutoff and report on the feasibility of reopening the project. Bissell found the canal between nine and twelve feet wide, no more than six feet deep, hosting two feet of standing water, and with slack water on both ends outside the levee.[21] After Bissell reported back to Grant, the general sent his chief engineer, Captain Frederick E. Prime, to investigate other places where canals might be cut. Prime determined that a route could be cut from the Mississippi through Lake Providence, 45 miles north of Milliken's Bend, into the Tensas, Black, and Red rivers to rejoin the Mississippi 50 miles above Port Hudson. This would serve several purposes. It would allow the army to bypass Vicksburg and would block the Red River supply route to the Confederates in Vicksburg. It would also position Grant's army in such a way that he could move on Vicksburg and transport troops south to Port Hudson to assist General Nathaniel Banks's Army of the Gulf in capturing the Confederate stronghold there. Grant placed General James B. McPherson in charge of the 400-mile Lake Providence route.

Both canal efforts (Lake Providence and Williams's cutoff on De Soto Peninsula) began in late January, and Grant placed Captain Prime in charge. When the captain arrived at Williams's cutoff he counted a number of problems. The Mississippi rose five to six inches daily, but the water entering the canal was not forceful enough to scour the sides of the ditch and change the course of the river. Instead, water oozed into the low-lying areas, spreading for several miles. Deepening and widening the canal to increase the channel flow was difficult because there was already two to three feet of water in the ditch, and Prime lacked heavy equipment to dredge the channel. Levees built one and a half miles northeast of the actual canal and designed to keep the area around the canal as dry as possible started to leak, as did the railroad embankment three miles northeast of the canal, limiting the space where the men working on the project could pitch their tents. Thirty-nine infantry regiments called the dreary swamps home. Sanitary conditions were abhorrent as the "seepage from the river . . . kept the ground very soft." Some soldiers "managed to get boards for our tent floors, but this was the exception and not the rule."[22]

What was critical for the success of the canal was not the depth, although it needed to be deep enough to allow heavy vessels through, but the width. The canal needed to be sixty feet wide for transports, rams, and ironclads to pass through. As the sides of the ditch were dug away the stagnant water in the canal spread, making it difficult to dig deeper than three or four feet. Constant rains raised the level of water in the canal, saturated the shelf of ground on both sides of the ditch, and left soldiers in a dismal state. Captain Henry G. Ankeny, Company H, 4th Iowa Infantry, wrote to his wife: "At camp near Vicksburg raining constantly, terrific thunder, camp overflowed. What we suffer will never be known outside these precincts. Work on the canal going on . . . Great deal of sickness prevailing in the army. Some new regiments have 300 sick. Many die. 72 left in my company."[23]

Colonel Bissell's regiment arrived on the scene in mid-February, and Prime put them to work extracting stumps in the present canal and in digging a channel 200 feet long to draw off the water between the canal and a plantation on the peninsula. Prime ordered work parties to dam both the entrance and exit to the canal and ordered the raising of the canal levees. Wooden frames were built to support the dams, and five hundred contrabands from surrounding plantations were assigned to dig a new entrance 600 feet upstream. The black laborers dug only four feet before hitting water, and the soldiers worked twenty-four hours a day to strengthen the levees protecting their campgrounds. As men frantically worked, the water in the canal rose to within seven feet of the water level in the Mississippi.[24]

To lower the water level around the new entrance to the canal, the engineers built a steam-powered sump pump to draw down the water as the black laborers completed digging a sixty-foot-wide entrance, four feet deep. Incessant rain in late February continued to hamper activity even though two steam dredges arrived from Memphis and started to clear an approach channel for the new entrance.[25] Then the weather broke. Now, with dredging operations underway, Bissell's engineer troops cutting out stumps from the expanding canal, and soldiers working around the clock, a sense of optimism began to spread as the water appeared to recede. Henri Lovie, a correspondent-illustrator for *Frank Leslie's Illustrated Newspaper*, predicted to his readership that "we will . . . be able to run our largest boats through the cut-off in less than two weeks."[26]

Confederate General John C. Pemberton and his staff observed the canal project with great interest and moved batteries into positions opposite the canal's exit. Pemberton, a Pennsylvanian who joined the Southern cause as a result of his marriage to a Virginian, was cognizant of Yankee determination and ingenuity. He informed the Southern War Department of the canal work and reported that if it was successful, he would need to fortify Grand Gulf, 46 river miles south of Vicksburg and only 23 miles over roads. Anticipating Federal suc-

Captain Frederick E. Prime, Corps of Engineers, ordered Grant's canal on De Soto Peninsula, opposite Vicksburg, closed and drained so that soldiers could deepen it. On February 9, 1863, soldiers completed building the levees that finally prevented water from entering the canal from the Mississippi River. *Frank Leslie's Illustrated*, March 28, 1863.

cess, Pemberton ordered two regiments under General John S. Bowen to Grand Gulf, where the river changed course from northeast to west and eddies made navigation problematic. Bluffs extending for 6 miles also afforded Bowen the opportunity to deliver plunging fire and extend his line so that additional guns would cover the entrance to the Big Black River, 7 miles south of Grand Gulf.[27]

Meanwhile, Captain Prime had begun work on a battery position to defend the exit of the canal. With most of the Missouri engineers employed in removing stumps and operating the dredges and Company D in building pontoons for transporting soldiers and supplies through the bayous, Prime assigned Company I, 35th Missouri Infantry, the task of building gabions and fascines to strengthen the canal's gun positions. Lieutenant Christian Lochbiler and his company made 30 gabions and 120 fascines.[28]

The better weather instilled in all the men working on the batteries and canal a false sense of security. The Mississippi continued to rise, and on the night of March 6 the mighty river broke through the upper dam and cascaded through the channel. The force of the water did not blow out the lower dam, but as the water pooled in the channel, the levee built parallel to the canal during the excavation started to leak. This was the same levee purposely cut to expel water during the heavy rains in February. Within hours of the initial breach, the crevasse in the levee started to leak water so forcefully that a 150-foot-wide opening emptied four to five feet of water over the campsites near the canal.[29]

To relieve pressure, the downstream dam was blasted out and attention turned to closing the gap. Men laid sandbags in the gap while one of the dredges was guided into the canal.[30] Water pressure knocked soldiers down into the mud, made visibility poor, raised adrenaline levels, and drenched clothing. Sticks and large branches cut arms, legs, and faces, and some men lost their breath as gallons of river water flew into mouths and noses. Once the dredge arrived it attempted to scoop earth to dump onto the sandbags, but all it was able to gather was mud that oozed from the iron bucket. Prime then ordered foraging parties to dismantle buildings on plantations, and he called upriver for a pile driver. Within eight days, wooden planks reinforced the levee and the dredge managed to place enough dirt and mud to close the breach. All this frenetic work left a gap to the east of the canal entrance and perpendicular to the levee.[31]

When the pile driver arrived, the engineers drove a post into this final gap and planned to attach an earthen-filled barge to the post. The latter, acting as a bollard and anchored in mud, gave way against the weight of the barge, almost capsizing the dredge. Prime decided to cut the mooring lines and let the current take the barge away. He then asked the engineers to alter the course of the water streaming into the gap by opening a runoff near the new entrance. This finally stopped the water from flowing into the canal, although water continued to leak out through crevasses in the levee. It had been ten days since the initial breach.

Now Prime was ready for the final push. The dredges began moving through the canal, widening the channel as they closed to within half a mile of the exit. Prime believed the steam-powered machines would accomplish the remaining work in the canal and felt that operations along the Yazoo River needed his attention, so he turned the command over to Colonel George F. Pride, Grant's chief engineer of military railroads. Perhaps Prime decided the canal was a forlorn hope. Confederate guns positioned opposite the canal's exit began to shell the dredges as they came within range. Prime's suggestion to Grant was to consider altering the course of the canal. Pride was more than capable of seeing the canal project to a successful conclusion, but he could not prevent the Southern batteries' nightly cannonades directed at the dredges. On March 22, Grant decided the canal was a failure, and although work continued for two more days. the dredges and soldiers were quietly withdrawn. He wrote to General Banks: "I have prosecuted that work [the canal], and would before this have had it completed to the width of 60 feet but for the heavy rise in the river breaking down the dam across the upper end. It is exceedingly doubtful if this canal can be made of any practical use, even if completed. The enemy have established a battery of heavy guns opposite the mouth of the canal, completely commanding it for one-half its length."[32] It would not be until 1877 that the river naturally broke through the De Soto Peninsula. A section of Grant's Canal can still be seen off Interstate 20 in Madison Parish, Louisiana. In 1863 the canal was more than five miles south from the bend in the river, which was opposite the city. Today the remains of the canal are only a mile and a half south of the river. The channel at this point is a mile wide and 100 feet deep during high water, and this portion of the mighty Mississippi passes one mile south of the city.

Grant was never convinced that the De Soto canal operation would succeed. He agreed with Sherman's assessment that the efforts there were "labor lost," but Grant did hold out hope that an avenue cut in the Lake Providence region, 75 miles north of the city, "bids fair to be the most practicable route for turning Vicksburg."[33] If a canal could be opened from the Mississippi into Lake Providence and a channel made between Bayou Baxter and Bayou Macon, the Union army and navy would have access to 200 miles of waterway leading to the mouth of the Red River, 150 miles south of Vicksburg.[34] General McPherson was given command of the operation, and by the time one of his division commanders, General John McArthur, reached Lake Providence, Grant's chief of artillery, Lieutenant W. L. Duff, had nearly completed the canal connecting the Mississippi and the lake.

When McPherson arrived he immediately investigated Bayou Baxter south, seven miles west beyond the lake, looking for a spot where his men could clear a watercourse between the two bayous. McPherson was a West Point–trained engineer. He had graduated first in his class in 1853 and worked on river and harbor improvements before the war. He was Grant's chief engineer for the Fort Henry, Fort Donelson, and Shiloh campaigns and was promoted to major general, United States Volunteers, when given command of the XVII Corps in December 1862. Now he had a decision to make. Option one was to send soldiers to dig a canal between Bayou Baxter and Bayou Macon, clear trees and debris from the passage, and build the proper levees. This task would prove arduous work because heavy rains had left area lowlands covered in two to three feet of water. If the water level in the lake continued to rise, the work of digging a canal, albeit just a two-mile one, would become more difficult and frustrating.

Option two was to open the levee at the entrance to the Mississippi, allowing water to overflow the lake, the bayou, and the surrounding area. Then, using

Bissell's Island No. 10 technique, McPherson would order steamboats and barges into the proposed bayou-connecting waterway, using the steamboats' capstans to haul out the trees and logs felled by the men on the barges. Finally, the trees would be sawed off below the surface of the water.[35]

Before McPherson made his decision he considered one more possibility. On March 1 he traveled north just over the Louisiana and Arkansas border to discover that the town of Ashton, Arkansas, on the Mississippi was just six miles east of Bayou Macon. This might provide the best option because blowing out the levee in Ashton would flood the countryside beyond with water deep enough for boats to reach the bayou. He asked Colonel Bissell if the plan seemed feasible to him, and when Bissell responded in the affirmative, McPherson ordered Bissell to open the levee in Ashton.[36] Furthermore, McPherson decided not to open the Lake Providence levee but instead to lift the steam tug *J. A. Rawlins* over the dam into the lake and from there move it into the watercourse being opened between Baxter and Macon, to assist in hauling out snags and logs. A detail from the 15th Iowa Infantry used ropes, tackle, and rollers to move the 3,500-ton vessel over the levee, through the Village of Lake Providence, and into the lake.[37]

When Bissell opened the levee at Ashton, so much water poured through that a ten- to fifteen-mile area was flooded—but it did not pool deep enough in the area between Ashton and Bayou Macon to navigate a large ship or ironclad. Now McPherson was ready to open the levee at Lake Providence. The levee was breached on March 17, and one day later the opening was scoured from 30 feet to 200 feet wide and the entrance was estimated to be 20 feet deep. In five days the level of the lake and the Mississippi were equal.[38] By this time, however, Grant was frustrated. Multiple attempts to safely extend his supply lines and move his army south of Vicksburg had washed out. The Mississippi River would not do the army's bidding. It would not be controlled. High water dampened efforts to open waterways, and the eventual drop in the water level would surely ground vessels trying to pass through the backwater, bayous, and swamps of Louisiana. He wrote to Halleck: "[On] the work of getting through Lake Providence and Bayou Macon, there is but little possibility of proving successful. The land from Lake Providence and also from Bayou Macon recedes until the lowest interval between the two widens out into a cypress swamp, where Bayou Baxter, which connects the two, is lost. This flat is now filled to the depth of several feet of water, making the work of clearing out the timber exceedingly slow, and rendering it impracticable to make an artificial channel."[39]

With four attempts to get at Vicksburg from the north and east all ending in failure, Grant needed to find a more reliable way to bypass the city, if one existed. He began to ponder an overland approach that would require marching his forces and supplies approximately 90 miles along the western side of the great river, while at the same time disguising his movements from the enemy. He also understood that his engineers and infantrymen would have to build that route.

When Grant decided to march south, Major General John A. McClernand's XIII Corps was camped at Milliken's Bend and in the best location to conduct a reconnaissance of the region. Sherman's XV Corps was at Young's Point, McPherson's XVII Corps was at Lake Providence, and Major General Stephen A. Hurlbut's XVI Corps was in Corinth. Grant was aware of a single wagon road, resting on the natural levee bordering two bayous, to the town of New Carthage 20 miles below Vicksburg. The road was through a cypress swamp, so Grant needed McClernand to determine whether an alternative method of moving supplies to New Carthage could be found.[40] McClernand selected Brigadier General Peter J. Osterhaus, commander of the Ninth Division, to oversee the operation. Osterhaus, born in Koblenz in Rhenish Prussia, had learned the building trade under his father's tutelage and, at the age of twenty, had entered the Prussian Army. He joined the revolutionaries in 1848 and then fled to the United States. His dream was to become a historian, but now he found himself in the role of pathfinder.[41]

The men chosen for the assignment were the 69th Indiana Infantry under the command of Colonel Thomas W. Bennett. Bennett, most likely a product of common school education, had become a professor of mathematics and natural sciences and a lawyer before the war.[42] Bennett's reconnaissance force would include two companies of the 2nd Illinois Cavalry, two mountain howitzers manned by a detachment from the 6th Missouri Cavalry, and Patterson's Kentucky Company of Engineers and Mechanics.

Forcing a passage south along Roundaway Bayou to New Carthage was a dangerous proposition. Everything was underwater except the levees along the bayou, and there were ample opportunities for the Confederates to stage an ambush or place rifle pits across the road in advance of a Union force. The route selected had to guarantee the safe passage of supplies, and the roadway had to guarantee it could handle heavy, continuous traffic.[43] A military wagon generally carried 3,000 to 4,000 pounds of cargo that included about 20 pounds of fodder and grain per day for each of the six horses or mules. Each infantry division of approximately 6,000 men required a minimum of 18,000 pounds of food per day, which meant each division needed more than twenty wagons of food for

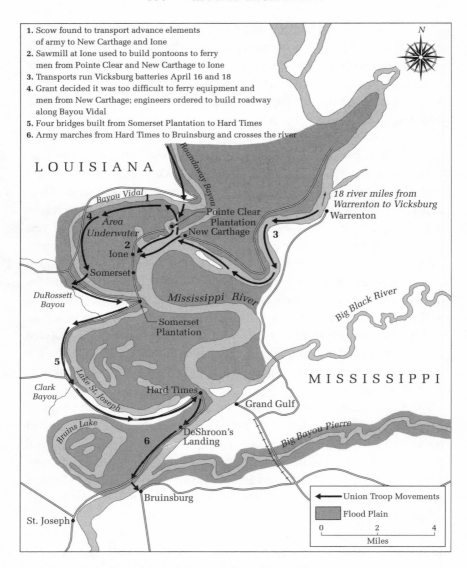

1. Scow found to transport advance elements of army to New Carthage and Ione
2. Sawmill at Ione used to build pontoons to ferry men from Pointe Clear and New Carthage to Ione
3. Transports run Vicksburg batteries April 16 and 18
4. Grant decided it was too difficult to ferry equipment and men from New Carthage; engineers ordered to build roadway along Bayou Vidal
5. Four bridges built from Somerset Plantation to Hard Times
6. Army marches from Hard Times to Bruinsburg and crosses the river

Grant Crosses the Mississippi and the Engineers Open a Supply Line

a three-day march. This did not include ammunition wagons, ambulances, a portable forge, wood for the forge, and, when available, a pontoon train. In addition, 100 to 130 horses were required to move a six-gun battery, ammunition, and supplies, consuming 3,600 pounds of fodder each day. Clearly, identifying the proper road was not simply a matter of finding a path through the woods.[44]

On the first day from Milliken's Bend, Bennett's command marched south-west toward the little village of Richmond. Patterson's engineers built a 200-foot-long bridge across Roundaway Bayou to Richmond, using boards salvaged from a nearby plantation. Patterson then moved south toward New Carthage. With this first leg of the supply line opened at Richmond, Osterhaus sent his division forward, and he personally led Bennett's vanguard toward Pointe Clear plantation where Roundaway Bayou and Bayou Vidal joined. This was only two miles north of New Carthage, but the area was entirely underwater, with houses submerged up to their roofs. Water rushed through porous levees, which prevented the engineers from building a bridge. That evening, contraband slaves told Osterhaus about a large scow hidden on Bayou Vidal several miles from their position. The scow was retrieved, and Patterson and his men converted it into a small gunship. They boarded up the sides with planks to a height of about five to six feet, cutting holes for oars and gun ports, and mounted a mountain howitzer in the bow.[45]

The scow, named the *Opossum* by Patterson's men, chased away the small Confederate force at New Carthage, so Osterhaus continued another mile and a half south to Joshua James's plantation, Ione, surrounded by twenty acres of dry ground. From this position the Mississippi levee ran unbroken for 41 river miles south to St. Joseph. Osterhaus was told that Confederate cavalry under Major Isaac F. Harrison was in the immediate area, along with two infantry regiments and a six-gun battery. It was imperative for the Union advance guard to hold the plantation so that it could become the staging area for a Mississippi River crossing.

On April 6, Grant ordered McClernand to move his remaining divisions to the vicinity of New Carthage. This opened the Milliken's Bend area for McPherson's corps moving from Lake Providence. Developments were dictating that commanders act with alacrity. Problems that required immediate solutions were confronting officers by the hour. There was no time to send a request up the chain of command and await a response. General Osterhaus appointed Major John W. Beekman of the 12th Ohio Infantry as acting engineer.[46] Beekman would organize the runaway slaves now flooding the Union lines into working parties. Patterson was to gather as many soldiers as possible with mechanical skills to join him at the sawmill on Ione plantation to build pontoons and boats. Men and equipment were ferried from Pointe Clear to New Carthage and Ione as fast as boats could be assembled.

Meanwhile, Grant decided the southern passage along the Louisiana side of the Mississippi was his final option to move his forces south. Crossing over the river onto dry land, he would lay siege to Vicksburg. Unlike the previous attempts at bypassing Vicksburg, to abandon this new plan was to end the campaign. No other options were left, and because Admiral Porter had agreed to run his ironclads and transports past the city's guns to meet Grant's army somewhere south of Vicksburg, the stakes were exceptionally high. Moving downstream, the warships would have the current with them, and Porter was confident his ships would make it by the Confederate guns guarding the river. But Porter then warned the general that his plan had better work because it would be impossible to return the 5,000-ton ships back upstream. "I am ready to cooperate with you in the matter of landing troops on the other side," Porter said, "but you must recollect that, when those gunboats once go below, we give up all hopes of getting them up again."[47]

Gunships, transports, and barges ran the batteries on April 16 and 18, so Grant rode to McClernand's headquarters at Pointe Clear to prepare for the crossing. When Grant arrived he did not like what he saw. First, he observed that narrow roadways and bayous hampered the supply line from Milliken's Bend. Next, the staging area around Pointe Clear, New Carthage, and Ione was too small to bivouac more than five divisions, and his entire army consisting of ten divisions was necessary for a successful amphibious landing. Finally, the navy transports would need to move twenty miles upstream to Warrenton, where the men, artillery, and horses would disembark; the transports would then return to New Carthage for the next trip. By this time, General Pemberton would be alerted to the enemy's landing and would rush men to Warrenton, only eight miles south of Vicksburg, and destroy Union soldiers on the east bank of the Mississippi before they could be reinforced by sufficient numbers. So, Grant decided to move his army farther south and strike at the Confederate fortifications at Grand Gulf.

Osterhaus reported that twenty-five miles to the south of Ione was Hard Times. The roads and extended area around Hard Times were dry, and it was only five miles from Grand Gulf. This sounded promising, but Grant remained skeptical. He asked his chief engineer, Colonel Wilson, to scout the area and assess the Osterhaus plan to determine whether an alternative route might be more feasible. The problem was that for Osterhaus's division, already on dry ground beyond New Carthage, moving overland to Somerset Plantation and then following the levee around Lake St. Joseph to Hard Times would make perfect sense. But moving seven more divisions and enormous amounts of supplies between Pointe Clear and New Carthage, where deep running water made bridging difficult, was a significant concern. Barges had managed to move material through Roundaway Bayou to Richmond and Pointe Clear, and there was a channel between Pointe Clear and New Carthage. But Grant was worried that the barges would eventually get stuck in the bayou because the Mississippi had finally begun to fall at a rate of six inches per day.[48] After the war Grant recalled: "I visited New Carthage in person, and saw that the process of getting troops through in the way we were doing was so tedious that a better method must be devised. The water was falling, and in a few days, there would not be depth enough to use boats; nor would the land be dry enough to march over."[49]

After his reconnaissance, Wilson confirmed McClernand's report that a possible route existed along a road that arched northwest then southeast along Bayou Vidal to Somerset Plantation. Sections of the road, however, needed serious bridging. Brigadier General Alvin P. Hovey's Twelfth Division was assigned the task of opening the causeway along Bayou Vidal with the assistance of his own pioneer corps under Captain George W. Jackson and Patterson's Kentucky engineers. Jackson was from Huntington, Indiana, where the Wabash and Erie Canal passed, and so was familiar with bridging and watercourses. On April 22 the work began. Under Captain Jackson's direction, soldiers were first assigned the task of cutting and hauling timber for the corduroyed roads and bridges. Patterson's men and additional infantrymen started work on the first bridge.

The bridge was constructed on a 100-foot-long flatboat anchored across the main channel bayou by a cable and chain on the south end and with a brace against a tree on the north end. Timber ties 6 or 8 inches in thickness were laid over the gunwales of the flatboat, on top of which rested 8-inch by 12-inch stringers (frames) supporting the floor planks. Men standing neck-high in water then began building four more sections toward each shore. The first section beyond the flatboat was made of 12-inch by 12-inch timber notched halfway into existing tree trunks, with additional planks attached from the flatboat to the notches. The remaining spans were trestle sections formed from four uprights secured at the top and bottom by square logs. The roadway was fixed in place by heavy beams pinned to the floor planks. The bridge was 362 feet long, but only 240 feet of it rested on trestles and was immoveable. Patterson feared that if the bayou rose or fell more than 18 inches, the connection between the floating sections and the stationary ones might render the bridge impassable.[50]

After completing the first bridge, Patterson's engineers marched south along the bayou and began work on a curved 550-foot bridge, using newly built 40-foot-long flatboats with piers and trestles on each end. The boats were anchored to a 2½-inch cable stretched from shore to shore and supported in the center by a tree. Some of the boats were fastened directly to the cable passing over their bows, and others were connected by short ropes.[51]

A third bridge, 150 feet long, was constructed across a slough. It rested on a center pier formed of logs placed crosswise and on trestles at either side of the pier. Men slogged through mud, stood in mosquito-infested water, and developed trench foot and blisters, yet by April 25 the work was completed: three bridges with a combined length of more than 1,000 feet and two miles of corduroyed road, "threading one of the most difficult regions that ever tested the resources of an army," were open for traffic. General Hovey described it as "the great military route through the overflowed lands from Milliken's Bend to the Mississippi River below Vicksburg."[52]

Of this feat, Colonel Wilson wrote: "With troops less capable and commanders less resolute and resourceful, we might have been beaten before getting within reach of the enemy." "When it is remembered that the bridges were built by green volunteers who had never seen a bridge train nor had an hour's drill in bridge building," he continued, "some conception may be had of the quality of the men and officers who carried through the remarkable work."[53] It is true that the army had not trained most of the men who worked on the bridge and road construction around Bayou Vidal as engineers, but these men had considerable experience as carpenters, mechanics, boat builders, and lumbermen before the war. The West Point–educated Wilson displayed some of his arrogance as a proud member of the Corps of Engineers: he clearly thought how extraordinary it was to see ordinary soldiers perform such skilled engineering work. Grant was more accurate in his praise: "the ingenuity of the Yankee soldier was equal to any emergency."[54]

As incredible as Northern engineering efforts were in providing Grant's army with a route south, more was about to be asked of them. With passage through Bayou Vidal completed, General Osterhaus ordered Colonel James Keigwin of the 49th Indiana Infantry to organize a combat patrol from Somerset Plantation to Hard Times. This patrol's primary purpose was to drive off a Confederate cavalry unit known to be operating in the region. Grant issued two orders: first, Colonel Wilson was to ferry across the Mississippi under cover of darkness to reconnoiter all roads leading from the river to the bluffs, and second, General McClernand was to ready his corps for an amphibious assault on Grand Gulf.[55]

Wilson's patrol confirmed Grant's fears. The area north and west of the Big Black River was inundated with water and had no practicable roads leading to the highlands. South of the Big Black the hills were swarming with gray coats.[56]

Keigwin and the 49th Indiana, 114th Ohio, a detachment from the 2nd Illinois Cavalry, and a section of the 7th Michigan Battery were also not having much luck. Four miles from Somerset Plantation, the Confederate cavalry they were looking for burned the bridge over Holt's Bayou.[57]

Keigwin detailed 100 soldiers from each regiment and under the direction of Colonel John Beekman, who had already demonstrated bridge-building skills, rebuilt the structure over the 80-foot-wide bayou in three hours. The patrol resumed the march for approximately one mile then came to another roadblock. The 120-foot bridge over Du Rossett Bayou was destroyed and the current in the stream lapped over the banks. The current was swift and the riverbed was quicksand.[58] Lieutenant James Fullyard led a fatigue party of men selected from the 49th Indiana and, with Lieutenant Francis Tunica of the Corps of Engineers, they began work on the bridge. Fullyard was yet another citizen soldier who could perform engineering duties in the field and whose improvisational ability was not limited by a particular methodology taught and practiced by West Point–trained engineers.[59]

Fullyard and Tunica decided to solve the problem of the quicksand by stripping half-inch weatherboarding from nearby plantation buildings and layering it crosswise and lengthwise at the base of their trestle structure. Once enough buoyancy was established, working throughout the night, they converted a trestle bridge into a floating bridge. Early the following morning, with a detachment of soldiers left behind to guard the new bridge, Colonel Keigwin's patrol continued its march toward Hard Times.[60]

Early on April 26, the combat patrol found Major Harrison's 400-man Confederate cavalry force deployed behind Clark Bayou a quarter of a mile away. Keigwin's men on the northern bank of Phelps Bayou eyeballed the Rebels and the two bridges they had burned to the ground. Not to be denied the roadway to Hard Times, Keigwin opened fire with his artillery, and then his infantry forded the bayous to attack Harrison. The numbers favored Keigwin, so Harrison's men mounted and hurriedly left the vicinity.[61] Now Captain William H. Peckinpaugh would have the honor of supervising the construction of the next two bridges.[62]

All the construction material for the two bridges was taken from neighboring barns. Large dry beams 50 feet long were used to support the flooring, which was kept in place by 6-inch by 6-inch blocks. Several of the banks were steep, so the engineers and pioneers dug an approach road by cutting down and tapering the embankments then corduroying the excavated area. This was to prevent the road from becoming a quagmire with the heavy traffic. Once the bridges were completed, Keigwin reported to Osterhaus that a practical road was opened from Somerset Plantation to Hard Times. In his official report Keigwin was effusive in his praise of his small command. He could not "speak in too high terms of all the officers and men in the detachment. They were ever ready to assist in all the labors of building bridges and to obey any command."[63] Finally, the men of the 49th Indiana and 114th Ohio returned to their brigades and prepared for the amphibious attack on Grand Gulf.

The bombardment of Grand Gulf began on April 29, 1863, as Admiral Porter's Mississippi Squadron attempted to soften the Confederate defenses on the eastern bank of the river. Instead, each member of the fleet took direct hits that damaged the vessels and inflicted seventy-five casualties. The defenders had three killed and nineteen wounded, and the infantry waiting in the rear for the Union infantry assault was untouched. Captain Shirk of the USS *Tuscumbia* invited two army officers to see the damage to his ship firsthand. A Colonel Warmoth wrote: "The boat was completely riddled. Torn to *smash*. Hog chains broken and a good many other things broken."[64]

Understanding Clausewitz's axiom that generals had to anticipate things going wrong on the battlefield, Grant, the quintessential Clausewitzian (even though he never read a word the Prussian wrote), had developed a contingency plan. His scouts had determined that the way was clear to march the army farther south toward DeShroon's Landing, and they learned from a runaway slave that there was a good road connecting Bruinsburg with Port Gibson on the east bank of the Mississippi. After a fitful night's sleep, infantrymen from the 24th Iowa and 46th Indiana marched south to DeShroon's Landing, where they were ferried across the river and touched the eastern shore of Mississippi in the midmorning hours of April 30. Grant's army was now on dry land.[65]

The first contact Grant had with a sizable Confederate army was on May 1 at Port Gibson, ten miles due east of Bruinsburg. Outnumbered three to one, the Southern forces under General Bowen fought a remarkable delaying action. They retreated north, burning two bridges behind them, one over the south fork of Bayou Pierre and the other over the north fork. A railroad bridge linking Port Gibson to Grand Gulf was also burned, but the Federals determined it was not necessary for crossing the river. Grant had accompanied General McPherson into Port Gibson and was in a hurry to pursue the retreating Confederates north, so he ordered Colonel Wilson, Captain Patterson's engineer company, and Captain Stewart R. Tresilian, commander of McPherson's pioneer corps, to begin work at once. The roadway of the suspension bridge over the south fork of Bayou Pierre was completely destroyed, and the decision was made to construct a floating bridge about twenty yards north of the old bridge. Plenty of buoyant materials were found by tearing down barns and cotton gins in the neighborhood.[66] Under time constraints, the pioneers and engineers completed the bridge by noon on May 2. Infantry officers did not think much of the structure, however, so they tried it out on an artillery piece drawn by a team of mules. Half way across, the span started to rock and then flipped over, dumping the gun and mules into the river. One infantryman said: "It was rather an expensive trial, but better than a column of infantry."[67] The next effort proved more successful. The bridge was a continuous raft 166 feet long with three rows of large mill-beams lying across the current in the bayou, with intervals between the beams filled by buoyant timber. The entire bridge was firmly tied together by a cross-floor or deck. The approaches to the bridge were over quicksand, so the men laid layers of logs and covered them with earth to prevent the animal teams from losing their footing on the logs.

Eight miles northeast of this new bridge was Grindstone Ferry and a suspension bridge over the north fork of Bayou Pierre. When the engineers arrived the bridge was in flames. The men in the vanguard and runaways from local plantations quickly extinguished the fire. A section of the bridge's roadway, side-truss, and string-pieces were destroyed. Colonel Wilson later wrote in his official report: "The charred parts were left undisturbed, and dispositions made to construct a new roadway over the remains of the old one so as to distribute all of the strain upon that part of the cross-ties just next to the stirrups of the suspension-rods. Timbers were lashed firmly to the suspension-rods by wire taken from the telegraph line, and rested on the charred cross-ties . . . The road covering . . . obtained from the farm houses near by, was . . . secured by side rails, spiked, and lashed to their places. The roadway was made perfectly secure by rack-lashing . . . 5 inch grass cable, passing around the new string-pieces and cross-ties and over the suspension-chain. The lashings were drawn taut by using rack-sticks 5 feet long and twisting around the suspension-rods." The new road rested approximately ten inches above the old road. It was a brilliantly improvised design. As historian Phillip Thienel observed, the engineers and pioneers built a "suspension bridge within a suspension bridge."[68]

Captain Andrew Hickenlooper, chief engineer of McPherson's XVII Corps, offered sublime praise of the resolute engineers. "The bridging of the South Fork of Bayou Pierre in four hours by Captain Tresillian [sic], a detailed engineer of the Third Division, and the complete reconstruction of the suspension bridge—nearly three hundred feet in length, and forty feet above the bed of North Fork—in a single dark and stormy night by a pioneer corps command[ed] by Captain Patterson, assisted by troops worn out by two days and nights of continuous marching and fighting . . . will ever remain as examples of what may be accomplished by intelligent direction of American soldiers."[69]

Like Tresilian and Patterson, Hickenlooper was one of those men, increasingly common as the war continued, who initially served as artillery or infantry officers but soon found themselves in the engineers. Fighting at Shiloh as captain of the 5th Ohio Battery, Hickenlooper was promoted by McPherson to chief of artillery and then chief of engineers. A civil engineer before the war, Hickenlooper observed that it was improbable "that in any other army ever organized could such bodies of men have been so quickly selected, efficiently organized, and rapidly qualified for the duties assigned as were the Volunteer Engineer Corps of the Union army, because in no other army of similar magnitude was there ever to be found such a versatility of first class talent subject to command at a moment's notice."[70]

In fifteen days, Captain Hickenlooper and his fellow engineers would be called upon to carry out another engineering feat: crossing the Big Black River. Before that, however, Grant's army would break from its supply base and, to the surprise and confusion of General Pemberton, march east to the Mississippi capital of Jackson. There the army would block any attempt by Confederate Joseph Johnston to rescue Pemberton at Vicksburg. Grant would then turn west, defeat the Southern army at Champion's Hill, and continue to pursue the retreating Confederates toward the Big Black River. Once the Big Black was crossed, the city of Vicksburg was next.

The battles of Champion's Hill and the Big Black River were decisive victories for the Army of the Tennessee. General Pemberton's Confederates retreated into the eight miles of fortifications surrounding the city in the north, east, and south. Grant's soldiers began the campaign in late February 1863, under deplorable conditions. Rain, hail, mud, and cold were their lot for several months as they tried to dig canals, move through swamps, and avoid falling prey to typhoid, typhus, and acute diarrhea; in April, as the weather warmed and the flooding abated, mosquitoes and snakes became the enemy. In May, for a period of seventeen days, the men fought five successful battles and marched about eighty-two miles, arriving at the last obstacle before Vicksburg.[71] Grant demanded no delay, but his concern was that his three corps of 44,000 men were packed within ten miles of each other. Located like points on an isosceles triangle, McClernand's corps was at the railroad bridge over the Big Black River; McPherson was five miles southeast of McClernand's position; Sherman was five miles northeast of McPherson and nine miles east of McClernand.

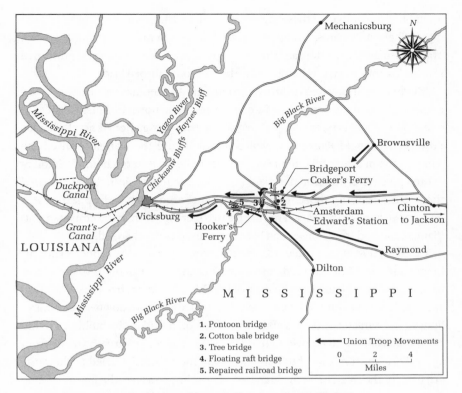

Engineers, Pioneers, and Infantry Build Four Bridges over the Big Black River

Grant imagined a massive logjam as the entire army tried to cross the only bridge over the river, a permanent wooden railroad bridge built on a masonry foundation. Of course, the retreating Southerners burned the entire bridge leaving just the foundation standing. The Big Black River was between forty and seventy yards wide, and at low-water stages the banks could be as high as twenty feet. The riverbed was mostly silt, sand, and soft clay, which meant that, unlike railroad bridges with foundations, road bridges with wooden trestle footings would be unstable and flash flooding would easily wash out these structures.[72] To cope with the unique geographical conditions, the eight crossing points that locals used from Bridgeport and Grand Gulf were by ferry. But ferrying 44,000 men, 38 artillery batteries, and all the wagons would take too long. Grant ordered the railroad bridge rebuilt and three other floating bridges constructed.[73]

Captain Christian Lochbiher's Company I, 35th Missouri Infantry, had hauled an India-rubber pontoon from Milliken's Bend to the Big Black. Now, along with the 127th Illinois and four companies of Bissell's Missouri engineers, his regiment built the bridge on the army's right flank at Bridgeport to allow the crossing of Sherman's corps. Bridgeport was located two miles north of the town of Amsterdam. At Amsterdam the river turned west for six miles, then turned south. One mile from this bend was the burned-out railroad bridge.[74]

McPherson discovered two roads around the Amsterdam area leading to the river. One was at Coaker's Ferry, a quarter of a mile north of Amsterdam, and one at Hooker's Ferry, two miles west of the town. Brigadier General Thomas Edward Greenfield Ransom, a civil engineer before the war, supervised the work on the bridge at Hooker's Ferry and completed it in ten hours.[75] Soldiers cut tall pine trees from the riverbank, stripped them of their limbs, and floated them to the opposite shore. The trees were anchored on shore by placing them on wooden tripods that raised the bridge two feet higher than the embankment. Planks collected from houses and barns were used as the roadbed.[76]

Construction of the bridge at Coaker's Ferry, supervised by captains Hickenlooper and Tresilian, required even more imagination than the Hooker's Ferry bridge. Working all night by the light of brushfires, soldiers from the 48th Indiana, 59th Indiana, 4th Minnesota, and 18th Wisconsin tore down several warehouses and hauled the planks and bales of cotton found in the buildings to the riverbank. Nails were recovered from the wood and a guy wire was made from horses' bridles. A raft was built and, with two men aboard, ferried to the western bank of the river; here, the men tied a sheer line to a tree, the line also being fastened to a tree on the eastern bank. The river was 102 feet wide. For each section, soldiers took two beams 34 feet long and laid them side by side 10 feet apart. The beams were joined together with 1-inch strips nailed 2½ feet apart. Uprights were nailed to the ends of each strip. Two cotton bales were placed on each cross-piece and pressed against the end uprights. The bales were kept in place by nailing strips crisscross in front of them. Additional bales of cotton were placed in the same way until they filled the entire 34-foot by 10-foot frame. The three sections were floated into the river, fastened together, and tied to the sheer line. The flooring was nailed in place, and at dawn Tresilian watched a 20-pound Parrot gun sink the bridge only 14 inches, leaving an excess of buoyancy of 16 inches.[77]

At the damaged railroad bridge, more drama unfolded. Under cover of darkness, lieutenants Tunica and Hains stealthily moved to the riverbank to determine the extent of the damage and to site a location for a new structure. Working with quiet resolve, members of the Kentucky engineers stripped the unburned sections of the bridge, dismantled several farmhouses, and, under sniper fire, constructed a 200-foot-long floating raft bridge about 150 yards

With the pontoon train miles to the rear of the Federal army's XVII Corps, its chief engineer, Captain Andrew Hickenlooper, and the engineer of the corps' Third Division, Captain Stewart Tresilian, devised an ingenious plan. They used cotton bales confiscated from a nearby ware-house to bridge the 102-foot-wide, 30-foot-deep Big Black River near Vicksburg. Tresilian observed that "a 20 pounder Parrott sunk the structure only 14 inches, leaving an excess buoyancy of 16 inches." Theodore R. Davis, *Harper's Weekly*, June 27, 1863.

northwest of the ruined railroad bridge.[78] Now, as described by Colonel Wilson of Grant's staff, "Counting these improvised bridges on the Big Black River and those at Milliken's Bend and Bruinsburg, there were five to six thousand feet of such bridges."[79]

On May 18, the Army of the Tennessee crossed the Big Black and formed on the outskirts of Vicksburg. The final phase of the Federal army's campaign to capture the Gibraltar of the Confederacy began. It is not an exaggeration to say that Grant's engineers, pioneers, and infantry regiments had built their way to Vicksburg.

The Confederate defenses surrounding Vicksburg ran in an approximate half-moon north to south for over eight miles. Captain David B. Harris of the Confederate Provisional Engineer Corps was responsible in early 1862 for lo-cating and supervising the construction of the first batteries at Vicksburg. The Confederate engineer most responsible for the defenses on the eastern side of the city, however, was Major Samuel Henry Lockett. Lockett graduated second in his class at West Point in 1859 and entered the Corps of Engineers. Born in Virginia, he grew up in Marion, Alabama, and joined the Confederacy at the start of the war. After serving with Braxton Bragg in Tennessee, he was ordered to Vicksburg in June 1862. On his arrival, he set to work surveying and mapping the terrain. Finally, he laid out a "system of redoubts, redans, lunettes, and small field works, connecting them by rifle pits so as to give a continuous line of defense."[80]

The ridges along the high ground, the deep ravines and gullies, were forested with magnificent magnolia trees and a dense undergrowth of cane. Lockett found a general line of commanding ground surrounding the city and decided to focus particular attention on the five major roadways entering Vicksburg. The defenses at their closest point to the city were three-quarters of a mile out, and at the farthest point, two miles.[81] Ten earthworks were built, one of which was the 26th Louisiana Redoubt. The structure had a six-foot-high parapet, twenty feet wide. On the exterior of the parapet was a ditch that served as a trench, ten feet wide and six feet deep, and in front of the trench was a rough palisade and glacis. Beyond the parapet was a two-foot-high firestep, with twenty-foot-wide terreplein, or horizontal surface, in the rear of the parapet.[82]

In November, Lockett was promoted to chief engineer of the Department of Mississippi and East Louisiana, which meant that he was responsible for the fortifications from Holly Springs to Port Hudson. He would no longer supervise the day-to-day construction of the Vicksburg defenses, and until April 1863, supervision of the remaining work was done by committee. Unfortunately, attention to detail lapsed, so when Grant's army crossed the Big Black River and Pemberton's forces occupied their fortifications, Lockett noticed a series of problems that he had failed to anticipate.

The redans and redoubts had not been occupied since their construction, and areas were now washed out and weakened by the winter rains. Along parts of the line the rifle pits were never finished, no obstructions had been laid, traverses were incomplete, and covered ways were nonexistent. Finally, there were only approximately 500 picks and shovels to be disbursed among 18,500 effectives spread out through eight miles of earthworks.[83] "They [the tools] were distributed to the different brigades according to the amount of work required," Lockett wrote, "and being much scattered along our lines were considered so precious by both men and officers that when not in actual use they were hidden for fear that they would be stolen by other troops, or ordered to some other part of the line by the chief engineer. They were entirely inadequate for the work, and the men soon improvised wooden shovels, using their bayonets as picks."[84]

Lockett's workforce consisted of twenty-six sappers and miners, eight detailed mechanics and firemen, four overseers for slaves, seventy-two slaves
(twenty were sick), three four-mule teams, and twenty-five yoke of draught
oxen. In addition, eleven engineer officers were under Lockett's command.[85] Although these officers' names appear in the *Official Records*, little else is known
about them. Captain Powhatan Robinson graduated from William and Mary,
Lieutenant Arthur W. Gloster was a railroad engineer, and Captain D. Wintter,
First Lieutenant E. McMahon, and Second Lieutenant F. Gillooly each commanded a company of sappers and miners.

The undermanned engineer force continued to rely on slave labor to dig new
sections of the fortifications and repair existing ones, and although the attackers marveled at the Confederate earthworks there was nothing unique or novel
about them. Brigadier General Francis A. Shoup, who commanded troops at the
26th Louisiana Redoubt, recalled that "the fortifications about Vicksburg were
a poorly run and poorly constructed set of earthworks. But there was no point
of the whole line which could not have been carried by a simple assault without
ladders or any sort of machines."[86]

On May 25, under a flag of truce, Pemberton sent a note to Grant proposing
a ceasefire so that dead and wounded Federal soldiers in front of Confederate
lines could be buried or receive medical assistance. Grant agreed. During the
lull in the killing, General Sherman introduced himself to Lockett, and soon
the two were sitting on a log enjoying a pleasant conversation. At one point
Sherman said: "You have an admirable position for defense here, and you have
taken excellent advantage of the ground." "Yes General," Lockett replied, "but
it is equally as well adapted to offensive operations and your engineers have not
been slow to discover it." Sherman agreed.[87]

Union engineers and pioneer detachments did adapt to siege operations
with aplomb. Since only a handful of army-trained engineers were present
in the siege lines, soldiers received little instruction in building approach
trenches, saps, and mines, but they took what information they were given and
combined it with their own ingenuity to successfully tighten the noose around
Vicksburg.[88]

Captain Prime was the army's chief engineer officer responsible for the entire siege operation. He made it clear that the engineering organization for conducting the siege works was deficient. He no doubt believed that trained military engineers familiar with the principles of Sebastien Vauban—considered
the father of siege operations for the brilliance he displayed at the siege of
Maastricht, Netherlands, from June 13 to 26, 1673—were the only ones capable

of supervising the complex business around Vicksburg. Prime estimated that he needed at least thirty officers for the task and acknowledged that those detailed from the staff of corps with some engineering experience would help. In his official report after the siege, however, he did not mention the three dozen volunteer officers in the Army of the Tennessee who performed engineer officer duties in constructing canals, roads, and bridges and in clearing swamps and bayous.[89] He also never mentioned Colonel Bissell's Missouri Engineer Regiment, which remained on the west side of the Mississippi, maintaining and repairing roads and bridges for the supply wagons, or Captain Patterson's Kentucky Company of Mechanics and Engineers attached to McClernand's XIII Corps.

The plan devised to cross no-man's-land between the two lines and bring Union soldiers within close range of the Confederate fortifications was to dig ten approach trenches, known as saps, from different locations along the line. Soldiers detached from their units to act as engineer pioneers were to dig zigzag trenches from the Union lines to the Confederate lines.[90]

Runaway slaves (paid ten dollars a month) and infantrymen were instructed by division engineers and pioneer officers on the techniques of digging saps, constructing sap rollers, fabricating gabions and fascines, and building batteries, parallels, magazines, and platforms for 30-pound Parrott guns. Several field guns were obtained from the navy, so pioneer soldiers had to haul these pieces from the river thirty miles away.[91] Captain Tresilian built wooden mortars by shrinking iron bands on cylinders of sweet gum trees and boring out the center to set off six- or twelve-pound shells. These innovative devices could throw a projectile 150 yards. Tresilian also guided soldiers in building twenty-two-foot-long scaling ladders. A rope was attached to the ladder so that men approaching fortifications on their bellies could pull the ladders behind and not be detected by enemy soldiers.[92]

The workhorse of the siege was the sap roller. As the approach trenches were dug in a zigzag pattern, using the contours of the geography to protect the excavators from enfilading fire, men remained vulnerable to enemy sharpshooters. To defend against this fire, sap rollers were pushed along in front of the workers to afford protection. The sap roller was a woven cylinder made from grape vines or cane, open at the top and bottom, with braces built on the inside so the roller would maintain its shape; this was pushed in front of the men in the trench for protection.[93] The solid cane offered excellent protection but was too heavy to maneuver over such difficult ground. Some men tried stuffing the rollers with cotton bales to provide more support, but weight remained a problem. Lieutenant Hains devised an innovative solution to the weight problem. He placed two empty barrels head to head, with fascines secured around their exteriors. He then tied smaller cane bundles between the fascines, and telegraph wire was wrapped around the exterior to hold everything together.

By the end of June, Prime reported that 1,200 fascines, 1,000 gabions, and six sap rollers had been constructed. In addition, pioneers had built 89 batteries and three magazines, dug twelve miles of trenches, hauled 370 wagonloads of lumber, and placed 220 field and siege guns. The approaches, deriving their names from the brigade or division commander who furnished the working parties, were built with unique characteristics to conform to the terrain in front of them.[94] For example, Thayer's Approach began near the crest of a ridge, ran down the slope toward the Confederates' 26th Louisiana Redoubt, then back up the ridge on which the redoubt was positioned. Under the supervision of Captain Herman Klostermann, who commanded the pioneer company of Major General Frederick Steele's First Division, a tunnel was dug from the north side of the first ridge that placed the pioneers at the base of the second ridge. From this ravine a trench was dug to the foot of the second spur, and then two traverse trenches were started up the slope. The trenches were six feet wide and six feet deep and were covered by a blinding of cane bundles across the top, affording protection against enemy rifle fire.[95]

Logan's Approach was begun on May 26, starting 150 feet southeast of Shirley House and 400 yards east of its objective, the 3rd Louisiana Redan. Captain Hickenlooper was in command of the operation, and the sap roller used was a railroad flatcar with wooden wheels that was stacked with twenty cotton bales. Loopholes in the cotton bales allowed men to use the flatcar as a rolling firing platform. On June 3 the sap reached a knoll, so two trenches, one left and one right, were extended from the original sap, and a Union battery was positioned there.[96]

At 75 yards from the redan, Confederate infantry were able to shoot directly into the approach trench, bringing work to an abrupt halt. Men from the 23rd Indiana then suggested building a tower behind the Union battery from which a sniper could harass the enemy so that work on the sap could continue. Hickenlooper agreed to the plan, the tower was built, and "Coonskin's Tower," named after Lieutenant Henry C. Foster, the tower's occupant, effectively shut down Louisiana sharpshooters. The work on the sap continued.[97] By June 16, Federal soldiers were within 25 yards of the redan. Now, with hand grenades being thrown into the Union trench and Union soldiers attempting to throw them back, a mining operation was begun under the immediate command of Lieutenant Russell of the 7th Missouri and Sergeant Morris of the 32nd Ohio.[98]

With drills, short-handled picks, and shovels, the men dug a gallery four feet wide and five feet high, at right-angles to the face of the fort's parapet. The main gallery was 45 feet long, and from the end of the main gallery two others were dug at 45 degrees on either side for a distance of 15 feet. The reddish clay soil was easy to dig and required very little bracing. Eight hundred pounds of power was placed in the main shaft and 700 pounds at the end of each of the lateral galleries. From each powder charge, two strands of safety fuse were laid to cover the possibility that one might fail to burn. The earth, which had been removed in grain sacks, was now carried back and deposited compactly, well braced by heavy timber, beyond the junction point of the three galleries. The rest of the entrance to the gallery was also packed with dirt.[99]

At 3:00 p.m. on June 25 the mine was exploded, leaving a crater thirty feet wide and fifteen feet deep. "At the appointed moment," Hickenlooper wrote, "it appeared as though the whole fort and connecting outworks commenced an upward movement, gradually breaking into fragments and growing less bulky in appearance, until it looked like an immense fountain of finely pulverized earth, mingled with flashes of fire and clouds of smoke, through which could occasionally be caught a glimpse of some dark objects—men, gun-carriages, shelters, etc."[100] Volunteers from the 31st and 45th Illinois Infantry ran into the crater, only to be met with withering fire from Confederates who had pulled back from the area because their commanding general sensed imminent danger. The falling debris also formed an artificial parapet commanded from a distance by Confederate artillery. With the Union assault a forlorn hope, troops were withdrawn to a new line beyond the range of artillery shells.[101]

The Yankees exploded another mine at Logan's Approach on July 1, and a gallery at Ewing's Approach was completed but never used. Southerners made a couple of attempts at mining and countermining; for example, at A. J. Smith's Approach they tried to blow up a sap roller but underestimated the distance and used a weak charge.[102]

The Army of the Tennessee finally had a chokehold on Vicksburg. Citizens of the town and the soldiers defending it all were suffering. Water was contaminated, disease and malnourishment were rampant, the dead went unburied, and the wounded went unattended. Men, women, and children lived on a diet of mule meat and boiled peas, the smell of human waste filled the air, and there was no hope of relief. There was no choice; on July 4, 1863, Pemberton surrendered to Grant. With sarcasm and exhaustion, Captain Patterson wrote to his wife: "What shall I say of today? I have had headache. You will see by the paper we have the place. Taking Vicksburg July 4 calculated to make one happy. They were reduced to [eating] mule meat. Poor fellows were glad to be captured."[103]

Some historians have argued that Vicksburg proved to be a costly white elephant for the North. As noted earlier, Albert Castel, Herman Hattaway, and Archer Jones all agree that the loss of Vicksburg did not deprive Southern forces of supplies from the trans-Mississippi because little was coming from that region before the loss.[104] Northern commerce along the river did not revive to prewar levels, Southern partisans threatened the safety of Northern river traffic south of the city after it surrendered, and the North just gained more territory to guard, siphoning resources from other operations.[105] These interpretations challenge President Lincoln's notion that, after the victory, "The Father of Waters again goes unvexed to the sea."[106]

Lincoln's point, however, was not to suggest that, after the collapse of Vicksburg, commerce was freely flowing along the Mississippi. The purpose of his letter to James C. Conkling on August 26, 1863, was to remind Democrats and Republicans alike that he was determined to emancipate the slaves, whether either party agreed with him or not, and to save the Union. Regarding slaves, Lincoln wrote: "Why should they [blacks] do any thing for us, if we do nothing for them? If they stake their lives for us, they must be prompted by the strongest motive—even the promise of freedom. And the promise being made, must be kept."[107] To save the Union, the war had to be won. "The strength of the rebellion, is its military—its army," Lincoln wrote. "That army dominates all the country, and all the people, within its range."[108] Destroy the army and the Union destroys the Confederacy and slavery. At Vicksburg, Grant had captured an entire Confederate army. Twenty-nine thousand soldiers surrendered, along with 172 pieces of artillery, 38,000 projectiles, 58,000 pounds of black powder 4,800 artillery cartridges, 50,000 shoulder weapons, 600,000 rounds of ammunition, and 350,000 percussion caps.[109]

Union commerce was limited along the Mississippi for the remainder of the war, but that was not the point. A Confederate army was captured and, strategically, the victory at Vicksburg forced the South to operate its other army groups in a dwindling territory. If Vicksburg had remained in Southern control, the North's ability to destroy the rebellion would be that more difficult. It would continue to delay other major actions, including operations along the critical Nashville and Atlanta corridor. With a presidential election coming in 1864, a stalemate favored the Peace Democrats in the North and threatened President Lincoln's second term. Lincoln's letter to Conkling in August 1863 was political commentary, not strategic analysis. He told his constituencies: "The signs look better."[110] The signs were much-needed victories, and Grant had delivered a huge one.

The capture of the city and 29,000 men was not the only important outcome of Grant's campaign. The Confederate defeat contributed to the surrender of Port Hudson, south of Vicksburg, and demoralized Southern citizens. Georgia Governor Joseph Brown, omitting any mention of Gettysburg, on July 17 urged his citizens not to despair over "the late serious disasters to our arms" in Mississippi and Tennessee. "The disastrous movement of Lee into Pennsylvania and the fall of Vicksburg," a Confederate congressman wrote, "the later [sic] especially, will end in the ruin of the South."[111] And a Texas sergeant in General Johnston's army commented after Vicksburg: "I have little hope in the future."[112] Across the Confederacy, the fall of Vicksburg struck like a death knell. In Richmond, chief of the Confederate Ordnance Department, Colonel Josiah Gorgas, wrote in his diary: "Vicksburg and Port Hudson capitulated, surrendering thirty-five thousand men and forty-five thousand arms. It seems incredible that human power could effect such a change in so brief a space. Yesterday we rode on the pinnacle of success—today absolute ruin seems to be our position. The Confederacy totters to its destruction."[113]

Vicksburg ensconced Grant as Lincoln's number one commander. The political machinations were over. The calls from other generals and the press to sack Grant because of his drinking abruptly ended. Soon to become a lieutenant general and commander of all the Union armies, Grant showed brilliance and tenacity that would bring about the collapse of the Confederate armies. In hindsight, Grant was the greatest American general of the nineteenth century and, one could argue, the greatest general in American history. Success at Vicksburg gave Grant the chance to prove his greatness.

Finally, Vicksburg tested the mettle of the Union soldier, not just as a warrior but also as an innovator and builder. The campaign showed that defeating the Confederacy on its own territory required moving supplies over long distances and controlling vast expanses. Control of railroads, roads, and bridges was the fundamental element needed to take the war to the South. By contrast, Southerners were unable to bring supplies to Vicksburg because of their inability to build additional light-draft boats to haul food from the productive plantations of the Yazoo River Delta. Additional food rotted on the Vicksburg wharves because of a lack of public storage facilities.[114]

With a limited number of trained military engineers to cope with the size and scope of the campaigns and armies, Northern volunteer soldiers were called upon to assume the role of engineers and pioneers. As a Federal infantryman recalled, "Every man in the investing line became an army engineer day and night."[115] Other generals besides Grant now understood the power they had in the volunteer soldier's ability to perform engineering tasks to help move the army into positions previously thought inaccessible.

Captain Prime, not always effusive in his praise of volunteer engineers, in his final report of the campaign had to admit his amazement and admiration for what these men accomplished. "Over a line so extended and ground so rough as that which surrounds Vicksburg, only a general supervision was possible, and this gave to the siege one of its peculiar characteristics, namely, that many times, at different places, the work that should be done . . . depended on officers, or even on men, without either theoretical or practical knowledge of siege operations, and who had to rely upon their native good sense and ingenuity."[116]

Vicksburg was a critical victory for the Union, and it was the most remarkable feat of engineering during the entire war. General Grant was a risk taker, and he had vision, determination, and moxie. He also had the men, who in previous lives were lumberjacks, railroad workers, mechanics, and machinists. These men understood the value of innovation and problem solving as civilians, and they transferred this knowledge to the problems faced as soldiers. Prussian Army Major Justus Scheibert, an observer attached to the Army of Northern Virginia for seven months in 1863, wrote of Vicksburg that the Union army owed its success to a navy, to heavy artillery, and "to engineering superior to their Confederate counterparts."[117] Grant had the men who engineered and built the Army of the Tennessee's victory at Vicksburg.

CHAPTER TEN

Gettysburg

Meade declined the challenge, and Lee resuming the retreat, crossed on the bridge of boats that had been thrown over the river at Falling Waters by the engineers—and a crazy affair it was, too.

Lieutenant Colonel Moxley Sorrel, CSA

During the third year of the war, the differences between Union and Confederate railroad management, engineering, and use of infantry as effective engineers or pioneers grew increasingly apparent. As operations became more complex and the logistical support for those operations became more critical, these differences highlighted one of the central reasons for Union success and Confederate setbacks.

In the spring of 1863, while the Army of the Tennessee drew closer to Vicksburg and operated with an unorthodox yet highly effective engineering organization, a thousand miles northeast in Washington, DC, Major General Henry W. Halleck suggested to Congress the establishment of a more orthodox engineer corps suitable to the responsibilities carried out by regular army engineering officers. Halleck found it almost incomprehensible that a first lieutenant, Cyrus B. Comstock, served as chief engineer of the Army of the Potomac. Halleck understood that more army engineers were needed in all theaters of operations. From Vicksburg each week, corps and division commanders requested that the War Department send additional engineers. In his report on the siege, Captain Prime wrote: "The engineer organization here, as in all our armies, was very deficient, if we judge either from the practice of nations wiser in the art of war than ourselves or from results. Thirty officers of engineers would have found full employment."[1]

In middle Tennessee, General Rosecrans used the 1st Michigan Engineers and Mechanics to repair railroads and bridges, but the engineers reported to and were supervised by regular army engineers. A significant part of the military culture was the understanding that engineering required the best-trained and most technologically skilled officers in the army. Halleck's proposal was now to give those skilled officers the recognition and ranks they deserved. The chief engineer of a field army, Halleck proposed, would hold the rank of colonel, and the chief engineer of an army corps, the rank of lieutenant colonel or major. These promotions were overdue. Furthermore, Halleck argued, at the start of the war, skilled engineering officers like Rosecrans and Meade offered to lead volunteers and were promoted to brigadier general. It was better for their careers that they become field commanders rather than remain as chief engineers of army groups with the rank of lieutenant or captain.

Halleck was still not willing to concede, however, that volunteer engineers should be placed on the same level as West Point–trained regular army engineers. Nonetheless, this professional ethos had started to wear down. Grant discovered that citizen soldiers, together with a small professional cadre of engineers, were capable of improvised solutions to the many logistical challenges faced by the army. According to historian Edward Hagerman, "The Civil War forced the regular officer . . . to reconsider his professionalism from a broader intellectual and organizational perspective. The changes in warfare that he had to bring under control required that he adopt a more open, flexible, and historical, and less static, mechanistic, and absolutist military world view."[2] For Halleck, before he could reconsider his professionalism from a broader perspective, the first step was to seek proper professional recognition for the role engineers had played in the first two years of the war and the role they would surely play as the war continued.

Senator Henry Wilson of Massachusetts took up Halleck's suggestion to provide proper promotions for engineers and to expand the Corps of Engineers without excessive costs to the already bloated wartime budget. Wilson introduced a bill out of the Committee on Military Affairs to merge the Corps of Topographical Engineers with the Corps of Engineers and to raise the authorized strength of the latter to 107 officers. At the time there were 48 officers in the engineer corps and another 40 in the topographical corps. The bill also included creating five colonels, ten lieutenant colonels, twenty majors, and thirty captains.[3]

Some senators and congressmen felt it was shortsighted to spend money on an expanded Corps of Engineers and not on additional infantry. Others, like General Sherman's younger brother, Senator John Sherman of Ohio, saw

Wilson's efforts as a deceitful way of expanding the regular army after the war. Wilson convinced his colleagues that he would revise the measure, and as a result, the bill passed in both the House and Senate on March 3, 1863. Three key changes to the bill were made to assuage the worries of Senator Sherman and his faction. First, no engineer officer could be promoted to a field grade rank without passing an examination given by three senior engineer officers. If the officer failed, he would be suspended for a year and then allowed to take the exam again. If he failed a second time, he would be asked to leave the army. Since suspension in wartime was too a great a risk for ambitious engineers, Sherman hoped most engineer officers would remain captains and lieutenants. Second, the act would be in effect only during the rebellion, after which the president would reduce the number of engineers to that authorized by law before the new act went into effect, from 107 to 55. Third, engineer officers promoted during the war would revert to their prewar rank or, in the case of West Point graduates in the classes of 1862–65, to the rank of second lieutenant.[4]

With the act operational in early April, the newly appointed commander of the Army of the Potomac, Major General Joseph Hooker, prepared for his spring offensive in northern Virginia. Hooker was ambitious, confident, and personable, although he had a political dark side. Lincoln had heard that when the army was under Burnside's command, Hooker was rather disingenuous in his dealings with his fellow general, often trying to thwart him. Now Lincoln placed Hooker in charge, and in a letter to the general, he was both sharp and direct. "I have heard, in such a way as to believe it, of your recently saying that both the Army and Government needed a dictator. Of course it was not for this, but in spite of it, that I have given you command. Only those generals who gain success can set up dictators. What I now ask of you is military success, and I will risk the dictatorship."[5]

Military success was not forthcoming. After launching an excellent plan to envelope Lee's army in the Fredericksburg area, Hooker lost his nerve. He was holding to a defensive position centered on Chancellorsville when Stonewall Jackson made his famous march onto the Union flank and rolled up a portion of Hooker's army, at the cost of Jackson's own life. The Army of the Potomac's engineers, under new leadership, did perform well in building fifteen bridges over the Rappahannock River between April 28 and May 4 and then dismantling all of them. Seven bridges were built by the 15th New York, six by the 50th New York, and two by the US Engineers. Two bridges built by the 50th New York were transported sixteen miles in nineteen hours from Franklin and Pollock's

Mill Creek Crossing up to Banks' Ford, and one bridge built by the 15th New York required the men to make extra trestles on the spot.[6]

Brigadier General Henry Benham of Cheshire, Connecticut, was now commander of the Engineer Brigade (volunteers). An 1837 graduate of the Military Academy, Benham had been in charge of superintending the fortifications in Boston and Plymouth harbor before being assigned to the Engineer Brigade. He was a skilled pontoon bridge builder, and part of his responsibilities included command of the engineers' depot in Washington, DC.

There were an estimated two hundred pontoons at the depot; some required repairs, and all needed to be arranged into bridge trains and made ready for delivery to places the commanding general designated. The new commander of the 15th New York was Major Walter L. Cassin. Captain George H. Mendell, a former member of the topographical engineers, was now commanding the Army of the Potomac's Engineer Battalion (regulars), having replaced Captain Duane when he was reassigned to the Department of the South after General McClellan's dismissal in November 1862.[7]

The chain of command among the engineers in the Army of the Potomac was awkward and confusing. The chief engineer for the army was Brigadier General of Volunteers Gouverneur Warren.[8] He had received his promotion in September 1862 and still retained the rank of captain of engineers. Benham technically reported to Warren. Benham, however, was promoted to brigadier general of volunteers in August 1861 and retained the rank of major of engineers. On paper Benham outranked Warren. It was also made more confusing by the fact that both Colonel William H. Pettes, commander of the 50th New York, and Cassin outranked Mendell. Yet Mendell was a West Point graduate and a member of the prestigious Corps of Engineers; Pettes and Cassin were not.

Fortunately, as the army moved toward Gettysburg, the engineers' orders were direct: on June 17, the Engineer Battalion and 250 men from the 50th New York were to place a pontoon bridge train in the Chesapeake & Ohio Canal at Georgetown and proceed to the area around Edwards Ferry on the Potomac River, thirty-two miles northwest of the capital. By the morning of June 21, the engineers had built a 1,340-foot pontoon bridge with sixty-four boats and three crib trestles.[9] By June 27 a second bridge was built, and that evening the entire army of 90,000 soldiers with horses, artillery, cavalry, and supply wagons crossed the river in search of Lee's army.

The lead elements of General Robert E. Lee's Army of Northern Virginia were in Chambersburg, Pennsylvania, before General Hooker's army started over the pontoon bridges. Using the South Mountains as a screen, Lee's three corps were separated by thirty-two miles and the army was operating fifty-five miles from its supply base in Winchester, Virginia. For historians, and perhaps even for General Lee, the specific geographical objective was unclear. Nevertheless, if his army continued into Union territory and turned toward Philadelphia or Baltimore, Lee would have to require his engineers to maintain, repair, or build bridges across several major rivers.

There were several reasons that Lee was invading the North in the summer of 1863, and one of them was that the Confederate railroad system had failed the army in the winter of 1862–63. That winter, Lee was certain the Army of the Potomac was preparing to cross the Rappahannock and begin a major spring offensive toward Richmond. As he contemplated his own strategic options at Fredericksburg in January and February, his men and horses were starving. Supplies carried by trains on the Richmond, Fredericksburg & Potomac Railroad provided little food and forage. The reason for this: no central control of the railroads.[10] The Wilmington & Weldon Railroad, a single-track road, ran due north and connected at Weldon with the Petersburg Railroad. This railroad directly connected to the Richmond line and the Richmond, Fredericksburg & Potomac. The east coast of North Carolina constituted a critical wealth of corn, bacon, grain, and fish, more than enough for Lee's army, but the wretched condition of the tracks and no cooperation or coordination between railroads left the Army of Northern Virginia cold and hungry.

Colonel William Wadley, Confederate coordinator of transportation, complained to James A. Seddon, the new secretary of war, about the problems on the Wilmington & Weldon. The State of North Carolina owned a controlling interest in the railroad, so Seddon made known to Governor Zebulon Vance that Lee's army was starving and depended on supplies from his state. Seddon asked Vance if more could be done to ensure a more efficient management of the railroad.[11] Wadley, for his part, recommended that mechanics be released from military service to work for the railroads and that the government pass an act making it "obligatory upon the railroads of the country to perform promptly Government transportation. The law, without allowing men and supplies, will be of no use, for without these the roads cannot exist." Wadley's frustration was palpable. To the highest-ranking general in the Confederacy, Adjutant and Inspector General Samuel Cooper, he wrote: "In every direction there is an accumulation of freight that is being wasted or damaged for want of protection, and a number of Government agents and messengers accompanying it in the character of protectors and forwarders would, I have not the least doubt, form a full regiment."[12] Nothing resulted from his sarcasm or his appeal.

In March, Seddon admitted to Lee that the lack of supplies for the Army of Northern Virginia was due to railroads, which were "daily growing less efficient and serviceable." One month later, Lee again wrote to Seddon pleading for help. The Army of Northern Virginia, plagued with scurvy and typhus, needed fresh food. With incredulity Lee noted that "I also learn that there have been 100 carloads of sugar and other supplies for this army detained at Raleigh and Gaston for more than a fortnight."[13]

In April, desperate for food, Lee sent General John Imboden's cavalry on an expedition to destroy the Baltimore & Ohio Railroad in western Virginia and to bring in beef cattle for the army. Imboden was successful, but the sustenance was too late for Lee to launch an offensive before Hooker's army struck him. The resulting Battle of Chancellorsville led Lee to believe that a Union summer campaign in Virginia would place his army on the defensive, but a move north over the Potomac River might lure Hooker out of Virginia and present Lee with an opportunity to destroy the Army of the Potomac. A move north would pose a threat to Baltimore, Washington, and Philadelphia, provide much-needed subsistence and forage, and give farmers in the Shenandoah Valley a chance to harvest their crops free from interruption by military operations.[14] Finally, Lee believed the Peace Democrats in the North were gaining strength, especially because the Emancipation Proclamation had uncorked racist emotions among Northern politicians, and that an invading army might agitate malcontents such as Ohio Congressman Clement Vallandigham to further action.

It was alleged that Vallandigham had encouraged Confederate President Davis and his cabinet to back Lee's plan to invade the North. In a *New York Times* editorial, Henery Reinish stated that, according to reliable sources, "Vallandigham . . . assured [Davis and his cabinet] that the North was ripe for revolution." "Mr. Vallandigham's representations were corroborated by the tone of the majority of the Northern journals," he continued, "who surely would not denounce the Administration so boldly except by assurance of having the masses strongly in their favor."[15] These comments might have reinforced Lee's belief that when his army marched through the Maryland countryside, people would greet him with open arms. To the general's surprise, they did not.

Confederate railroad management and transportation issues played a major role in Lee's decision to invade the North in the summer of 1863. The Confederate Congress and president finally attempted to address the transportation problem with a railroad regulation act in May, giving the government authority to set through-freight schedules and to enforce them at government discretion.[16] The law was never invoked in 1863, however, because individual rights and private enterprise trumped Southern nationalism and military logistics. President Davis did try to preserve a Railroad Bureau by appointing Wadley's protégé, Frederick W. Sims, to the position as head.

The new head inherited from his retired predecessor little more than the framework for an organization. The bureau had no control over quartermasters in the field, and both General Bragg's Army of Tennessee and General Joseph E. Johnston's army in Mississippi had their own railroad bureau. The officer responsible for railroad traffic through Richmond, Major D. H. Wood, reported to the quartermaster general, and Sims and his bureau had no representative in the major railway hub of Atlanta.[17] The quartermaster bureau had eighty-eight clerks; the subsistence bureau, thirty-six; and ordnance, twenty-four. The engineer bureau had three clerks and the railroad bureau had two.[18]

Sims worked the bureaucratic system well. He got his bureau transferred to the Quartermaster Department where he could better manage competition among staff officers for supplies. He could use the department to assist him in moving engines and rolling stock to the east from sections of Mississippi abandoned by the Confederate military. Yet, even in his efforts to salvage this abandoned equipment, private railroad superintendents demanded the rescued trains. Confederate army use of locomotives competed with private railroad use, and no one took the time to make the necessary mechanical repairs to the engines. The locomotives were run into the ground, owners lost their valuable machines, and they had no redress with the Confederate government. In a letter to the quartermaster general in October 1863, Sims asked rhetorically: "Is it any wonder that transportation is deficient? Is it not rather a wonder that we have any transportation by rail at all?"[19]

Sims continued to lobby for a more centralized bureau as his administrative skill and energy brought about some improvements to the railroad transportation system. Nonetheless, individual interests continued to work against him and damaged the Confederacy as much as Yankee guns and bullets. Army officers often seized rolling stock for their own purposes. In November 1863, at a major railroad convention held in Macon, Georgia, owners made clear that the purpose of the conference was to consider raising the rates that railroads charged the Confederate government for transportation. Finally, out of necessity, General Lucius B. Northrop, head of the Commissary Department, maintained his own railroad representatives in certain areas of the Confederacy and in August 1863 named the president of the Georgia Central Railroad, Richard R. Cuyler, as head of transportation for southern Georgia.[20]

The question must be asked: If Lee's army had been well fed in the winter of 1862–63, would he have embarked on a move north to risk all in the spring?[21] Perhaps he would. We can only speculate, but we know for certain that one of Lee's major reasons for going on the offensive was to feed his starving men and animals. So, the Army of Northern Virginia's march into Maryland and Pennsylvania in June 1863 uncovered a significant logistical problem faced by the Confederacy. This supply problem, however, was not the only one confronting Southern armies. Gettysburg revealed two other weaknesses operating against Lee's army as it marched north: poor maps and poor engineering.

In February, Stonewall Jackson had asked his topographical engineer, Jedediah Hotchkiss, to prepare a small-scale map of northern Virginia, central Maryland, and south-central Pennsylvania. The map was also to include Baltimore, Philadelphia, and Washington, DC. When finished, the map would encompass more than 1,000 square miles of territory. The scope of the project was vast, with Union territory the most significant portion, and time was limited, so Hotchkiss had to rely on Pennsylvania county maps. On a 38-inch by 42-inch sheet of heavy watercolor paper, he penciled a grid consisting of thousands of square centimeters. He superimposed a similar pencil grid on a county map, then transcribed the map on a smaller scale onto the heavy-paper grid. It was a long process and the project was incomplete at the time Jackson was mortally wounded at Chancellorsville.[22]

This incomplete map was the one that General Lee used during the Gettysburg Campaign. The map itself was a work of art. On cream-colored paper, red pencil lines identified roadways, blue marked rivers and streams, and black, in impeccable handwriting, listed towns, mills, blacksmith shops, major topographical features, and every rural resident's name. More nuanced features such as mild declivities, small hillocks, woods, road surfaces, and fording sites did not appear. The salient and eventually famous landmarks of the battle— Seminary Ridge, Culp's Hill, Cemetery Hill, Little Round Top—also did not appear on the map.[23]

These omissions were costly for Lee and his army. On July 1, two regiments were lost from Confederate Brigadier General Joseph R. Davis's brigade when his men jumped into an unmapped twenty-foot-deep railroad cut and were captured by Colonel Rufus R. Dawes's 6th Wisconsin Regiment. As Lee planned to envelope Meade's left flank on July 2, his engineer, Captain Samuel R. Johnston, insisted that he had ridden to Little Round Top and found it unoccupied. In fact, the area was swarming with General John Buford's Union cavalry. It was probably Warfield Ridge or Houck's Ridge that Johnston had found to be empty.

When Johnston was ordered to guide Hood's and McLaws's divisions to the southern end of the battlefield in preparation for the early afternoon attack on the Union left, he had to backtrack and countermarch because he had no map and misunderstood the area's topographical features. Finally, the map of Gettysburg made by Hotchkiss that accompanied Lee's and Ewell's written reports of the battle indicated a keen awareness of the ground north of the town, but to the south, the map does not include accurate illustrations of the round tops or the undulating and wooded areas east of the Emmitsburg Road, including the terrain around Weikert's and Trestle's farms, Rose's Woods, the Wheat Field, and the valley between Houck's Ridge and Little Round Top.[24]

The lack of adequate maps placed the Army of Northern Virginia at a distinct disadvantage, especially when a knowledge of local geographical features was so crucial to a movement's outcome. Perhaps Lee expected he would receive local intelligence to fill in whatever gaps there were in his maps. He had received this type of information in the past, but he had always fought in friendly territory. Gettysburg highlighted the need for an invading army to have technically skilled men and materials at a general's disposal. This explains why Lee's artillery chief said after the battle: "Not only was the selection of ground about as bad as possible, but there does not seem to have been any special thought given to the matter. It seems to have been allowed to select itself as if it were a matter of no consequence."[25]

The major misadventure of the campaign, however, came when Lee's army, after the three-day battle, attempted to escape across the Potomac and back into Virginia. Complete failure in the river crossing would have cost Lee his entire army and the war. Disaster was barely averted, and the episode revealed the Confederacy's vulnerability when it came to engineering.

Conflicting testimony makes it difficult to determine exactly what happened at Williamsport and Falling Waters between July 4 and 14, yet evidence suggests that the operation to build a pontoon bridge over the Potomac River did not go well and left Lee frustrated with his engineers. On the night of July 6, General Lee learned from General Imboden, who was leading the army's wagon train and ambulances, and from General Stuart, whose cavalry was acting as a rearguard for the train, that Federal cavalry had partially destroyed the Confederate's unguarded pontoon bridge at Falling Waters. Moreover, the river at Williamsport had risen because of the unremitting downpours of the past several days and was now not fordable.

There were about twenty engineer officers with Lee's army, including members of his own staff, his corps commanders' staffs, and some divisional staffs. There were no engineer soldiers; instead, men were detached as pioneers from brigades and regiments. Consequently, no one was responsible for the pontoon train once the Army of Northern Virginia had crossed the Potomac in June. When the bridge was disassembled it was left behind. Many of Lee's men had forded the river in June when it was only about three feet high in certain sections, and he fully expected his army would do the same on the return trip. The lack of concern about the pontoons also suggests that Lee believed his army could ford any river if and when they turned east toward Baltimore or Philadelphia.[26]

On the morning of July 7, Lee allegedly ordered Major John Harman, General Ewell's quartermaster, to begin work on the bridge. General Imboden, however, recalled that "General Lee expressed great impatience at the tardiness in building rude pontoons at the river," and frustrated, called in Major Harman to supervise the construction. That was July 9.[27] Perhaps construction went so poorly on July 7 and 8 that General Lee decided to send Harman to the rescue. History does not record who might have been responsible for any blunders in the first attempt to build the bridge.

We do know that one day after Harman arrived at Williamsport, also ordered to Williamsport to work on the bridge were Major John G. Clarke, Longstreet's chief engineer; Captain Summerfield Smith, assigned to Clarke, and his pioneers; Lieutenant Henry Herbert Harris and his pioneers; and Captain Justus Scheibert, a former engineer in the Prussian Army and a recent immigrant to America.[28] Harris recorded in his diary that Major Clarke, Lieutenant Colonel William Proctor Smith, Captain Samuel Johnston (Lee's chief engineer), and Captain Henry T. Douglas (of A. P. Hill's staff) consumed some time drawing the dimensions of the pontoons necessary to support the bridge. The effort was becoming a bridge by committee.[29]

According to Harris, work on the pontoons began in earnest on July 11. Five days had passed since Lee ordered work on the bridge begun. The pioneers tore down warehouses along the canal in Williamsport, and a local sawmill and lumberyard were used to construct the boats. Oakum was picked from old ropes and forced into the seams of the pontoons, hot tar from the quartermaster wagons was used to caulk the boats, and they were now ready to float the six miles to Falling Waters, the selected site of the bridge. Meanwhile, Harman supervised the operation of a flatboat, guided by a wire strung across the river, to ferry the wounded and prisoners.[30]

By July 13, the twenty-six pontoons, at least fifteen new boats and the others from the old train, were strapped together using trestlework. Heavy cables anchored the bridge on each shore, and wooden boxes filled with stones acted as anchors. The latter did not work well, as some were not heavy or stable enough to prevent the bridge from swaying.[31] Fortunately for Lee's army, the river level had begun to drop at Williamsport, from eleven feet to just over four feet, which made fording the river possible. The pioneers dug the approaches to the pontoon bridge, and early on the morning of July 14, the Army of Northern Virginia began its trek across the river to home—a retreat described by Longstreet's chief of staff, Colonel Moxley Sorrel, as "a crazy affair."[32] Colonel Alexander, Lee's artillery chief, later recalled that "at last, not long after sunrise, we came to the pontoon bridge. It had a very bad approach and [was] on a curve—a bad location and several wagons, caissons, etc., had gone into the river during the night."[33] Captain Smith was ordered to stay behind and break down the bridge, saving those sections that were serviceable and destroying the rest. Most of the bridge was destroyed.[34]

The need for well-trained engineer troops was evident to Lee both before and after the battle. Yet, Army regulations had made it clear that officers of engineer troops would neither assume "nor be ordered on any duty beyond the line of their immediate profession, except by special order of the president." Adjutant General Samuel Cooper's General Orders No. 60, dated June 26, 1863, went so far as to state the following: "Officers of engineers will not be required to give other supervision to the fatigue parties or laborers employed in the construction of works than is necessary to indicate, in a clear and distinct manner to those directing the labor, their plans and the character of the work to be done."[35] The officers communicated their plans to the supervisors and then the officers could walk away. There was no room for improvisation and ingenuity unless the ideas came from the top. Engineering regulations in the Confederate army stood in stark contrast to Union Captain Duane's *Manual for Engineer Troops*. Confederate regulations provided three pages on the construction of fortifications, a section on siege operations, and fifteen pages of report formats; the section on fortifications included standing orders to keep the grass mowed and the wooden floors swept. Confederate engineering regulations compared unfavorably with Duane's 265 pages with sections on pontoon drill, rules for conducting a siege, military mining, and construction of batteries.[36]

On May 22, 1863, before Gettysburg, Lee had received a directive from the War Department that proposed uniting pioneer companies from various divisions to form a permanent engineer regiment. After the Gettysburg Campaign, the army commander wrote to Secretary of War James A. Seddon that a "regiment of engineer troops would be very desirable to serve this army, but, from my experience of the past campaigns, I do not think that the duties specially assigned to such troops would authorize the withdrawal of so large a body of the best men from the ranks of the army at this time." Chancellorsville and Gettysburg had cost Lee 40,884 soldiers, and creating special engineer troops from the remainder, he believed, would diminish the army's fighting power. Despite the combat performance of the 1st Michigan Engineers and Mechanics who fought at Murfreesboro (and Lee was probably unaware of the Michigan regiment's combat role), Lee believed it was impracticable to get an engineer regiment into combat.[37]

The supposition that limited manpower resources explained why the Southern high command could ill afford to take men away from the front and assign them to logistical support roles did not take into account other factors. General officers had large personal staffs, although Lee was an exception; the Quartermaster and Commissary departments were well staffed, and the government could have nationalized the railroads but chose not to do so. The Corps of Engineers was understaffed, first, because President Davis had the elitist notion that only West Point graduates could fill the role; second, because there were just not enough men who had developed engineering or mechanical skills before the war; and third, because it was not a priority. Fighting, not logistics, would determine the outcome of the war. Disastrously for the South, however, logistics very often determined the outcome of the fighting.

By the end of the summer of 1863, the 1st Regiment of Engineers did begin to take shape. Men conscripted into the army who possessed basic mechanical or trade skills were trained as pontoon bridge builders or instructed in making gabions and chevaux-de-frise.[38] It was not until March 1864, however, that the regiment was filled out, because finding men with mechanical ability was difficult and construction of pontoon trains took time.[39] Therefore, Lee's army continued to depend on pioneers carved from divisions and on impressed slave labor. Virginia had passed impressment legislation in October 1862 and only then, out of respect for states' rights, did the Confederate government begin developing an overall impressment policy in the spring of 1863. By November, backed by General Lee's support, engineer officers could impress slaves to work on engineering projects.[40] In the spring of 1864, the Engineer Bureau asked the Conscription Bureau to draft into service a force of 20,000 slaves, which the Engineer Bureau would organize into "gangs" of one hundred, "groups" of eight gangs, and "directorates" of three groups.[41]

The Confederacy's attempt to establish an engineering regiment, the difficulty of building a pontoon bridge at Falling Waters, poor maps, and the chaos and obstructions faced by the Railroad Bureau in moving vital supplies to the front—all contrasted significantly with Union railroad and engineering operations in the spring, summer, and fall of 1863. When Union General George G. Meade replaced General Hooker as commander of the Army of the Potomac on June 28, Union supply lines supporting his 90,000 men were in a precarious state. Ewell's corps had torn up track from Harrisburg to Carlisle, Pennsylvania, and Stuart's cavalry had destroyed bridges and rails along the north- and south-running Northern Central Railroad, the B&O to Frederick, Maryland, the Columbia-Wrightsville railroad bridge over the Susquehanna River, and railroad track west of York and east of Hanover Junction, Pennsylvania. For Meade's army to avoid delay in pursuing Lee, before the Army of Northern Virginia crossed the Susquehanna at Harrisburg—leaving Meade trapped behind the river and Lee free to move on Philadelphia—a steady flow of supplies had to reach the Army of the Potomac from Baltimore.

Herman Haupt, with authority extended by the War Department to manage all railroads in Maryland and Pennsylvania, decided to use the twenty-nine-mile Western Maryland Railroad running from Baltimore to Westminster, Maryland. The decision by the War Department to give Haupt complete control of the operation was a critical one. Railroad presidents, generals, quartermasters, and ordnance officers all answered to Haupt. Even if the South had had a man with Haupt's abilities, that person would not have been given the authority to conduct such an operation.

Haupt analyzed the problem, studied the maps, and determined that from the terminus of the railroad in Westminster, wagon roads would connect with Union troop locations in Taneytown, Uniontown, Union Mills, Littletown, and Manchester.[42] Yet, the single track was abraded and without sidings, water stations, and turntables. It had no telegraph lines and could run only three or four trains a day. Haupt would need to run thirty trains a day.[43]

Haupt had already built railroad bridges, designed ferries to carry rolling stock and engines from Washington to the docks at Aquia Creek, Virginia, near the mouth of the Potomac River, and organized the Construction Corps. Now he was asked to open a vital supply line for Meade's army in its desperate struggle with the invading Confederates. First, from Baltimore, Haupt wired Adna Anderson, chief of railroads in Virginia. Anderson was one of Haupt's protégés, and he did not hesitate to call upon him for assistance. Anderson arrived within a day of receiving the telegram with 400 members of the Railroad Construction

Herman Haupt sits atop his makeshift raft made of India-rubber pontoon cylinders that were inflated through a nozzle using a bellows, ca. 1862 or 1863. Library of Congress, Prints & Photographs Division, Civil War Photographs, LC-DIG-ppmsca-10341.

Corps. Second, Haupt ordered essential supplies. Trains to Baltimore brought tools, equipment, lanterns, water buckets, and split wood.

Anderson dispatched repair teams to open the Northern Central to Hanover Junction as the fighting began at Gettysburg. Haupt ran trains in convoys to Westminster. By July 3, the Western Maryland was moving 1,500 tons of supplies daily, and returning trains were bringing out thousands of wounded men to York hospitals. By late in the afternoon of July 4, the Northern Central was open from Baltimore to Hanover Junction.[44]

The 50th New York Engineers and Engineer Battalion had been detailed to build pontoon bridges over the Potomac during the army's initial entry into Maryland. They had dismantled the bridges and sent two of them back to Washington, while the other pontoon bridge followed the army north to near Gettysburg. During the battle, the engineers were used to guard the railhead at Westminster, escort the wagons to and from Gettysburg, and march prisoners to the provost guard's collection point.

The Union army's effort in general, and Haupt and Anderson's in particular, to establish a supply line during the Battle of Gettysburg proved vital to the success of the campaign. The greatest logistical challenges for the Army of the

Potomac, Army of the Tennessee, and Army of the Cumberland, however, lay ahead.

After Gettysburg, Southern strategy shifted to the defensive: hold on for as long as possible and do the most damage possible. If major transportation and commercial hubs such as Petersburg, Atlanta, Charleston, and Wilmington could remain in Confederate hands until November 1864, then a war-weary Northern public might give the White House to a Democrat, who would probably recognize the Confederacy. Confederate independence could still be won. President Lincoln was aware of this possibility as early as the summer of 1863, and for the president and his commanders, the only strategic solution was to take the war to the South and destroy the Southern people's will to fight. To execute this strategy, engineers would be required to open roads and build bridges deep in enemy territory, and others would need to repair and keep the railroads in operation over extended supply lines. After Gettysburg, while two great armies recovered from the devastation and slaughter that would thereafter be known by the names Slaughter Pen, the Peach Orchard, and the Valley of Death, Union hopes would turn west to General William Rosecrans and his Army of the Cumberland.

CHAPTER ELEVEN

Chattanooga

It was as fine a thing as was ever done.

Captain Alfred Hough to his wife, describing the
Union army's operation at Brown's Ferry,
Chattanooga, October 1863

After the Battle of Stones River in late December 1862, which pushed the Confederates south, out of Tennessee, and fired Northern hope that victory could be won in the western theater, Major General William Rosecrans began organizing for a major offensive into northern Alabama and Georgia. Rosecrans would prepare very slowly and deliberately, however, frustrating everyone in Washington.

Rosecrans's main fault was that he lacked political acumen. The occasional sarcastic remark in his correspondence with General Halleck was not appreciated by the chief of staff. Rosecrans—like General McClellan, a former engineering officer—did have a keen mind and considerable skill when it came to organizing his army and thinking imaginatively about logistics. For the first six months of 1863, under his leadership, the Army of the Cumberland adopted new techniques and innovative practices that would benefit not only his army but also those of General George Thomas and General William T. Sherman in their 1864 and 1865 campaigns. Between January and June 1863, Rosecrans's engineers and pioneers would develop new mapping technology, design a new pontoon system, and introduce the use of fabricated truss bridges to rest on original masonry piers. Historian Philip L. Shiman argues that it was Rosecrans who institutionalized these innovations, establishing a foundation for "the continuing development of his organizations and technology."[1]

In early January 1863, along the Nashville & Chattanooga Railroad at Murfreesboro, Tennessee, Rosecrans ordered the building of a forward supply base large enough to withstand a siege with a force of at least 60,000 men. Captain James St. Clair Morton's pioneers, with depots for the commissary, quarter-master, and ordnance departments and the help of four steam-powered saw-mills, warehouses, blockhouses, magazines, railroad spurs, and mutually sup-porting lunettes, constructed Fortress Rosecrans.[2] Once the fortification was completed and his supply lines were secure, Rosecrans believed his army was capable of a greater movement toward Chattanooga and Tullahoma, Tennes-see, where Braxton Bragg's Confederate army was lurking. Halleck was eager for Rosecrans to attack Bragg because the Union War Department feared that Bragg's army would be sent to rescue Pemberton at Vicksburg. Nonetheless, Rosecrans was deliberate in his preparations, which angered Halleck.

While Fortress Rosecrans was under construction, Rosecrans gave top pri-ority to mapmaking, asking Captain William E. Merrill of the Corps of Engi-neers to reorganize the army's topographical department. Merrill, unlike his predecessor Captain Nathaniel Michler and other corps members such as Cap-tain J. C. Duane, did not resent volunteer engineers and believed that the North-ern war effort depended upon the abilities of these civilians. Consequently, Merrill began to detail volunteer officers from various brigades and divisions to his topographical staff, and with prismatic compasses and portable drafting kits, he trained these men in the rudimentary art of mapmaking.[3]

Merrill then devised a process for managing topographical information. Mapmakers assigned to brigades sketched out areas and passed their informa-tion to division topographers; these topographers compiled the data with their own information and transmitted the material to corps topographers, who eventually passed it on to army headquarters. Merrill also adopted a system de-signed to create "information maps" in which individual topographers would take county maps and add additional geographical information; these maps were forwarded to headquarters and used to revise existing maps, which were then reissued to general officers. The challenge here was to print and dissemi-nate the new maps as quickly as possible.[4]

The army first used photography as a means of reproducing maps, but pho-tography proved problematic. The photographic equipment was bulky, lenses were inadequate for taking sharp, clear pictures at close range, and sunlight was needed to develop the photographs. After studying this problem, Merrill de-cided to procure lithographic presses, and the results were clear, legible maps. The image of the map was traced on a specially prepared flat stone that was then inked and pressed onto the paper. The drawback to this mapmaking method was that the stones were too heavy to transport into forward field positions and the preparation process was cumbersome.[5]

The most ingenious technique for reproducing maps was devised by Captain William Margedant of the 10th Ohio Infantry, an experienced photographer before the war. Margedant, while serving with Rosecrans during the western Virginia campaigns in 1861, had demonstrated how to copy maps using photographic chemicals. So perhaps at Rosecrans's suggestion, Merrill now adopted this creative method. A map was drawn in black ink on tracing paper on top of a piece of paper treated with silver nitrate. In the sunlight, the treated paper stayed white under the black ink, while the remainder of the treated paper turned black. The resulting black maps, as they were called, had white rivers hand-colored in blue and white roads colored in red, to avoid confusion. These maps had the benefit of quick revision, although the process was expensive. Margedant's mapmaking technique was enhanced and proved valuable to the Union army throughout the Atlanta Campaign and the remaining months of the war.[6]

During the Tullahoma Campaign between June 23 and 28, 1863, Rosecrans skillfully maneuvered Bragg's Army of Tennessee into a precarious position with its back to the Elk River. Only Forrest's Confederate cavalry prevented the Army of the Cumberland from springing the trap, by blocking Northern mounted infantry from destroying the bridge over which Bragg would escape. As a result, Bragg managed to slip across the river and destroy all the bridges over the Elk, setting the stage for the Chickamauga Campaign.

The topographical maps produced by his engineers enhanced Rosecrans's operation around Tullahoma. Yet, during the heavy rains in the six days of complex flanking movements around Bragg's army, Rosecrans observed that his engineers and pioneers had a miserable time trying to move the pontoon train along muddy and viscid narrow roads. The Army of the Cumberland used Russian-type canvas pontoons. The canvas was stretched over a wooden frame, lashed in the center, then wrapped around the stern and bow, pulled tightly, and lashed down again. The wooden frames were not as heavy as the Army of the Potomac's French pontoons, but they were cumbersome and required special wagons. Rosecrans had designed a prototype pontoon, built by his engineers, with two sections that were to be joined together by pins. Unfortunately, he did not have the facilities to make more of these frames, so he established an engineer shop in Nashville that would eventually build what became known as Cumberland pontoons.[7]

With Bragg's army in Georgia, Rosecrans hesitated to pursue the enemy until his supply line back to Murfreesboro was secure, which meant rebuilding the turnpike and railroad bridge across the Elk River. John B. Anderson's civilian construction corps, Morton's pioneers, fatigue parties drawn from infantry regiments, and the Michigan Engineers were available to do the work. The Michigan "Wolverines" were assigned the railroad bridge. Using three existing stone abutments, the engineers needed to build a span 470 feet long and 50 feet high. The new bridge's twenty wooden bents and stringers were cut from the surrounding woods, but railroad spikes, rails, and sawed planks for the flooring had to be hauled in from Nashville.[8] Because of the delay in receiving nails and other materials, it took the engineers six days to build the bridge, one day less than it took Lee's engineers to build a pontoon bridge across the Potomac.

Rosecrans recognized that the makeshift railroad over the Elk River could collapse if a thunderstorm suddenly dropped three or four inches of rain and produced freshets in the rapidly running river. Thus, in September, Rosecrans contracted with the McCallum Bridge Company to build a railroad bridge at Bridgeport, Alabama, after the Michigan Engineers had completed the Elk River bridge. The lumber for the Bridgeport bridge would come from Cincinnati and Louisville. Two months later, Construction Corps superintendent Anderson would arrange shipment from Louisville of six framed bridges. The practice of building bridge sections offsite and moving them to the proper locations was in its infancy, but this was a remarkable technological achievement. In May 1862, Herman Haupt had experimented with prefabricated truss structures called shad belly trusses; he made spans 60 feet long, amounting to approximately 1,000 feet of bridge in all. Now Rosecrans was asking private companies to do the same.[9]

Unlike the administration of military railroads in northern Virginia, the operation of the railroads in Tennessee lacked centralized control. John Anderson had clashed on numerous occasions with Colonel Innes over the use of civilian repair crews, the use of trains, and who should get replacement parts for track repairs. In August 1863, the ambitious Innes finally got himself appointed military superintendent of all military railroads within the department. He was now in charge of all aspects of running the railroads, including freight rates, repairs, and expenditures, and he would also have to manage recalcitrant officers like himself.

The operation of the railroads was not Innes's only headache. Bridge building continued unabated as Rosecrans's army readied itself for the September offensive against Bragg, and the Michigan men now had to get the Army of the Cumberland across the Tennessee River. First the engineers had to select the

Haupt's experimental shad belly truss being tested by Construction Corps workers sometime in 1862 or 1863. Eventually, railroad flat cars carried these prefabricated sections to bridge construction sites. Library of Congress, Prints & Photographs Division, Civil War Photographs, LC-DIG-ppmsca-10353.

best places to cross the river, and then they had to gather the material and build the bridges. The four sites selected were at Bridgeport, Shellmound (east of Bridgeport), Battle Creek (north of Bridgeport), and Caperton's Ferry (south of Bridgeport). Pioneers attached to various divisions built two of the bridges and one ferry. At Shellmound, Major General Joseph Reynolds's Fourth Division constructed a floating bridge from captured boats and bridge-building material. Brigadier General John Brannan's Third Division pioneers built rafts to ferry men over the river, and Brigadier General Jefferson C. Davis's First Division built a pontoon bridge at Caperton's Ferry.[10]

At Bridgeport, a detachment of Michigan engineers under the command of Lieutenant Colonel Kinsman Hunton, along with men from Morton's Pioneer Brigade, had the most difficult task among the four sites because they had only about sixty-two pontoons but 1,200 feet of river.[11] Hunton's companies started work on a trestle bridge from the west bank of the river to an island, and the pioneers built a pontoon bridge from the island to the east bank.

First Michigan engineers under the command of Lieutenant Colonel Kinsman Hunton skillfully built a combined trestle and pontoon bridge from the western bank of the Tennessee River to Long Island near Bridgeport, Alabama, in the autumn of 1863. The men built a trestle section because they had only sixty-two pontoons available to span 1,200 feet of river. Captain James St. Clair Morton's Pioneer Brigade used many of the pontoons to cross from Long Island to the east bank of the river. Library of Congress, Prints & Photographs Division, Civil War Photographs, LC-DIG-ppmsca-33490.

The timber for the trestle bridge was cut by infantrymen working in nearby woods, cut planks for flooring were delivered by train from army-run sawmills, and additional flooring was secured by stripping area barns and houses. The lumber might have come from a captured sawmill operated by pioneers at Scottsboro, twenty-eight miles west of Bridgeport, and from a sawmill at Anderson, Tennessee, or perhaps from an Ohio bridge company. Within a day, the engineers had anchored the trestles for one-third of the distance, and stringers and planks were laid as quickly as possible. Men worked through their exhaustion so that in four days the entire bridge was finished. But the problems were not over. In the middle of the afternoon of September 2, as General Philip Sheridan's infantry and artillery crossed the Tennessee, part of the bridge collapsed, throwing supply wagons and animals into the river. The tired engineers waded into the river and, over the next six hours, made repairs to the bridge. Wagons continued to cross, but close to midnight, seven bents fell as a result of wash at the bottom of the piles, and the bridge was closed.[12] The engineers worked twelve hours the following day to correct the problem, and the remainder of Sheridan's wagon train and cavalry crossed the river.

Rosecrans was relieved that his army was now on the eastern bank of the Tennessee, but was not satisfied that in the unfortunate circumstance of a retreat, the Bridgeport bridge would have to carry a large portion of his army back to the western side. So he ordered the Michigan engineers to remain at Bridgeport to construct more than fifty pontoons and use them to build a second bridge, parallel and close to the first. In hot and humid weather, men in long-sleeved cotton shirts, wearing a variety of hats to protect them from the sun, toiled another week to finish the bateaux and begin construction on the bridge. Rosecrans understood the demands he placed on the Wolverine regiment, and he understood that more engineer soldiers were needed to carry out the multiple tasks essential for the Army of the Cumberland's operations against Bragg's army. These responsibilities included corduroying roads, building supply warehouses, repairing track, and constructing pontoon trains. In addition, the engineers were conducting reconnaissance, making and reproducing maps, operating engineer shops in Nashville, and operating sawmills throughout southeastern Tennessee. The Pioneer Brigade, under the command of Captain Morton, theoretically provided additional manpower, but the structure of this organization made the pioneers an unreliable resource. Most corps and division commanders resented the Pioneer Brigade, and many men detached with the pioneers resented the assignment.

Men on pioneer duty were not under the control of their division or brigade commanders, they did not have to drill, they were careless with their tools, and they did not maintain the same discipline enjoined when they were part of their regiments or brigades. Brigadier General William B. Hazen bitterly complained to headquarters: "The whole pioneer concern [is] a stench in everybody's nostrils, and no one seems disposed to use them." "As it is now," he continued, "the pioneers get no drill, very little control, no sympathy, but the contempt of everybody."[13]

Throughout the summer and early fall, Rosecrans implored the Engineer Bureau and Congress to transform the Pioneer Brigade into an official organization. Congress, however, would not authorize a permanent Pioneer Brigade. Meanwhile, with his army southeast of Chattanooga, Rosecrans would attack Bragg's army, seeking the great victory for which Lincoln, Halleck, Stanton, and the Northern public were hoping.

The Confederate high command did not want to wait for Rosecrans to attack. Beauregard, Johnston, Bragg, and Lee agreed that the fastest way to bring peace and an independent Confederacy was to bring the Yankees to the negotiation table. A decisive "Napoleonic victory" followed up by another invasion north would bring about the desired results. The question that remained was where to concentrate manpower for the decisive blow. Lee, a powerful voice among the president's military advisers, suggested to Davis another move north in the fall of 1863, confident that given a second chance he would defeat the cautious George Meade. This time, however, Davis rejected Lee's suggestion, and feeling pressure from what historians Thomas Lawrence Connelly and Archer Jones called the Western Coalition, he decided that Bragg's Army of Tennessee would be reinforced and would defeat Rosecrans in northern Georgia. This would be followed by a move against Grant's communications network. General Beauregard, the most vocal supporter of the plan to strike Rosecrans, reasoned that troops from both Mississippi and Virginia could reach Chattanooga within five to seven days. After two weeks of mulling over his options, Davis agreed to send assistance to Bragg, and five brigades from James Longstreet's corps were selected for the journey south. Longstreet's men made the 500-mile rail trip via Lynchburg and down the Tennessee Valley over the Virginia & Tennessee Railroad in eleven days. They arrived without supply wagons or artillery.[14] Longstreet's artillery, under Colonel E. P. Alexander, arrived in Dalton, Georgia, twelve miles from Chickamauga early in the morning of September 25, five days after the battle was over.[15]

Quartermaster General Andrew R. Lawton and Major Frederick W. Sims, chief of the Confederate Railroad Bureau, were in charge of moving Longstreet's men and equipment. Transporting men from Lee's to Bragg's army was a Herculean task. It became even more arduous when it was discovered that Union General Ambrose Burnside had entered undefended Knoxville, Tennessee, on September 3 and now blocked Longstreet's intended route. With the Knoxville road cut, Sims decided the only alternative was a circuitous 800-mile route using at least ten different poorly maintained railroads. Part of Longstreet's two divisions would travel to Raleigh, Charlotte, and Branchville, South Carolina, before heading east to Atlanta. Others would move south through Goldsboro, Wilmington, Charleston, and Savannah before moving northwest to Atlanta. The new routes required eight separate transfers because of unconnected track, incompatible gauges, and deteriorated tracks. G. Moxley Sorrel, Longstreet's chief of staff, commented: "Never before were so many troops moved over such worn-out railways, none first-class from the beginning. Never before were such crazy cars—passenger, baggage, mail, coal, box, platform—

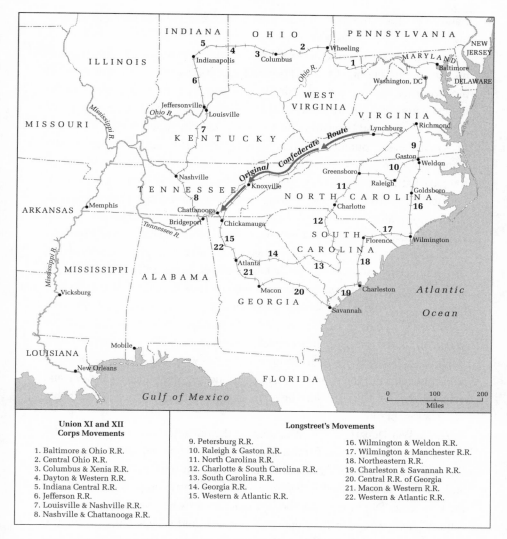

Union and Confederate Troop Movements for the Chickamauga and Chattanooga Campaigns

Union XI and XII Corps Movements

1. Baltimore & Ohio R.R.
2. Central Ohio R.R.
3. Columbus & Xenia R.R.
4. Dayton & Western R.R.
5. Indiana Central R.R.
6. Jefferson R.R.
7. Louisville & Nashville R.R.
8. Nashville & Chattanooga R.R.

Longstreet's Movements

9. Petersburg R.R.
10. Raleigh & Gaston R.R.
11. North Carolina R.R.
12. Charlotte & South Carolina R.R.
13. South Carolina R.R.
14. Georgia R.R.
15. Western & Atlantic R.R.
16. Wilmington & Weldon R.R.
17. Wilmington & Manchester R.R.
18. Northeastern R.R.
19. Charleston & Savannah R.R.
20. Central R.R. of Georgia
21. Macon & Western R.R.
22. Western & Atlantic R.R.

all and every sort, wabbling on the jumping strap-iron—used for hauling good soldiers."[16]

Although only half of Longstreet's infantry and none of his artillery arrived in time for the battle, the transfer of Longstreet's men from Virginia to Georgia was the most successful Confederate railroad operation of the war. On September 19, Longstreet leaped from the halted train at a little flag station called

Catoosa Platform and rode off to find Bragg and assist in bringing about a crushing blow to Rosecrans's army.

At Chickamauga, Rosecrans was both unlucky and lucky. Unlucky because his misunderstanding of the location of his units among the thick woods led him to order an unnecessary realignment of troops. Instead of strengthening his defensive line, which was the intended purpose, he inadvertently created a huge gap, and Longstreet hit the exact point left open by the Federal army's mistake. As Confederate infantry rushed through the hole, the Army of the Cumberland's right wing, fearing it would be enveloped, panicked and fled back toward Chattanooga.

Luckily for Rosecrans, his left wing, under the command of General George Thomas, remained on the field and, anchoring around Snodgrass Hill, held fast to his embattled position until nightfall. After dark, Thomas and his weary but stalwart soldiers, pursued by Bragg's Confederates, followed the rest of the Union army back to Chattanooga. Now it was the South's turn to besiege a Union army, forcing it to surrender or starve to death. Bragg's men occupied the heights surrounding three sides of the city and blocked access to the Tennessee River north. The Army of the Cumberland's only available supply route for its 51,000 tired and hungry men was over crude wagon roads to Bridgeport, sixty miles away. Under the best of circumstances this would have been an undesirable connection to the supply base, but with horses and mules weakening and dying by the day, the situation the Army of the Cumberland found itself in became untenable.

At the War Department in Washington, the telegraph clicked and clattered out an ominous message from Assistant Secretary of War Charles A. Dana: the Army of the Cumberland had been driven from the field at Chickamauga and was now besieged by Confederate forces in Chattanooga. Dana's telegram made it clear: "No time should be lost in rushing twenty to twenty-five thousand efficient troops to Bridgeport."[17] It was a bold suggestion, but Stanton convinced the president, Halleck, and several cabinet secretaries that it could be done. It was agreed that the XI and XII corps of the Army of the Potomac would be sent south under General Joseph Hooker. President Lincoln immediately issued a blanket order authorizing Hooker to take military possession of all railroads and their equipment "which may be necessary to the execution of the operation."[18]

The next morning, Stanton's transportation advisers, Samuel M. Felton, Thomas Scott, John W. Garrett, and William P. Smith, were called to Washington to recommend a feasible route for Hooker's two corps. These four men

wielded enormous power in the railroad industry, and each had important con-
tacts with other railroad executives in the North. Stanton had chosen wisely.[19]
They agreed to divide the operation into three sections. D. C. McCallum would
be responsible for the initial stage, supervising Hooker's embarkation from his
base in Culpeper, Virginia, to Washington. From there, Garrett would manage
the two corps' movement west to Jeffersonville, Indiana. Scott would oversee the final leg of the trip from Jeffersonville to Louisville and then along the
Louisville & Nashville Railroad to Stevenson and Bridgeport, Alabama. Two
days after Dana's telegram arrived in Washington, two trains, fifty-one troop
cars, and another four cars carrying field artillery rumbled through the nation's
capital on its way to relieve Rosecrans's starving and demoralized soldiers.[20]

On this 1,200-mile trek they used approximately 400 miles of track that had
been severely damaged by Lee's army or Morgan's and Forrest's cavalry and
subsequently repaired by Haupt's Construction Corps or the 1st Michigan Engineers and Mechanics. The first trains rolled into Bridgeport eight days after
departing Washington, DC. Hooker's entire force of approximately 16,000 men,
10 batteries, 3,000 horses and mules, 570 wagons, and 150 ambulance wagons
completed the journey in eleven days.[21] It was an astonishing feat of railroad
management, supported by skillful repair work to tracks and bridges, as well
as a tactical victory for the Union army during the siege of Chattanooga. Hooker's men could guard the Nashville & Chattanooga Railroad from Confederate
saboteurs. At least now the Army of the Cumberland was assured that supplies
would get from Nashville and Murfreesboro to Bridgeport unmolested.

Yet Hooker could not move on to Chattanooga because he barely had enough
food and fodder for his own men and animals, and they, too, would starve inside
the city. So before any attempt was made to reestablish the initiative and drive
Bragg's army off the mountains surrounding the city, a new supply line had to
be opened. The army needed an alternative to the treacherous sixty miles of
roads through narrow mountain passes and ankle-deep mud, contested by
Rebel sharpshooters who killed exhausted men and frightened the horses and
mules trying to haul the wagons. The situation was desperate.

Now, just as with the canal dug around Island No. 10, the bridges built during
the Peninsula Campaign, and the floating roads and bridges built on the Mississippi River floodplain at Vicksburg, Northern ingenuity and innovation would
turn the tide at Chattanooga. Historian James McDonough would write of the
Chattanooga Campaign that "once in a great while . . . the awful drama of war
narrows to a very small focus . . . and the story of a great struggle takes a decisive turn through the execution of a simple, daring plan—and the fearful mo-

mentum of war begins to swing from one army to the other. So it was at Brown's Ferry."[22]

Inside Chattanooga, Rosecrans had ordered the Michigan engineers to start building pontoons. With typical ingenuity they got an old steam engine to run and created a makeshift sawmill, cutting timber found in the town and logs obtained from across the river. Couriers on horseback brought in supplies of nails, and cotton was used as caulk.[23] Rosecrans also requested from the War Department that he be allowed to convert some of his infantry regiments into engineers under a law authorizing the formation of "veteran volunteer engineers." The War Department did not respond.[24]

Charles Dana was concerned both that Rosecrans was micromanaging everything and that he would try to pull his army out of Chattanooga over the sixty-mile wagon road instead of working to open a second supply route and break the siege. Dana wrote to Stanton: "General R[osecrans] insists on personally directing every department, and keeps every one waiting and uncertain till he himself can directly supervise every operation." Regarding a pullout from Chattanooga, he warned Stanton that given the conditions, the road for the only way out was almost impassable. "The returning trains [supply wagons] have now for some days been stopped on this side of the Sequatchie [river], and a civilian who reached here last night states that he saw fully five hundred teams halted between the mountain and the river, without forage for the animals and unable to move in any direction . . . And if the army is finally obliged to retreat, the probability is that it will fall back like a rabble."[25]

Halleck, Stanton, and Lincoln were disturbed by this news and thought that the only way to extract the Army of the Cumberland from its hazardous predicament was to remove "Old Rosy" before the army collapsed. Halleck issued an order on October 16 that created the Military Division of the Mississippi, and General Grant was made its commander. Grant's first decision was to replace Rosecrans with General George Thomas.

Rosecrans may have had a plan to establish a second supply route, the "Cracker Line," but to Grant's puzzlement he did not carry it out.[26] Historian Albert Castel argued that Rosecrans's new chief engineer, Brigadier General William "Baldy" Smith, suggested laying a pontoon bridge at Brown's Ferry on the same day that Rosecrans received word that he was relieved of command, so Rosecrans did not have the opportunity to approve or execute Smith's plan. Perhaps this was true, but if so, it took Rosecrans almost five weeks—from September 24 when the siege began to October 19 when he was fired—to develop some attempt to break the Confederate stranglehold. Smith claimed that when he arrived on October 10 he tried unsuccessfully to convince Rosecrans that the army simply could not exist unless the Tennessee River was opened.[27]

For the officers and men, the waiting and suspense were terrible. Captain Alfred L. Hough wrote to his wife in early October: "To live, and to go to sleep knowing that a hundred or so cannon are looking one another in the face, and may at any moment open on each other, that is our daily life. But it must soon end, a fight or a fall back by one side or the other must take place before many days."[28] Grant showed no signs of waiting. Prepared both to open the Tennessee River and to fight the Rebels, he rode with his staff and cavalry escort the sixty miles along Walden Ridge and Sequatchie Valley roads in driving rain and perilous conditions, arriving at Thomas's headquarters on October 23 in the late evening. Everyone on Thomas's staff knew that Grant had been thrown from a horse just weeks before and had badly sprained his ankle. So, with the injury and the miserable weather, they were shocked when Grant, wet and cold, limped through the doorway of Thomas's command post, refused dry clothes, and immediately asked to hear about plans to open the Cracker Line. Grant's demeanor signaled that he was there to manage the situation rather than allowing the situation to manage him.[29]

The scene was tense. Grant and Thomas had a cool relationship. Perhaps this was because, after Shiloh, when Grant was demoted to second in command behind Halleck, Thomas was given temporary command of the Army of the Tennessee, and Grant suspected Thomas had something to do with the change. Whatever the feelings, both men acted as professionals, and Grant listened intently as Baldy Smith presented his idea (with Thomas's approval) of opening a supply line that would avoid Confederate artillery positioned atop Lookout Mountain and the Southern sharpshooters at the Raccoon Mountain bend in the Tennessee River. As Smith put forward his proposal—floating pontoons from the city south toward Chattanooga Creek, then around the U-shaped bend in the river north to Brown's Ferry, about two and a half miles from Lookout Mountain, where a bridge would be constructed—Grant sat up in his chair and took notice.[30] Now supplies could be transported by river to Kelley's Ferry then hauled by wagon through a pass in Raccoon Mountain to Brown's Ferry, on to Moccasin Point, and over another pontoon bridge into Chattanooga. Simultaneously, Hooker would move his two corps along the Nashville & Chattanooga Railroad into Lookout Valley and tie into the Brown's Ferry position.[31] The following morning Grant scouted Brown's Ferry and approved the plan.

After listening to Smith's ideas the previous evening, Grant also telegraphed instructions to General Sherman to leave Iuka for Nashville and then move

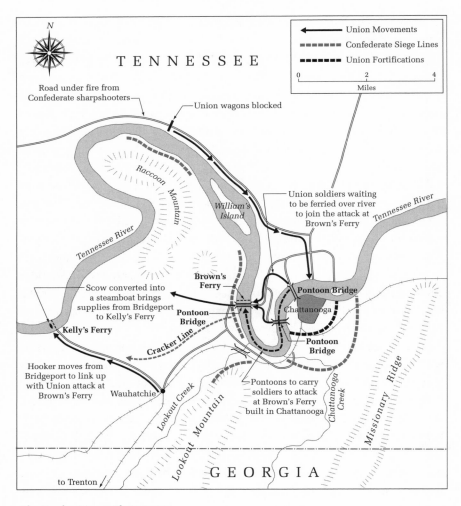

The Cracker Line at Chattanooga

rapidly into Chattanooga. With Sherman was General Grenville Dodge's Second Division of the XVI Corps. Grant grasped quickly that with the additional manpower from Hooker and Sherman, as well as the likelihood that he would need to supply Burnside's command in east Tennessee, a single-track railroad was not optimal. Remembering that Dodge was a railroad builder before the war, Grant sent a second message to Sherman ordering him to halt Dodge and his 8,000 men at Athens, Alabama; their task was to repair the Central Alabama Railroad below Nashville, extend it to the Tennessee River at Decatur, then repair the Memphis & Charleston between Decatur and Stevenson. This would give the army two roads as far as Stevenson over which to run supplies.[32]

Dodge first assigned men to guard against Confederate raiding parties, then others gathered food and forage from the surrounding countryside. Once these detachments had started their work, Dodge recruited blacksmiths from men in the ranks and assigned them to set up portable blacksmith shops to make the tools necessary for railroad and bridge building. Axmen were put to work cutting timber for bridges and fuel for locomotives, and mechanics worked to repair whatever engines and cars could be found. To support Dodge's efforts, Grant ordered General McPherson to send eight engines and rolling stock from Vicksburg to Nashville. He arranged to have rails taken from unused track and inactive locomotives and cars sent to Nashville. John Anderson, who succeeded his former replacement, Colonel Innes, was also directed to furnish rolling stock and as much bridge material as possible.[33]

In forty days, Dodge and his men were to lay or repair 102 miles of track and rebuild 182 bridges and culverts. This was Northern ingenuity at its finest. Now—much like McCallum and Haupt's development of the military railroad management system in Virginia and Stanton and his transportation board's acumen in moving Hooker's army south—Grant, Dodge, and Anderson had serendipitously established the foundation for efficient management of military railroads in the western theater of operations.

Dodge was not the only one with a unique engineering story to tell during the siege. In Bridgeport, after Hooker's arrival in early October, Captain Arthur Edwards, an assistant quartermaster from Detroit and, before the war, a shipbuilder in Lake Erie, prepared to convert a scow into a steamboat. The scow, mounted with an engine, boiler, and stern-wheel, would carry and tow supplies to Kelley's Ferry, nine miles west of Chattanooga, where the food and forage would be unloaded and transported overland to the city. The captain retained the services of a master mechanic named Turner and carpenters, who framed the boat and set it on blocks about six feet above the water level of the Tennessee.[34]

Suddenly the water began to rise. Before the planking was finished and the caulk and pitch were applied to the bottom, the water was within sixteen inches of the planks. Using pig iron left behind by Confederates, Turner started to weigh down the hull to prevent its being swept off the blocks and broken apart by the current. Another quartermaster, Lieutenant Colonel William Gates Le Duc, suggested to Turner and Edwards that carpenters should cross-timber the blocks, but Edwards pointed out the futility in trying to keep pace with the rising water. Le Duc reminded Edwards that Thomas's starving soldiers needed the steamboat and that if the planking got wet, it would be another two to four weeks before it would dry and the caulk and pitch could be applied. Edwards understood the predicament but believed nothing could be done except wait for the river to run its course.[35]

Le Duc had an idea. Extra pontoons were floated alongside the steamboat, and carpenters bored two-inch holes in the bateaux, gradually filling them with water. When the pontoons rode low enough, they were pushed under the steamboat as the blocks were hammered out. As soon as the steamboat was secure atop the several pontoons, the carpenters plugged the holes and other men bailed out the water. Within six hours the steamboat was safely resting on the pontoons, riding on top of the rising water. The hull was caulked and pitched, and three weeks from the start of the project the USS Chattanooga was launched. On October 30, under the direction of Le Duc and a soldier named Williams who had steered a steam ferry between Cincinnati and Covington (in the years before Roebling's suspension bridge was completed), the steamboat landed at Kelley's Ferry with 40,000 rations and 39,000 pounds of forage.[36]

Back in Chattanooga, immediately after Grant approved Baldy Smith's plan to build a pontoon bridge at Brown's Ferry, Smith spoke with Captain Perrin V. Fox of the 1st Michigan Engineers and Mechanics: the army needed fifty pontoons in two days. Already, the Wolverines had built most of the boats with unseasoned lumber and a limited supply of nails. Fox now needed to collect equipment and tools and assemble a bridge team train, because infantry would ride in the pontoons and two flat boats. Despite having to scramble for quartermaster wagons to haul equipment, all was finally ready. Late on the night of October 26, Fox, his men, and members of the assault team set off for Brown's Ferry.[37]

The assault team was made up of hand-picked men from the brigades of brigadier generals William Babcock Hazen and John Basil Turchin. At 3:00 a.m. each of the pontoons was loaded with twenty-five men and five oarsmen, and they quietly pushed off, gliding over the water so as not to be heard by Confederate pickets lining the western bank. The river was approximately 800 feet wide, and heavy fog provided the cover the Union force needed. At around 4:30 a.m., signal fires were lit on Moccasin Point to orient the boats to their position, nine miles from the jumping-off point. At about 5:00 a.m., the second section of the flotilla started to drift toward the middle of the river, and the audible sound of the oars pulling hard to redirect the boat to the eastern bank alerted Confeder-

Yankee ingenuity personified. Converted from a scow in just twenty-four days and launched on October 24, 1863, the *USS Chattanooga* delivered desperately needed supplies to Major General George Thomas's besieged Union army in Chattanooga. The improvised steamboat helped open the "Cracker Line." National Archives photo no. 111-B-672 (Brady Collection).

ate pickets. Hurriedly the men sprang onto the western shore, taking the Confederate pickets by surprise.[38]

Immediately, the heavy "wooa" sound of a Minié bullet was heard ripping over the heads of the men in the boats. The Union assault teams now began to row toward the rifle fire. Infantrymen on Moccasin Point were ferried across the river to support the initial assault wave, and once the west bank was secured (in less than an hour), Fox's men began building the 870-foot-long bridge. It took fewer than eight hours to complete. Reinforcements were moved across the new bridge to protect the bridgehead from Confederate counterattack.[39]

General Longstreet, in overall command of Confederate forces at the bridge, could muster only 4,000 men, while 5,000 bluecoats established firm control of the area. Longstreet, at the Battle of Wauhatchie in Lookout Valley, would also try to stop Hooker's two corps from linking up with the Army of the Cumberland at Brown's Ferry, and failed. As Captain Alfred Hough would write to his wife, the operation at Brown's Ferry "was as fine a thing as was ever done."[40]

With the Cracker Line opened, supplies flowed into Chattanooga and Grant once again took the strategic initiative, which in late November culminated in the battles of Missionary Ridge and Lookout Mountain. Bragg's Army of Tennessee was driven from Chattanooga, and this set the stage for the start of Sherman's Atlanta Campaign in the spring of 1864. From December 1863 to March 1864, in both the eastern and western theaters, Union armies prepared for major offensives. Grant was called to Washington in March to receive his commission as the army's only lieutenant general and became commander-in-chief of the United States Army. After deciding to operate his headquarters in the field alongside the Army of the Potomac, he promoted Sherman to command the three army groups of the Division of the Mississippi: Thomas's Army of the Cumberland, McPherson's Army of the Tennessee, and John Schofield's Army of the Ohio.

Perhaps because he learned from Rosecrans the importance of engineering, Thomas would take the time between Chattanooga and Atlanta to refurbish his combat engineer organization. In May 1864, Congress would authorize Thomas's Pioneer Brigade as an official engineer unit and would create the First US Veteran Volunteer Engineers. Merrill, Thomas's chief engineer, trained soldiers from each army brigade as mapmakers and made sure the engineer shops in Nashville were repairing locomotives and rolling stock, producing portable sawmills to cut lumber at bridge sites, and building new, lighter pontoons. The original idea for the new pontoons was Rosecrans's, but the improved design was Merrill's. Instead of using pins to hold the sections together, Merrill substituted hinges so that the frame folded and did not come apart. Merrill then dispatched a Captain O'Connell of the Pioneer Brigade to Nashville "with a detachment of pontoniers, to build a train of such boats, giving him the authority to make any additional improvements that he or *any one else* could suggest."[41]

Because of the daunting task facing the Michigan engineers in repairing railroads and building a backup line to transport supplies to Nashville, the Engineer Regiment of the West consolidated with the 25th Missouri Infantry to form the 1st Missouri Engineers. In addition, a pioneer company from the 59th Illinois Infantry, under the command of First Lieutenant Chesley A. Mosman, constructed a bridge over Running Water Ravine to complete the direct line from Bridgeport to Chattanooga. Trestles 16 feet in height and layered atop each other rose from the riverbed, with four 116-foot-high tiers completing the 780-foot-long structure. During the construction, Mosman wrote: "I have a hundred men at work on the bridge, mostly soldiers from Colonel William

First Lieutenant Chesley A. Mosman and the pioneer company of the 59th Illinois Volunteer Infantry built this four-tiered 780-foot-long Nashville & Chattanooga Railroad bridge over Running Water Ravine, near Whiteside, Tennessee. The first train crossed the bridge on January 14, 1864, about six weeks after construction started. Library of Congress, Prints & Photographs Division, Civil War Photographs, LC-USZC4-3925.

Grose's brigade who get one dollar a day extra. Two more bents (poles) were erected today. The pioneers have rigged up a windlass, a device for raising or hauling objects. With a windlass and tackle they raise bents in place."[42]

Sherman's offensive into Georgia required that railroads and bridges be in sound condition. When he finally began his campaign against General Joseph Johnston's army, Sherman's initial point of contact with the enemy would be 300 miles from his main supply base, and the distance would grow longer as his three armies, comprising more than 110,000 soldiers, moved farther southeast toward Atlanta. Daniel McCallum, called to Nashville to assess the condition of the railroads, determined that the number of cars and locomotives was insufficient for such an ambitious operation. Furthermore, he reported, track conditions were poor. John Anderson was relieved from command of military railroads. In his place, Adna Anderson was appointed general superintendent of transportation and maintenance, and W. W. Wright as chief engineer of construction in the Military Division of the Mississippi.[43]

Adna Anderson was replacing Herman Haupt, who had resigned from the military railroad department in September.[44] Fortunately for the Yankees, Anderson was an excellent railroad engineer and a skilled manager, and he inherited a remarkable operation. The Construction Corps, now a permanent organization, was well organized into five divisions: bridges and carpentry, track, water stations, masonry, and trains. Interchangeable bridge trusses were stockpiled, and a simple barge, first used during the Peninsula Campaign, was available for transporting eight loaded freight cars by water. Under Adna Anderson's leadership, the United States Military Railroad (USMRR) in the western theater would develop into an even larger organization, just as resourceful and proficient as the operation in the eastern theater.

During January and February 1864, both Union and Confederate soldiers huddled by campfires, wrote long letters home, and tried to recover from emotional and physical wounds sustained in the hard fighting of the past year. Generals sat in their headquarters, reorganizing their division and corps commanders and contemplating their next move and that of their enemy. In Richmond, the Davis government continued to wrestle with railroad presidents and states' rights governors about the use of railroads for military purposes. Engineers continued to direct slave labor in building fortifications in northwestern Georgia, in Atlanta, and around Petersburg and Richmond. Conversely, the Army of the Potomac and the Military Division of the Mississippi prepared to go on the offensive. Union engineers would play a major role, as they did at Vicksburg and Chattanooga, in determining the outcome of those operations. Union engineers, pioneers, Construction Corps workers, and sometimes details of infantry readied the rails for the next offensive. Locomotives were collected and repaired, rolling stock was gathered and made, and supplies were stockpiled. The next big push was about to come.

CHAPTER TWELVE

The Red River and Petersburg

We cannot train them for every possible encounter . . . because we
cannot anticipate what those encounters will be like. Instead we have to
develop them to be the kind of people who can sort it out for themselves
once they get there.

Colonel Barney Forsythe, on the Cadet Leadership Development System

At the corner of Third Street and Vine sat the majestic Burnet House, Cincinnati's handsomest hotel. The *Illustrated London News* called the neoclassically styled, golden-domed, five-story building "the finest hotel in the world."[1] There, in March 1864, generals Grant and Sherman met to plan the spring campaigns, which they believed would finally bring the Confederacy to its knees. Sitting in Parlor A, Grant revealed his strategy, calling for simultaneous Union offensives along the entire Confederate line. The Army of the Potomac would attack Lee's army in northern Virginia; Sherman would strike at General Joseph Johnston's forces along the Chattanooga-Atlanta corridor; the newly formed Army of the James, under Ambrose Burnside, would pressure Lee from the south or shift its concentration to coastal North Carolina; and Nathaniel P. Banks's Department of the Gulf would prepare a combined operation with the navy to capture Mobile, Alabama.[2]

While Grant and Sherman worked to ready their forces for a coordinated pincer movement against the Confederacy's two major armies, Union forces under Nathaniel Banks, in the Department of the Gulf, launched a successful campaign against Southern fortification along the southeast coast of Texas, with the hope of threatening the French in Mexico. In March, Banks was also fully engaged with Confederate forces from the trans-Mississippi region for control of the Red River, the wealthy cotton plantations along the river, and the Trans-Mississippi Department's headquarters at Shreveport, Louisiana.

During the Texas coast operations, Banks's engineers were the 1st and 2nd Regiment Engineers, Corps d'Afrique. Formed from the Louisiana Native Guards, the Corps d'Afrique was made up of property-owning freeborn creoles and blacks and, later, freedmen from refugee camps. Much of the work done by the engineers included building fortifications and digging trenches and latrines. It was the hard, backbreaking work that white soldiers preferred not to do. Banks demanded that white troops respect the work of his soldiers of color, and he believed that black soldiers were better at performing the hard duty of "throwing up defensive earthworks . . . unwillingly performed by white troops." A captain of the 53rd Massachusetts wrote to his wife about the black engineers in a racist fashion typical of the period: "The[y] can be put into the unhealthy localities in the department and not suffer like white men."[3]

This, of course, was far from true. During the siege of Port Hudson in the summer of 1863, Corps d'Afrique troops died at an alarming rate as a result of typhoid and dysentery. The engineers' presence, however, aided in the capture of the Confederate stronghold. The 3rd Regiment Engineers had built a bridge over the Tunica River at Bayou Sara Road, which allowed Banks's forces to approach Port Hudson from both north and south. On May 27, the 1st and 3rd Engineers participated in the first assault on the city and lost 37 killed, 155 wounded, and 116 captured. The 1st Regiment's commander, Captain Andre Cailloux, one of the first black officers in the Union army, was one of those killed.[4]

In the months following the Port Hudson and Texas coast operations, a number of soldiers in the Corps d'Afrique deserted, primarily because of poor treatment by the white soldiers and officers and the cruel conditions under which they lived. By the fall, Banks had purged the corps of all the black officers, and their white replacements acted as overseers and martinets. The Bureau of Colored Troops—established on May 22, 1863, under the command of Major Charles W. Foster, a Lincoln Republican from Ohio—attempted to recruit other soldiers, but most of the regiments remained understrength. Two additional engineer regiments, the 4th and 5th Engineers, were organized in time for the Red River Campaign. The 4th would be stationed in New Orleans, and the 5th would join the 3rd Engineers and take part in two of the most remarkable and least-remembered engineering feats of the war.

In brief, in the spring of 1864, General Banks had intentions of capturing Mobile Bay, closing off the Confederacy's last major port in the Gulf of Mexico. His plans were interrupted, however, when General Halleck insisted that Banks move along the Red River and defeat Lieutenant General Richard Taylor's trans-Mississippi forces. Grant, promoted to lieutenant general and given command of all Union armies at the start of the Red River Campaign, agreed with Banks that Mobile Bay, not Taylor's army, should be the focus. Yet, out of deference to Halleck, the commanding general did not call off the Red River effort. He did, however, give Banks a time limit: either Banks complete his objectives by April 25 or the campaign would be halted. This time constraint placed additional pressure on Banks, which almost led to a disaster, averted only by the creative skill of the engineers.

Banks's plan called for three movements. First, elements of Thomas Edward Ransom's XIII Corps, William Buel Franklin's XIX Corps, a cavalry division, and four infantry regiments of the Corps d'Afrique would march northwest along Bayou Teche and Vermillion Bayou to Alexandria, Louisiana.[5] From the confluence of the Mississippi and Red River to Alexandria was at least 70 river miles. Second, General McPherson would send a detachment of soldiers from Vicksburg down the Mississippi, join Admiral Porter's fleet (13 ironclads and 7 light-draft gunboats) at the Red River, and move to Alexandria. Third, General Frederick Steele, operating in Arkansas, would send forces from Little Rock. All three groups would converge at Shreveport. The 3rd and 5th Regiment Engineers, Corps d'Afrique, built pontoon bridges both at Vermillion Bayou and at Cane River as Banks made his way north.

At Alexandria the problems began. Low water made it only just possible, taking ten days, for the fleet to pass the double rapids above the city. Then, moving forward, Banks's men clashed with Confederates in the first major engagement of the campaign at Sabine Cross-Roads (April 8), and Banks was driven south toward Pleasant Hill, suffering a significant loss of men and material. After another fight at Pleasant Hill on the following day, Banks decided to abandon his attempt to capture Shreveport. Confederate partisans, 1,000 cavalrymen, obstructions in the river, demanding terrain, and Steele's failed attempt to pressure Shreveport made the decision an easy one, but getting Porter's fleet safely back to Alexandria was a monumental undertaking. The water level of the Red River continued to drop. Taylor divided his Rebel forces between harassing Banks in Alexandria and obstructing the river 25 river miles south of the city at Snaggy Point.

By April 25, Banks's situation was dire. His 25,000-man force could fight its way back to the Mississippi, but Porter's fleet, twelve ironclads and eight gunboats, was trapped above Alexandria, and the possibility of losing all the vessels, along with the valuable cotton they carried, to a small Confederate force was both real and alarming. Just as the hopes of the Federal government and nation were resting on Grant and Sherman's spring offensives, an announcement that the navy had lost twenty ships and the army was soundly defeated by a smaller force in the Red River Campaign would devastate Union morale and boost their enemy's. Furthermore, members of Lincoln's own party, led by the politically ambitious and untrustworthy Secretary of the Treasury Salmon P. Chase, were attempting to win the Republican nomination for president in the fall election. An overwhelming loss in Louisiana would be a severe blow to Lincoln's reelection efforts.

On April 25, Colonel Joseph Bailey, acting chief engineer for General Franklin's XIX Corps, suggested a unique plan for rescuing Porter's ships. Colonel Bailey was born in Ohio, received a common school education, and before the war was employed as a civil engineer and lumberman. Remembering a technique used by his fellow Wisconsin lumbermen to funnel logs downriver to the mills, he proposed to build wing dams, both to deepen the water and to shoot the boats over the rapids. Porter's flotilla needed seven feet of water to prevent grounding, and the water over the rapids was only three feet deep. Franklin thought the idea impractical, but the fleet was still 300 miles from the Mississippi, and Banks's army had just three weeks of half rations remaining and virtually no forage for the animals. Franklin conferred with Banks, who gave him permission to begin.[6]

What happened next made Bailey's project controversial. In his report dated May 17, 1864, Bailey mentioned the technical aspects of the dams and the regiments involved in constructing the dams, and, as if as an afterthought, he wrote: "In addition to the dam at the foot of the falls, I constructed two wing-dams on each side of the river at the head of the falls."[7]

The first of the two dams was constructed above the second set of rapids, closest to Alexandria where the river was 758 feet wide and the current at ten miles an hour. On the north bank of the river, the left wing of the dam was built of felled trees laid in the river, aligned with the current, their branches locked and trunks tied together. On the south bank, the right wing of the dam consisted of a crib filled with stones and scrap iron and placed in such a way that there was a 150-foot gap between the two wings. Finally, transport barges filled with rubble were sunk in the gap. It took eight days and nights to complete the dam, and everything was ready for May 8. The idea was that the barges would be hauled away and the rising water behind the dam would explode through the opening with such force and enough water to funnel Porter's ships through. Sailors stripped side armor from the boats and brought ashore anchors, chains, ammunition, and guns to be transported by wagons downstream to reunite with their ships later.

The pressure was too much for the sunken barges as they were pushed aside by the great weight of the water, and the sudden cascade sent vessels behind the dam bouncing, dipping, scraping, and skating into the river with enough water to float them into the deeper part of the river. It was a great success, but ten boats, not yet ready for the break in the dam, remained upstream beyond the first set or upper falls. It was for this reason, Bailey wrote, that a second dam was built.

Bailey thanked a number of officers for their determination and zeal while working on the dams, including Colonel George Dorgue Robinson, commander of the 97th Regiment Colored Troops, and Lieutenant Colonel Uri Balcom Pearsall's 99th Regiment Colored Troops. Before April 4 these units had been designated the 3rd Regiment Engineers, Corps d'Afrique, and 5th Regiment Engineers, Corps d'Afrique. They remained engineer troops by a different name. Others mentioned by Bailey included the 29th Maine, 116th New York, 24th Iowa, 16th Ohio, 27th Indiana, and 19th Kentucky.[8] Bailey received the "Thanks of Congress" and a presentation sword from the Navy Department for an engineering feat that saved the fleet during the Red River Campaign.[9]

Colonel George D. Robinson, commanding the 97th Regiment, US Colored Troops (3rd Regiment Engineers), reported a slightly different version of Bailey's story. Immediately after Bailey received permission to begin his dam project, Robinson, an 1861 graduate of the University of Michigan, was ordered to meet with him. The two men, along with Lieutenant Colonel Pearsall, commander of the 99th Regiment, US Colored Troops (5th Regiment Engineers), walked to the river bank and discussed where to build the dam.[10] Pearsall suggested building two dams, one at each set of rapids, but Bailey insisted that a dam at the lower rapids would be sufficient. Robinson then set his men to work removing barricades from the town, cutting timber, and constructing a battery for six guns along Bayou Rapids Road. When the upper dam was built, Robinson mentioned that the dam was Pearsall's idea and it was the men of the 97th and 99th who built it. Robinson wrote: "The plan for building two dams across Red River, which from necessity was finally adopted, was originally proposed by him [Pearsall], and the success of the dam was in my opinion, mainly due to his efforts."[11]

Pearsall, in his report, was more emphatic. During the building of the first dam, it was men of the 97th and 99th who built the crib section and men of the

29th Maine, 110th and 161st New York, and the XIII Corps' Pioneer Brigade who built the log section. Then, when the center section gave way on May 9, Bailey told Pearsall, who had already suggested a second dam, to build a dam at the upper rapids. The upper dam that saved ten gunboats and ironclads in Porter's fleet was Pearsall's idea and was built primarily by US Colored Troops.

Pearsall was familiar with wing dams. Born in Owego, New York, he attended common schools and Oxford Academy in Owego and worked for his father in the lumber business; his father had built the first dam on the Susquehanna. At sixteen, Pearsall moved to Wisconsin to work for his uncle in the same trade. Now, at twenty-four, he used his knowledge and creativity to save more than half of Porter's boats. His men built two-legged trestles for a "bracket dam." Because of the swift current, "some pieces of iron bolts (size one-half inch) were procured and one set into the foot of the legs of each trestle; also one in the cap pieces at the end resting on the bottom up stream . . . The trestles were fastened as soon as they were in position by means of taking 'sets' and driving the iron bolts above referred to down into the bottom." The conditions were dangerous. Standing in four feet of water, some men were swept away by the current and rescued downstream. Finally, planks were placed horizontally on the trestles to form the dam, and the design worked as imagined. The remaining boats in the fleet were pulled through the funnel on May 12, and the entire armada was now safely below Alexandria.[12]

Colonel Bailey deserved credit for originally suggesting the idea of the wing dams to Franklin and Banks and for the design of the lower dam. He was also responsible for the entire operation. In addition, on May 18, as the Union army approached Simsport near the Mississippi River, Taylor's Confederates were threatening to catch Banks's troops with their back to the river.[13] To his sole credit, Bailey came up with a unique solution. Mooring twenty-two steamboats side by side and then nailing planks over the bow of each, Bailey built a floating bridge. During the next two days, the army and its wagon train crossed unimpeded. Bailey merited the encomiums he received. Wickman Hoffman, an adjutant with General Franklin, recalled that "we crossed the Atchafalaya by a novel bridge constructed of steamboats. This, too, was Bailey's work."[14] John Merwin of the 161st New York recorded in his diary on May 19: "No bridge, but Colonel Bailey is equal to the occasion once more, and has lashed twenty-two steamers together bows on and using the boat bridges or gang planks has formed a bridge that covers the muddy waters from shore to shore, over which we cross in safety."[15]

There is no doubt that Bailey performed at a high level of efficiency during the Red River campaign and was responsible, on two occasions, for contributing to the army's salvation. The historian for the XIX Corps, Richard Irwin, summed up how most white soldiers and sailors felt about Bailey: "At Simmesport [sic] the skill and readiness of Bailey were once more put to good use in improvising a bridge of steamboats across the Atchafalaya."[16] It was not surprising, then, that a recommendation from a former politician such as Banks combined with support from the Department of the Navy earned Bailey a "Thanks of Congress." It was also not surprising, given the treatment of black soldiers by their white comrades and the disparity between white officers and white officers commanding black soldiers, that with the exception of Bailey's brief comments, Robinson's and Pearsall's efforts went unrecognized.

Many white officers would have been scornful of and cynical about white men leading black troops, but this did not stop Pearsall and Robinson from expressing their opinions. Pearsall thought it was incredible that naval authorities reported that Bailey was the only person who believed the dam project practicable. Pearsall wrote that he had suggested the idea of wing dams days before Bailey, when the army was in Grand Ecore. "I beg leave to state that the project of building a dam across Red River, although difficult, could never have been pronounced impracticable by any man who followed a similar avocation [lumbering] in civil life."[17] Colonel Robinson was more direct: "If the thanks of Congress are due to any one for the final success of this dam I believe they are due to him [Pearsall] as much as to any one else." He also did not equivocate when he spoke of his men. "In conclusion, I would say that the organization of colored engineers is regarded as a complete success by all who have witnessed their operations."[18]

While blacks and whites in the Union army attempted to work out a cooperative relationship in their fight for the common cause of victory, the Confederate army and government made it perfectly clear that there would be no such codependent relationship in their armed forces. In the spring of 1864, when Colonel Thomas M. R. Talcott, commander of the newly formed 1st Confederate Engineers, proposed that free blacks be impressed to form pioneer companies, Secretary of War James A. Seddon reminded Talcott that free blacks and slaves worked on projects as impressed laborers, not as soldiers.

Some slaves and free blacks had developed technical skills before the war. Slaves on large plantations had learned blacksmith and carpentry skills, and many could operate and repair machinery. Free blacks also worked as carpenters, ironworkers, and toolmakers and could have been important additions in remedying the South's dearth of engineer troops. In the months before Grant's Overland Campaign and Sherman's Atlanta Campaign, the Confederate military attempted to strengthen its engineering organization, but African American craftsmen were not a part of that calculus. As a result, the South enjoyed only isolated success. The Confederacy had skilled engineers, and the army did well with what they had on hand. Some of the men in the ranks also made good engineer soldiers and pioneers. But they were not enough.

As the war progressed and the value of engineers became more apparent, the South simply did not have the skilled manpower to meet the army's needs. Brigadier General E. P. Alexander, for example, envied the size and quality of the Army of the Potomac's engineer units. He believed they were worth the equivalent of a corps to Grant and Meade.[19]

By the spring of 1864, the Army of Northern Virginia did have a functioning engineer regiment, although it was a meager asset. The 1st Confederate Engineers commander was Colonel Talcott, a railroad engineer before the war. Second-in-command was Lieutenant Colonel William W. Blackford. Blackford had studied engineering at the University of Virginia, served briefly as a civil engineer, and then joined the army as a cavalry lieutenant. Both men were good engineers, but recruiting and training additional officers was a significant problem. Whatever these additional officers learned was from the few military engineers in the regiment or from books such as Dennis Hart Mahan's *A Treatise on Field Fortifications* and Captain J. C. Duane's *Manual for Engineer Troops*.[20]

Some of the men recruited as engineers were among the veteran infantry, but because division and brigade commanders were reluctant to give up their experienced soldiers, some of the engineer troops were conscripts and the training they underwent was not rigorous. In his Civil War diary, Lieutenant Henry Herbert Harris recorded how the engineers guarded Yankee prisoners and worked on repairing corduroy roads. It was not until April 19,1863, that companies began pontoon drill on a millpond. Two days later he wrote: "Took the company out near the pond where two other companies were pontooning and looked on for a while." The next day he "went out to see the pontooners exercise . . . Hunted two ducks without killing either and fished a while without getting . . . a nibble. In the afternoon I busied myself on the company clothing account, which occupied me until nearly ten o'clock."[21] There was no school for pontooniers in the Confederate army. The Army of the Potomac did have such a school, near Belle Plain, Virginia, started in 1862. General Benham, who established the school, reported on January 25, 1864, that it made a vast difference in the engineer troops of the Union army. Practice was constant. For example, on March 6, 1864, the Volunteer Engineer brigade received the following general order: "The Ponton drills of this command will be resumed at once. These will be a drill by successive Pontons on Monday, Wednesday & Friday mornings . . . On Tuesday & Thursday . . . there will be pontoon drill by battalion under the direction of a Field Officer of each regiment when practicable."[22] Virtually every company in a regiment could maintain a bridge and install one. Only five companies from the eleven-company 1st Confederate could build a bridge.[23]

Also in April, Confederate Colonel Talcott ordered all company commanders to classify "as artificers those that they know to be such."[24] This was done to identify those men in the regiment who demonstrated ability as skilled craftsmen and artisans. According to the regimental roster, one soldier in ten was promoted to artificer. This did not compare favorably with the 15th New York Engineers where one in four was similarly ranked.[25]

The Engineer Bureau attempted to recruit the 2nd Confederate Engineers, but only two companies were raised, G and H. As it turned out, this regiment did not exist as a separate command. It had no commanding officer, and G and H companies were placed under the command of the 1st Engineers. An additional eight undersized companies were formed and spread throughout the southeast, reporting to local commanders.[26] Finally, the 3rd Confederate Engineers was raised and served with the Army of Tennessee for the remainder of the war. The 4th Confederate Engineers was organized in late 1864, and its members were scattered over the expansive Trans-Mississippi Department until General Kirby Smith surrendered to Union forces on June 2, 1865, in Galveston, Texas—the last major Confederate army to do so.[27]

In the spring of 1864, before Grant's Overland Campaign began, Lee appointed Major General Martin Luther Smith as chief engineer of the Army of Northern Virginia. An 1856 graduate of West Point, Smith had served as a topographical engineer before resigning his commission to join the Confederate army. In 1863 he had supervised the Vicksburg defenses and now, on Lee's staff, he would be instrumental in laying out the fortifications around Petersburg and Richmond. In Richmond, Captain Alfred L. Rives was appointed acting head of the Engineer Bureau in Jeremy Gilmer's absence. Rives was a skilled engineer, and he served as department head for the rest of the war. Despite the shortage of equipment and money and the lack of centralized control, Rives managed well with what he had. During the Petersburg Campaign, in cooperation with the Railroad Bureau, he assigned engineers to repair forty miles of track destroyed by Union cavalry along the Richmond & Danville Railroad. This was a crucial line connecting Lee's army with essential supplies from North Carolina and South Carolina. In addition, at Burkesville the line connected with the Southside Railroad, the latter running east to Petersburg.

At perhaps the apex of Confederate railroad engineering, men took up rails from the York River Railroad and forwarded them to the Richmond & Danville construction sites, while others, according to Corporal Charles Venable, "began cutting hewing and delivering the cross ties as rapidly as possible. The road bed was gone over and prepared so that track laying would be practicable, and by judicious distribution of our forces we soon had the repairs well in hand."[28] Newly promoted Colonel Rives wrote to his wife on July 24 that a large crew was working hard and that "we hope to have the trains running again & the road better than ever in a fortnight."[29] Unfortunately, the success enjoyed by the engineers in July would not be repeated, and for the rest of the war Confederate railroad operations steadily declined.

Local carriers refused to lend their skilled workers to the army. State railroad companies had the manpower but not the iron; the Confederate Railroad Bureau had some iron but not the workers. Disintegrating tracks, cars, and engines and increasing demands by the government to use the railroads did not elicit assistance from private companies. Furthermore, since the government had no central control of the lines and equipment, some private companies increased their fares and decreased their service. Sly business deals made some railroad companies a fortune, while Confederate armies operated with tenuous supply lines.

Although the Confederate railroad network collapsed, the fortifications designed and built by Confederate engineers during Grant's Overland Campaign and around Richmond and Petersburg exacted a high price from the Yankees in human life and in frustrating Grant's efforts to draw Lee's army out into the open.[30]

From his headquarters in Culpeper County, Virginia, in early 1864, General Grant had started to prepare for his spring offensive against the Army of Northern Virginia, which he hoped would end the war.[31] His plan was to move south and outflank Lee's army. The Corps of Engineers began preparations for the campaign under new leadership. Brigadier General Totten, stricken with pneumonia in the early spring, died on April 22, 1864, at the age of seventy-six. His replacement was sixty-six-year-old Richard Delafield, who prior to his appointment had supervised harbor fortifications for the city of New York. Delafield was in the first class at West Point (1818) to be assigned class ranks. He graduated first in his class and went on to a distinguished career in the engineers, including an assignment by then Secretary of War Jefferson Davis to observe the operations of European armies in the Crimean War.[32]

Delafield proved to be a competent chief engineer, but like several of his colleagues, he was too old for fieldwork. Fortunately, Grant had twenty-four of the eighty-six officers in the regular Corps of Engineers at his disposal (Sherman had nine).[33] Two of the oldest engineer officers with Grant were forty-year-old James C. Duane, chief engineer, and Nathaniel Michler, chief of mapping, both with the Army of the Potomac. Others, like Cyrus B. Comstock and John Parke, were in their thirties, and several, including Francis U. Farquhar, George L. Gillespie, and Peter Smith Michie, were recent graduates of West Point.[34]

Thirty-three-year-old James St. Clair Morton, who commanded General Rosecrans's Pioneer Brigade in the previous year, was appointed chief engineer of Major General Ambrose Burnside's IX Corps, which until May 24 was formally part of the Army of the Ohio, and thus Morton reported to Grant rather than Meade. In a similar fashion, Grant created the Army of the James, which also reported to him and operated with the Army of the Potomac under the command of Benjamin F. Butler. Farquhar and Michie served with this army, but all the engineers in Grant's combined forces operating against the Army of Northern Virginia and Richmond reported to Major General John Barnard, whom Grant appointed to his headquarters' staff and named chief engineer of the armies in the field.[35]

The United States Engineer Battalion was now under the command of Captain George H. Mendell, and Brigadier General Henry W. Benham directed the Volunteer Engineer Brigade, made up of the 50th New York Engineers, eight companies of the 1st New York Engineers, and the 15th New York Engineers.[36] In the spring of 1864, the 15th Engineers were detached to the engineer depot in Washington, DC, to repair and build pontoons. Benham, because of his demonstrated bridge-building skills, was detached from the army in the field to supervise construction work at the depot. Because of his new assignment, Ira Spaulding, commander of the 50th, and Edward W. Serrell, commander of the 1st, now reported directly to Duane rather than to Benham.

Duane divided the 50th New York into three units of three companies each and designated two companies, under the command of Ira Spaulding, as a reserve. Each unit carried fourteen pontoons, except the reserve unit, which carried twenty-four, and each unit was assigned to a corps. Only on rare occasions, and only for several days, was Mendell's regular Engineer Battalion attached to a corps.[37]

As the engineers continued to gather equipment and train for the upcoming offensive, Grant collected several thousand wagons to carry the vast quantities of supplies necessary to press the attack against Lee. The depots in Washington and Alexandria shipped supplies by train along the Orange & Alexandria Railroad to Culpeper. Grant's plan was to cross the Rapidan River at Culpeper and establish bases on the river, which would allow him to operate on Lee's right flank. With waterborne supplies, Grant would not have to worry about Rebels cutting the O&A. Unlike Sherman, whose lifeline was the railroad, Grant anticipated that his wagon trains would be more than adequate for his movements.

Grant was tenacious in attacking Lee during the Overland Campaign, but at every turn, especially at Spotsylvania, North Anna River, and Cold Harbor, the Army of Northern Virginia had anticipated the Federals' movements. The results were Union frontal assaults with ghastly consequences, earning Grant the sobriquet "the Butcher." Grant lost 55,000 men killed, wounded, or missing, and it was too much. He needed to move on Lee's flank before the Southern general had time to prepare strong defensive positions. "Without greater sacrifice of human life than I am willing to make all cannot be accomplished that I had designed outside the City [Richmond]," Grant wrote.[38] His challenge was to quietly disengage his 110,000 men from Lee's forces, transfer his army south to seize Petersburg, and then turn north to operate against Lee's remaining rail line into Richmond.

The plan Grant decided upon was both bold and complex because it required moving his army over the James River and doing so undetected. Grant's army would first extricate itself from the Army of Northern Virginia by a wheeling maneuver in which units on the right flank would pull out successively and march across the rear of adjacent units to the roads they would take to the bridges assigned for their crossing. The army train, which included 2,800 cattle, was sixty-two miles long. A bystander watching the train pass would see the tail pass thirty-one hours after the head.[39] To move the entire army across the James River, the engineers would first have to bridge the Chickahominy River.

There were sixty wooden pontoons and sixteen canvas pontoons (in all, about 1,400 feet of bridge) traveling with the Army of the Potomac. Lieutenant Colonel Ira Spaulding of the 50th New York Engineers was placed in charge of the Chickahominy crossing, and on June 12 he selected three sites: Long Bridge, fifteen miles southeast of Cold Harbor; Jones' Bridge, five miles east of Long Bridge, and Windsor Shades, four miles east of Jones' Bridge.

Major George W. Ford commanded the engineers responsible for the bridge at Long Bridge. Accompanied by the V Corps, Ford quickly discovered that the bridge had burned and little was left. He would need to clear the debris off the remnants of the old bridge and cut down the abutments. Broad swamps bordered the approaches, and the new bridge had to be placed near a narrow passage formed by the roadway to the old bridge. When Ford, with a cavalry escort, crossed the river, he realized that the land mass he had seen from the river bank

On May 24, 1864, during Grant's Overland Campaign, men of the 50th New York Volunteer Engineers, under the direction of Major Wesley Brainerd, built two standard canvass pontoon bridges, both approximately 100 feet long, over the North Anna River. Alfred Waud, Library of Congress, Prints & Photographs Division, Civil War Photographs, LC-DIG-ppmsca-21161.

was actually an island between the main channel, 100 feet wide, and the south branch of the river, 60 feet wide. To make matters more troubling, Confederate pickets opened fire at the engineers and cavalrymen from the south bank, killing one engineer. The Confederate soldiers were soon driven off, and in three hours the two pontoon bridges were completed and the V Corps crossed. By 5:30 p.m. on June 13, the engineers had dismantled the bridge and started moving toward the James.[40]

While Major Ford's engineers were working at Long Bridge, Major Edmund O. Beers moved his pontoon train to Jones' Bridge. Here, as at Long Bridge, there were two branches of the river to be bridged, the island between being about 800 feet wide. Lieutenant Mahlon B. Folwell (William Folwell's brother) and his company built a canvas pontoon bridge over each branch, and Captain Asa C. Palmer bridged both branches with wooden pontoons. The bridges over the north branch were 60 feet long, and those over the south branch were 40 feet. The following day, Beers and his men constructed a permanent bridge over both branches, and by 10:00 a.m. on June 14, the VI and IX corps crossed.[41]

The final bridge built over the Chickahominy was a testament to Northern engineering and innovation. Captain Walter V. Personius and his Company G were headed for Windsor Shades when Colonel Spaulding found the marshes and swamps in the area were too extensive to construct the proper approaches. The next closest potential place to cross was at Cole's Ferry, ten miles southeast of Windsor Shades. Topographical engineers had scouted the area ahead of the bridge builders and discovered an old farm road leading to the disused Cole's Ferry. A final possible site for a crossing was at Barrett's Ferry, twenty-five miles south of Cole's Ferry and a mile and a half from the James River, well south of the site picked to cross the James.[42]

The lower Chickahominy flowed through flat bottomlands, and in mid-June considerable rain had raised the water level, which widened the river at Cole's Ferry to 1,200 feet. This presented a considerable problem for Captain Personius because he did not have enough pontoons for a bridge that long. Early on June 14, a messenger was sent to find Major Duane and report on the precarious situation faced by Personius. When Duane received the news he ordered Colonel Spaulding to collect sufficient material, ride to Cole's Ferry, and get the work done. Meanwhile, Personius, following the army maxim "he succeeds who hustles while he waits," began to build a wharf of boats on each side of the river and a large pontoon raft, which ferried squads of cavalry and wagons to the south bank. Other men built rafts of four boats each, with material on each raft for making connections when the additional pontoons arrived.[43]

Major Beers arrived from Jones' Bridge at 1:00 p.m. and immediately began unloading his equipment to aid in constructing the bridge of rafts begun by Personius. In addition, Captain Peirce, assistant chief quartermaster, along with men from the US Colored Troops, started to build a timber approach about 250 feet in length on the north shore. By 5:00 p.m., after a twelve-mile march from Charles City Court House, Major Ford arrived on the south bank of the river, bringing together all the "land pontoons" the army had in its possession—with the exception of the eight canvas pontoons held by Lieutenant Folwell with General Phil Sheridan's cavalry, and those were now on their way to Cole's Ferry. This was turning into a remarkable operation, which required ingenuity, teamwork, cooperation, and deft management.[44]

Two of Ford's captains, James H. McDonald (Company K) and Asa C. Palmer (Company D), started working on different sections of the bridge. McDonald constructed the south abutment, and his men tacked on their pontoons to those already placed by Personius's men. Palmer set to work with another detail of US Colored Troops on a 200-foot raised corduroy approach road over wetlands. With Captain Von Brocklin's eight canvas boats attached to McDonald's pontoons and, at nightfall, with Lieutenant Folwell's eight canvas pontoons added to the rafts from the north end, all the material was used and the bridge was still short in the middle by 30 feet. Personius and Folwell had attempted to lengthen the spans with additional balk, but this still left the bridge short in the center of the river.[45]

It was a wild situation. Darkness fell. Men slept while standing. Mosquitoes bit. Soldiers cursed. Some urinated in the river. All drank river water. Major Ford, taken sick at Long Bridge, was too ill to walk. Fourteen miles south, lead elements of Grant's army were about to cross the James River. Waiting impatiently to cross the Chickahominy before moving to the James was the Army of the Potomac's sixty-two-mile wagon train, 2,800 head of cattle, and an army division of about 5,000 men. Nerves were on edge.[46]

The engineer officers all agreed that it would be necessary to detach the bridge from the north shore and connect it to the southern section. Then the men set to work constructing additional cribs and corduroy to make up the 30-foot difference. They did it by 3:00 a.m. on June 15 and the bridge was opened to traffic. The total length of the bridge was 1,240 feet, and the length of the timber and corduroy approaches was about 450 feet. The engineers had completed the bridge in an impressive fourteen hours. By late on June 16, the trains, men, and cattle had crossed without delay or incident. The bridge was then dismantled and towed to the James River where the pontoons met up with the wagons, which had gone overland.[47]

Three days before army wagons started across the bridge at Cole's Ferry, General Godfrey Weitzel, chief engineer of the Army of the James, directed his assistant, Lieutenant Peter S. Michie, to reconnoiter specific sites for the crossing of the James River. The site chosen was at Douthart's house, midway between Wilcox's Landing and Weyanoke Landing, on a neck of the river that ran north to south. The bridge would hit the opposite shore between Windmill Point and Fort Powhatan. The site was well beyond the observation of Confederate cavalry trying to discover the location of Grant's army, and constructing the approaches would not require the man-hours that would be required over the extensive marshlands at the two other possible sites. The difficulty would come in the actual bridge building. The width of the river was close to 2,000 feet. The currents were strong and the tides rose and fell about four feet. Because the army wanted to keep the area open to river traffic, a draw would need to be built into the center of the bridge, where the water in the main channel reached depths of 80 to 90 feet.[48]

On June 13, without official approval from Grant's headquarters, Weitzel

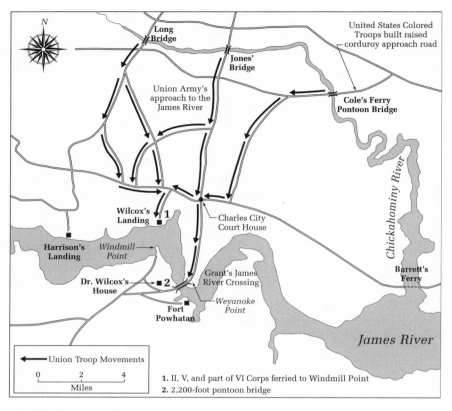

Crossing the James River, June 1864

sent Michie back to the site to begin work on the approaches, and by nightfall a detail of soldiers had cut and trimmed 1,200 feet of timber in logs averaging six inches in diameter and 20 feet long. On the southwestern shore above Fort Powhatan, the engineers cut 3,000 feet of timber, formed rafts, and floated them to the bridge site. Cypress logs, almost three and a half feet in diameter, were used in building approaches in the shallow part of the river. By midmorning on June 14, the approaches on both sides of the river were completed. On the southwestern side, a ramp was built, ruts and gullies filled over, and a roadway constructed to connect with Petersburg and City Point Road. On the northeastern side, trees were cleared and a 150-foot pier constructed over a small marsh. Grant had now approved the location of the bridge, and he directed General Benham at the engineer depot at Fort Monroe to send all available pontoons as quickly as possible to the bridge site. Finally, the regular battalion of engineers was ordered to the site on June 14 to construct the bridge, under the direction of Major Duane.

On the morning of June 14, Weitzel paced the riverfront waiting anxiously for the pontoons that Benham had promised to arrive. Finally, after his patience was worn thin by the delay and the recognition that Grant's lead elements were due to appear anytime, Weitzel sent a dispatch boat down the river to look for the steamer towing the pontoons. At noon, after being en route for twenty-two hours and traveling at three miles per hour, the steamer arrived with approximately 155 pontoons, enough for 3,100 feet of bridge.[49] Now there was excitement and chaos at the bridge site. Pontoons were taken off their towlines and bobbed freely in the river. Men from Captain George Mendall's United States Engineer Battalion jumped into the four-foot-high water, trying to push some of the boats on shore, while other men, slipping on underwater rocks and getting their feet encased in ooze and mud, started building an abutment that reached out into deeper water.

Once Duane had established control over the chaos of the unmoored pontoons, he ordered Mendall's men to steer fifty boats to the south bank of the river. Men of the 15th New York who had arrived with the pontoons from Fort Monroe and of the 50th New York—except Major Wesley Brainerd's battalion—were assigned to building the pontoon bridge from the north side. Brainerd was instructed to report to II Corps headquarters to support General Hancock's separate movement across the James. Grant expected that at least seventy-two hours would be required to get his entire army across the James River, and by then Lee might learn of his grand flanking movement and beat him to Petersburg. It was decided, therefore, that Hancock's II Corps would not wait for the pontoon bridge to be completed; instead, the men would be ferried across the river. The operation began on June 14, but by late afternoon the landing was progressing at a painstakingly slow pace. A wharf would help matters, and an old one, nearly destroyed by fire, was found.[50]

Sometime after 7:00 p.m., Brainerd was summoned to the headquarters of Third Division commander Major General David B. Birneys, and the general informed the engineer that he could have 800 infantry to rebuild the wharf to speed up the ferry crossings. Brainerd and his engineers worked until 3 a.m. trying to sink new pilings two feet beneath the riverbed. Then just before sunrise, pontoon boats arrived from General Benham and these were used to finish the new wharf.[51]

Meanwhile, work on the pontoon bridge proceeded with alacrity. The 50th New York Engineers worked for ten hours on the structure, and by 11 p.m. the

In one of the most incredible feats of military engineering of the Civil War, men from the 15th and 50th New York Volunteer Engineers constructed this 2,200-foot-long pontoon bridge and the approach road (foreground) across the James River, near Weyanoke Point, on June 14, 1864. Built in just eight hours, the bridge consisted of 101 pontoons and included a draw section in the middle to allow continuous river traffic. The schooners helped to anchor the bridge against the strong currents in the river. Library of Congress, Prints & Photographs Division, Civil War Photographs, LC-DIG-ds-05461.

remarkable bridge was finished. Because the current changed direction during the day, the pontoons had to be anchored both upstream and downstream. To provide anchorages in the main channel, the pontoons were fastened by guy wires to one of three schooners anchored above the bridge and one of three schooners anchored below the bridge. To permit passage of vessels upstream and downstream, a draw section made up of 100-foot-long rafts was built into the 2,200-foot-long bridge. At 1:00 a.m. on June 15, Benham received word from Meade to open the bridge to traffic.

Bridge trains and surplus artillery from three corps crossed first, and then at 6 a.m., the men of the Army of the Potomac began to cross—the V, VI, and IX corps, the Third Cavalry Division, the Army Headquarters, and finally the sixty-two-mile-long wagon train. The combat units crossed in fifteen hours.[52] Astonishingly, the only problem came three hours after the bridge was open to infantry traffic. An upstream schooner slipped its anchor, drifted into the bridge, and carried away a part of it. The damaged section was restored in several hours, and the march of Grant's army continued.[53]

The sight of this movement of the army across the James River mesmerized eyewitnesses. Private John H. Westervelt, 1st New York Volunteer Engineers, observed: "This is the first time I have seen anything like an army cross a pontoon bridge and I can assure you it is well worth seeing. From sunrise till 12M it was one steady stream tramp, tramp an a roar like a R road train all the time."[54] A correspondent of *Harper's Weekly* reported: "As we approach the pontoon bridge we see distinctly huge bodies of infantry, cavalry, horses, artillery, and wagons moving across the bridge. They extended across the entire length of the bridge, and can be seen wending along from far away up the east bank of the James, enveloped in a dense cloud of dust, while on the western bank is a part of the great body which has already effected its crossing."[55]

When the last animals crossed the bridge in the early evening of June 18, General Benham "breathed free again." Finally feeling a sense of great accomplishment, he described it as "the most successful effort on a large scale with pontoon bridging that has ever occurred in our country, if it does not rival those in any other land."[56] Benham's praise was merited.

The engineers had taken Grant's vision of a never before attempted grand turning movement and made it a reality. Unfortunately, costly mistakes on the eastern outskirts of Petersburg, a failed attempt to interdict Southerners' rail supply, and skilled Confederate generalship by P. G. T. Beauregard prevented the advance elements of Grant's army from capturing Petersburg in June 1864. Instead, a siege began that would last until April 1865. Before beginning the work of laying out and building fortifications and trenches, army engineers had built 8,678 feet of bridges. Now, beginning in July, the Army of the James, which occupied the northern portion of the siege lines, deployed troops from Richmond to Petersburg from its position on the Bermuda Hundred peninsula. The army's tactical movements required the engineers to build a number of bridges on the James and Appomattox rivers, including a 1,320-foot pontoon bridge across the James at Aiken's Landing built by the 1st New York Engineers. The engineers even attempted to dig a canal at Dutch Gap similar to the one dug around Island No. 10, but heavy Confederate shelling made it impossible to complete the canal.[57]

In the months after the army crossed the James, the siege of Petersburg required the engineers to build signal towers, make abatis, gabions, and fraise,

Union soldiers, with uncanny efficiency, systematically ruined Southern railroad tracks in June 1864 during Lieutenant General Ulysses S. Grant's Overland Campaign. This scene of destruction was repeated over and over again during General William T. Sherman's Atlanta, Savannah, and Carolina campaigns in the fall and winter of 1864–65. Alfred Waud, Library of Congress, Prints & Photographs Division, Civil War Photographs, LC-DIG-ppmsca-22456.

build magazines and bomb-proofs, construct parapets, lunettes, revetments, and platforms for guns, make maps, and draw plans for forts.[58] A significant amount of time was also spent corduroying roads and making covered ways—sunken roads wide enough to allow the passage of wagons and artillery, connecting two or more fortifications.[59] Yet, of all the work that engineer soldiers did during the eleven-month siege, the effort that became most daring—and very dangerous—was mining. Both sides employed this tactic, yet neither side enjoyed the success hoped for, although the first attempt made by the Federals might have worked had it not been for a mismanaged assault.

Lieutenant Colonel Henry Pleasants of the 48th Pennsylvania Infantry believed he and his men, many of them professional miners, could dig a mine beneath the 125 yards of soil separating Union from Confederate soldiers. By doing so they would punch a gaping hole in the Confederate trenches, which would then be exploited by thousands of Union infantry. With this break in the trenches, the entire Confederate line would collapse and lead to a crushing Confederate defeat and an end to the war. Approval for the mine had to work its way up the army's chain of command, and as it did so, Pleasants started the work. Sergeant Henry Reese of Company F was named foreman of the project, and improvisation and hard work were the essential tools used to tunnel under the wary yet unsuspecting Confederates.[60]

Pleasants wanted to be certain he had the correct distance from the mine entrance to the Confederate lines, and to accomplish this he triangulated using a theodolite. Historian Earl J. Hess explains the work: "His crew laid out short lines at five different locations within the Union position and created the other two sides of an imaginary triangle from each end of these lines to Pegram's Salient. The degrees of the angles at the sides were noted and the distance to the salient deduced from this and the length of the third, base line." Confederate guns were 133 yards from the Union front line.[61]

For ventilation, the miners first dug vertically from the gallery to the surface, creating a shaft twenty-two feet deep. Next the men constructed a square wooden pipe and laid it along the base of the tunnel, with a small iron furnace added to the bottom of the shaft. Finally, a partition, with a door, was build along the outside edge of the shaft so that fresh air entered the wooden pipe and exited deep within the gallery, then the heat from the furnace drew the exhaled air up the shaft. To make sure Confederates observing the Union line did not detect unusual activity, several campfires were maintained as decoys to draw attention away from the mineshaft.[62]

By the end of June, Pleasants had the mine ready. His men had dug two wings each about 36 feet long and packed these chambers with a total of 8,000 pounds of explosives. The tunnel was 510 feet long, and although Southern soldiers had dug two countermines to determine whether rumors about a Yankee mine were true, most Confederates had no idea what was about to happen. At about 4:45 a.m. on June 30, 1864, the mine exploded leaving a hole in the ground 200 feet long, 60 feet wide, and 30 feet deep. The blast killed or wounded more than 300 men. John Haley of the 17th Maine descried the scene: "Earth and heaven were rent by an explosion that would have done credit to several thunderstorms."[63]

What happened next, however, was the result of incompetence. Instead of pouring through the breech as planned, Union soldiers got to the crater carved out by the explosion and just stopped. They were uncertain what to do next, largely because their commanding officer, General James Ledlie, was sitting in a bombproof shelter drunk, and his soldiers had not been trained for the assault. A black regiment had prepared to attack after the explosion, but General Meade replaced them with white soldiers at the last minute. Confusion ensued as Union reinforcements, including the African American troops originally trained for the assault, rushed to the crater. By this time Confederate Briga-

Lieutenant Colonel Henry Pleasants of the 48th Pennsylvania Volunteer Infantry supervised the delivery and placement of powder kegs into the Union mine shaft that ran under Confederate lines at Petersburg, July 1864. Alfred Waud, Library of Congress, Prints & Photographs Division, Civil War Photographs, LC-DIG-ppmsca-21357.

dier General William Mahone managed to regroup his forces, stem the Union assault, and bottle up several thousand Northern soldiers inside the crater. By the time the fighting ended, the South suffered 1,491 casualties and the North, which lost a golden opportunity to break the Confederate lines, lost 3,798 men, killed, wounded, or missing.[64]

After the mine explosion at Pegram's Salient, both sides mined and countermined throughout the remainder of the campaign with no success. Confederate engineers Hugh T. Douglas and W. W. Blackford led efforts to dig mines under Federal lines. For example, on August 5, a mine detonated under Gracie's Salient to no effect because the explosive charge was only 850 pounds of powder and it exploded short of Union lines. The goal was to threaten Yankee mining efforts, but the attempt fell far short of the mark.[65]

The most creative device used to assuage Confederate soldiers' fears that the ground was about to erupt was a simple earth auger with which two men could bore a hole about three and a half inches in diameter to a depth of twenty feet. The hole was filled with water, which the clay soil retained, and if the water disappeared everyone would know the Yankees were beneath them.[66]

As Southern soldiers continued to resist assaults from Grant's massive army above and below ground, they clung to the belief that Southern independence was still possible. The Richmond *Sentinel* reported that "the war has continued for more than three years, the United States being weaker today than when hostilities commenced while the Confederate States are infinitely better prepared than ever before to resist the attacks of the enemy."[67] Perhaps if the war could be prolonged by forestalling Union offensives around Petersburg and northern Georgia, even with the loss of Mobile Bay to the Union navy on August 5, it might be possible that the Northern public would grow tired of the slaughter and elect someone to the White House in November who would be willing to negotiate a peace with slavery intact.

Lincoln certainly thought this bleak outcome was distinctly possible. On August 23, the president wrote: "This morning, as for some days past, it seems exceedingly probable that this Administration will not be re-elected. Then it will be my duty to so co-operate with the President elect, as to save the Union between the election and the inauguration; as he will have secured his election on such ground that he can not possibly save it afterwards."[68]

Despite the remarkable efforts to cross the James River and gain the flank of Lee's army and the attempt to blow a massive hole in the Southern trench line at Pegram's Salient, the Union army had reached an impasse with its enemy. How long would the Northern public continue to tolerate this stalemate and the senseless killing and maiming that accompanied it? In just three months, citizens would get their opportunity to answer that question in the November election. Grant's simultaneous offensives along the entire Confederate line brought great pressure to bear on all Southern forces and resources, but the Union army had yet to achieve the desired results. Time was running out. It was true that the South could no longer expect to win the war, but it still had an opportunity to win its independence. If the Southern armies could just hold out a little longer, Lincoln might be ousted, the emancipation proclamation rescinded, and their hope for a permanent Southern Confederacy might just become a reality.

CHAPTER THIRTEEN

Atlanta and the Carolina Campaigns

If his army goes to hell, it will corduroy the road.
General Joseph Johnston, CSA, on Sherman's march through the Carolinas

Despite President Lincoln's dour prediction, in the summer of 1864 there was still time to bring about the victory the citizens of the United States needed to persuade them that the war was worth continuing and that Lincoln was the man to continue to lead them. With Grant bogged down in Virginia, the best hope was for Sherman to take Atlanta—a difficult enterprise. It was a formidable task to move 100,000 soldiers and fight through the mountains of northern Georgia, with a dangerous enemy blocking the way and with the army's supply line along a single-track railroad from Nashville to the front lines.

For two years of war, Union engineering and railroad management had helped accelerate the advance made by Northern forces as they cut their way into Tennessee, Mississippi, Alabama, and Louisiana. Delays at forts Henry and Donelson, Nashville, Island No. 10, and Vicksburg, and the loss of an army at Chattanooga and a flotilla at Red River, all would have had a significant effect on the country's morale and, with the exception of the Red River fiasco, would have delayed Sherman's preparations for the Atlanta Campaign.

Yet, skillful and innovative engineering and the development of sound railroad management and policy, along with determined generalship and fine soldiers, brought the Military Division of the Mississippi to the gates of northern Georgia in the spring of 1864. Beginning on May 7, when Sherman's forces broke winter camp and started toward Dalton, Georgia, the Federals' formula for success would again have to be innovative engineering, skilled generalship, and determined soldiers.

Sherman faced a difficult logistical challenge because he would need to rely on single-track railroads extended over 300 miles for the sole source of supplies for his 100,000 men. His goal was to move 130 cars carrying supplies to the front daily. When Colonel McCallum arrived in Nashville to head railroad operations, he discovered that the 50 locomotives and 537 cars available were woefully inadequate to meet Sherman's needs. McCallum estimated that the campaign would require 200 locomotives and 3,000 cars along with sophisticated maintenance and construction facilities. The line itself would require constant attention because the track, according to McCallum, "was laid originally on an unballasted mud road-bed in a very imperfect manner, with a light U-rail on wooden stringers, which were badly decayed and caused almost daily accidents by spreading apart and letting the engines and cars drop between them."[1]

McCallum first set about recruiting able assistants: Adna Anderson, superintendent of the transportation and maintenance department, and William W. Wright, chief engineer in charge of the Construction Corps. Then, with authority from Secretary of War Stanton and General Sherman, McCallum made the chain of command clear to both men. Anderson and Wright would report directly to McCallum, but in the latter's absence, they would take orders only from Sherman or his corps commanders, McPherson, Schofield, and Thomas.

The organizations established under Anderson and Wright were derived from the blueprint drawn up by Herman Haupt in the eastern theater in 1862, but they were expanded to address the scope and complexity of railroad operations in five states: Mississippi, Alabama, Kentucky, Tennessee, and Georgia. The management structure for Anderson's transportation department consisted of masters of transportation who moved over certain sections of road to "see that the employes [sic] attended properly to their duties while out with their trains." At stations, a dispatcher made sure locomotives were in good order, the superintendent of repairs maintained the roads, and the master machinist managed repairs of locomotives. All managers were independent of each other and reported directly to the general superintendent.[2]

Wright's Construction Corps was organized into six divisions, and each division was made independent and equipped with tools, camp equipage, and field transportation. The division was under the command of a civilian engineer and was divided into sections, the largest of which comprised the track layers and the bridge builders. There was a foreman for each fifty men (a gang) and a sub-foreman for each ten men (a squad).[3]

Once Anderson and Wright started to build their respective workforces and requisition tools from the quartermasters, McCallum turned his attention to finding the additional locomotives, cars, shops, and machine tools necessary

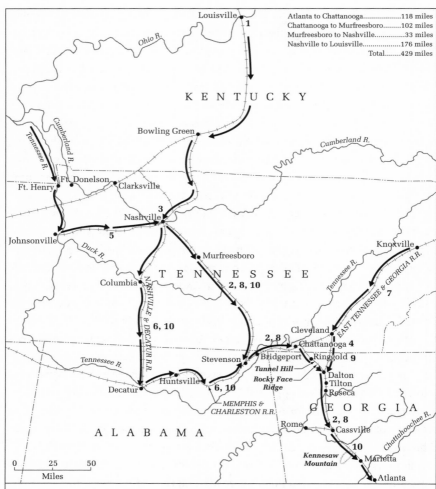

Atlanta to Chattanooga	118 miles
Chattanooga to Murfreesboro	102 miles
Murfreesboro to Nashville	33 miles
Nashville to Louisville	176 miles
Total	429 miles

Union Army Supply Lines for the Atlanta Campaign

1. Central depot Louisville
2. Single track from Atlanta to Nashville
3. Locomotive repair shops in Nashville
4. Sawmills and rolling mills in Chattanooga
5. USMRR built railroad from Johnsonville to Nashville to shorten supply route from Tennessee River
6. Nashville & Decatur R.R. and Memphis & Charleston R.R. used as auxiliary line of supply to Chattanooga

7. East Tennessee & Georgia R.R. served as auxiliary line
8. Siding constructed at eight-mile intervals along the railroad from Nashville to Atlanta
9. Ringgold became advance supply depot
10. Extra rails, spikes, and crossties left at certain locations to facilitate quick repairs

Sherman's Supply Line during the Atlanta Campaign

to sustain the planned spring offensive. To add gravitas to this endeavor, the secretary of war, on behalf of President Lincoln, wrote to every locomotive manufacturer in the country requesting or ordering their assistance. "In order to meet the wants of the military departments of the Government you will deliver to his [McCallum's] order such engines as he may direct, whether building under orders for other parties, or otherwise the Government being accountable to you for the same." Stanton reminded the railroad presidents that the need to supply the army in Tennessee was urgent and rendered "the engines indispensable for the equipment of the lines of communication, and it is hoped that this necessity will be recognized by you as a military necessity, paramount to all other considerations."[4]

The Union government's approach to the problem of acquiring additional locomotives and the private sector's response to the government's request stood in stark contrast to how private companies in the South responded to the Confederate government's needs. In places such as Taunton, Massachusetts (Mason Machine Works), Philadelphia, Pennsylvania (Norris Locomotive Works), and Manchester, New Hampshire (Manchester Locomotive Works), fifty-three engines and hundreds of cars were built at a startling rate and driven to Nashville. General Superintendent Anderson ordered underutilized locomotives and cars operating around Memphis to be transferred to the Atlanta operation. In April, 15 engines and 120 cars were taken from the Louisville & Nashville Railroad; in May, 2 engines and 68 cars were impressed from the Kentucky Central Railroad.[5] By June, 77 additional locomotives and 1,051 cars brought the totals to 124 engines and 1,488 cars available for supplying Sherman's armies. In addition, extensive machine and car shops were built at Nashville and Chattanooga, those at Nashville large enough to hold 100 locomotives and 1,000 cars at once.

So, as bacon, pork, salt beef, bread, flour, cornmeal, peas, hominy, mixed vegetables, coffee, sugar, and vinegar were loaded onto cars each day, the Construction Corps repaired, maintained, or rebuilt track to carry the supplies to sustain 100,000 men and 35,000 horses. To put things in perspective, if Sherman had had to depend upon animal-drawn wagon trains, he would have needed approximately 39,000 wagons and 220,000 horses to haul the necessary supplies twenty miles a day. It would have been impossible. Instead, approximately 130 railroad cars, each carrying ten tons, were shipped forward from Nashville to Chattanooga every day.[6]

The Nashville & Chattanooga Railroad, 151 miles long, was Sherman's main artery. About 115 miles of track were relaid, and sidings were put in at intervals of 8 miles.[7] Each siding could hold an eight-car freight train. Eventually, 19 miles of new sidings were added to the road, and forty-five new water tanks were built.[8]

Three other railroads would serve as auxiliary lines during the campaign, and in the early spring these also needed the attention of the Construction Corps and army pioneers. The first was the line between Nashville and Decatur, completed by General Dodge, which connected with the Memphis & Charleston east of Stevenson, Alabama. The next was the newly built military line running west from Nashville to the Tennessee River at Johnsonville, Tennessee. Finally, the East Tennessee & Georgia Railroad, from Chattanooga to Knoxville, was opened after the Construction Corps built the Tennessee River bridge at Loudon.[9]

The maintenance of all these single-track lines would determine how quickly Sherman could press his offensive against Johnston's army defending northern Georgia and, ultimately, the grand prize: Atlanta. By the end of April everything was ready. Quartermaster General Meigs reported to Sherman that enough food was accumulated at Nashville to feed 200,000 soldiers for four months, and sufficient grain to feed 50,000 animals for the rest of the year.[10] On May 4, 1864, General Thomas's Army of the Cumberland, the center of Sherman's three-prong attack, marched out from Ringgold, Georgia, prepared to begin the most arduous and arguably the most critical campaign of the war.[11]

Sherman's new chief engineer, Captain Orlando Metcalfe Poe, had spent most of his time before the campaign restructuring the army's engineer organization, which he deemed inadequate to meet the demands of 100,000 men moving in three separate armies. Poe was an interesting fellow. He graduated sixth in his 1856 West Point class and originally wanted to be an artillery officer. He believed that in a system where promotion came only through seniority, his best chance of promotions was in the artillery. He soon learned, however, that artillery openings were limited, so the young second lieutenant decided to switch to the topographical engineers.

After graduation, Poe worked for three years with Captain George Meade, doing survey work in the Great Lakes Region. Meade admired the lieutenant's determination and creativity, and by 1860 Poe had completed several projects, which included building nineteen weather stations on the five Great Lakes.[12] When the war began, Poe served on George McClellan's staff, organizing the defenses around Washington, DC, commanding the 2nd Michigan Infantry, and leading his regiment at Fair Oaks and Second Bull Run. Unfortunately, Republicans did not look favorable upon the apolitical Poe because of his connection

with McClellan. In the spring of 1863 when Poe was recommended for a brigadier generalship of volunteers, Congress never confirmed the appointment.
After the Emancipation Proclamation and fearing Democratic Party fallout, the
Radicals were intent on purging the military of political generals who were also
strong Democratic voices. Poe, unknowingly, was considered connected with
these men because of his association with McClellan.

So Poe went from being an unconfirmed brigadier general of infantry back
to a lieutenant of engineers in the regular army. Yet his reputation among his
fellow officers and the men in his command was solid, and soon he was made
a captain of engineers for the Army of the Ohio. He was serving in this position
when Sherman made him chief engineer.

Poe's performance throughout the summer of 1864 only confirmed the
faith others had placed in him. In restructuring the engineer organization,
he first took stock of what he had. The huge Army of the Cumberland had two
field-tested engineer units: the 1st Michigan Engineers and Mechanics and the
1st Missouri Engineers. The Michigan boys, after building a magazine in Chattanooga, 70 feet by 100 feet with 20-foot-high walls, and a railroad spur to the
magazine, were assigned the critical task of building railroad blockhouses to
help prevent Confederate cavalry and partisan raiders from destroying track
and, especially, bridges. The blockhouses, designed by Captain William Merrill with the help of Lieutenant Colonel Kinsman Hunton of the Michigan engineers, were two stories high. The first-story walls were forty inches thick,
and the walls of the second story, set diagonal to the ground floor, were made
twenty inches thick for better stability. Each blockhouse could hold a garrison
of twenty men.[13] In total, 105 blockhouses were built. The 1st Missouri, along
with McCallum's Construction Corps, focused on completing the newly built
military railroad running west from Nashville to the wharves on the Tennessee
River at Johnsonville.

Thanks to the efforts of General William Rosecrans the year before, the Army
of the Cumberland had a regiment of volunteer engineers made up of enlisted
men who had served with the army's Pioneer Brigade. These men from the
13th, 21st, and 22nd Michigan Infantry regiments and the 8th Ohio Infantry had
worked extensively on building blockhouses, operating sawmills, and building
fieldworks around Chattanooga. Captain Merrill, who was commissioned as a
colonel, led the 1st Veteran Volunteer Engineers.[14] The army also had a pontoon
train made up of the new Cumberland "hinged" bateaux, with additional pontoons and equipment held in reserve in Nashville.

Colonel George P. Buell and the 58th Indiana Infantry were placed in charge of bridging operations. The Indianans, Michiganians, and Ohioans were typical infantry regiments in the Union army, with many of the enlistees having mechanical and carpentry skills acquired before the war. With the shortage of engineer troops, these men were transferred to the engineers to perform the vital tasks of building roads, bridges, and fortifications critical for the success of Sherman's campaign. Both the Army of the Tennessee and the Army of the Ohio also faced a shortage of engineer troops, and they, too, would turn to infantry units for help.[15]

The Army of the Tennessee, with the exception of a well-established pioneer organization and a pontoon bridge train, had no engineer troops, so on Poe's suggestion, Sherman transferred the 1st Missouri Engineers to this army. Conversely, the Army of the Ohio had an Engineer Battalion, which had been established at Chattanooga when Grant was planning a move on Knoxville and east Tennessee.[16] Poe also had nine regular army engineers attached to the various field armies who were responsible for laying out fortifications for strategic points, conducting topographical surveys, and reconnoitering enemy positions.[17]

Sherman invaded Georgia in the late spring and summer of 1864. He planned to use a series of flanking maneuvers to force Confederate General Joseph E. Johnston to abandon his fortified positions and withdraw toward Atlanta. These movements would require maintaining a supply line over a tenuous single-track railroad from Sherman's advance positions to Chattanooga and Nashville. It would also require moving more than 100,000 soldiers over three major rivers: the Oostanaula, the Etowah, and the Chattahoochee.

The first of Sherman's flanking movements came in mid-May when the general's engineers determined that enemy forces were stretched across the Western & Atlantic Railroad, four miles northwest of Dalton, Georgia, on a north-south axis along Rocky Face Ridge. The Confederate commander hoped the Yankees would launch a frontal assault against a strong position, but Sherman would not oblige. Instead, he planned to have Thomas and Schofield demonstrate against Johnston's front, while McPherson's Army of the Tennessee would move twenty-five miles southeast through Snake Creek Gap and occupy the area around Resaca and hold the railroad and telegraph. This forced the Confederates to abandon their stronghold on Rocky Face Ridge and move into their defenses north and east of Resaca, north of the Oostanaula River.

Buell's pontoon train had covered approximately fifty miles from Chattanooga through Snake Creek Gap in about forty-eight hours and had stopped

to allow the men to eat and rest for no more than three hours during this grueling and often uphill march. Now in a valley east of Horn Mountain and west
of Resaca, the men of the 58th Indiana had finally halted and started to settle in
for a longer break when orders arrived to get the pontoons to Lay's Ferry on the
Oostanaula. Exhausted and ornery, Buell's command spent four hours moving
about in dense thickets on unidentified roads until they reached the east bank
of Snake Creek, one mile from where it flows into the Oostanaula.[18]

Sherman had attempted to get around Johnston's right flank with no success, and on May 15 he decided to gain a lodgment on the eastern bank of the
river, with the hope of cutting off the expected Confederate retreat from Resaca. Brigadier General Thomas Sweeny of County Cork, Ireland, a regular army
officer, was selected by McPherson to establish the bridgehead. Sweeny's soldiers described the hot-tempered general as speaking three languages, "English,
Irish-American, and Profane," and claimed he was most eloquent in the last.[19]
Now with an infantry escort, Buell's men assembled the Cumberland pontoon
boats and, after stretching on the canvas covers, paddled them down Snake
Creek to the Oostanaula and Lay's Ferry. Eventually, two pontoon bridges
were built to accommodate Sweeny's division and McPherson's corps, allowing
these forces to gain the Confederates' flank. This forced Johnston to evacuate
his formidable defensive position and move further south toward Atlanta.

Colonel Buell's 58th Indiana Infantry, with pontoon train in hand, conducted
an amphibious operation much like the one at Lay's Ferry on July 8, when the
regiment, launching the canvas pontoons filled with infantry down Soap Creek
into the Chattahoochee River, had debarked the soldiers to establish a bridgehead. Buell's men now threw two pontoon bridges over the river so that the entire Army of the Ohio could cross. The 58th Indiana then took up the pontoon
bridge, and the Engineer Battalion built a more permanent trestle bridge to
replace it. General Schofield, commander of the Army of the Ohio, wrote: "My
thanks are due to Colonel Buell and his regiment for the admirable manner in
which they performed their important part."[20] In his official report, Poe noted
"that whenever it was deemed necessary to use a bridge for a greater length of
time than forty-eight hours the pontoon bridges were invariably replaced by
wooden trestle bridges constructed from materials at hand, either by engineer
troops or the pioneer forces."[21]

One pioneer force that performed admirably throughout the campaign yet
received little recognition in Poe's reports was that of the Army of the Tennessee, under the command of Captain William Kossak, an aide-de-camp to Sher-

man and now an engineer officer. Kossak had distinguished himself at Vicksburg by tirelessly constructing mines and trenches against the Confederate defenses.

Now at Atlanta, Kossak's pioneers would build seven trestle bridges on the Chattahoochee River, each about 350 feet long, for Sherman's wagon trains to cross. The bridges were constructed from trees cut from the riverbank, and five of the bridges were double-tracked to accept two-way traffic. The efforts were impressive. Although in his report Poe singled out only Captain Reese and lieutenants Wharton and Twining, both West Pointers, all of his engineer and pioneer officers and men, most of whom were volunteers, performed extraordinarily well. Poe's experience with Congress in 1863 might have explained his reluctance to focus his report on others and not himself. After the campaign he wrote to his wife, Nell, that "not even a Congress can sever my name from its official connection with them [the history of the Atlanta Campaign]."[22]

Poe's subordinates laid out fifty miles of infantry and artillery parapets, repaired many miles of roads, and built six bridges averaging eighty feet long over Peachtree Creek, five bridges over the Flint River, and numerous smaller bridges. In addition, Poe's topographers made surveys of all the routes traveled by infantry columns, and he drew a map on a scale of four inches to one mile illustrating the siege of Atlanta. In all, 4,000 copies of campaign maps were issued to officers to facilitate military operations.[23]

On the Confederate side, engineers performed well at times, despite being plagued by a lack of engineer soldiers and, in some cases, poor planning. For example, most of the defensive positions at places such as Hardee's Salient, the Kennesaw Mountain Line, and the Chattahoochee River Defense Line were constructed in advance of the Confederate army's occupation, as were the fortifications surrounding the city of Atlanta. The latter, with redans and lunettes, posed a formidable obstacle to Sherman's three armies.

Yet, as archaeologist Robert J. Fryman points out, although the Atlanta defenses were carefully designed and built in 1863, a critical flaw still existed. Using archaeological analysis, old maps, and hand-held GPS units, Fryman discovered that the forts themselves were designed to be approximately 1.25 miles from the center of the city, and the woods were cleared for a distance of 1 mile from the forts. In December 1863, Colonel J. F. Gilmer, chief of the Confederate Engineer Bureau, visited Atlanta and approved what he saw.[24] The Army of Tennessee's artillery preferred the accuracy and strength of the bronze smoothbore 12-pounder Napoleon field gun, which fired solid shots and shells about 1,680 yards. As Fryman notes, "The construction of Atlanta's defensive perimeter at an average distance of 1.25 miles would have provided more than adequate protection for the city's buildings and infrastructure had the opposing Federal forces been armed with identical field artillery."[25] They were not. By 1863, Gilmer should have been aware of the capability of the Union army's artillery and planned accordingly. Sherman had weaponry far more powerful than the Confederates anticipated.

Sherman's three armies used 3-inch ordnance rifles, 10-pounder and 20-pounder Parrott rifles, and eight 4½-inch rifled siege guns, which could fire an average distance of 4,160 yards. This meant that during the siege, the Federal artillery could bombard Atlanta from distances beyond the reach of Confederate defensive weapons.[26]

The Confederate army fared much better when it came to the use of the railroads. Commanding General Joseph Johnston had ordered Major Stephen W. Presstman, acting chief engineer of the Army of Tennessee, to advise the Engineer Bureau in Richmond that Johnston's army wanted duplicate bridges (trestlework) built for every railroad bridge on the Western & Atlantic between Atlanta and Dalton, Georgia. The process of getting the work done was well coordinated, and the A. L. Maxwell, Jr., and Company foundry completed the trestles as requested. Johnston, however, never used the duplicate bridges. He anticipated the problem of replacing original, badly deteriorated trestles, but he did not anticipate how quickly Sherman's army would move against his flanks. Johnston, fearing he would be cut off from his supply base in Atlanta, began a retrograde movement, destroying railroad bridges and track.[27]

Johnston understood that Sherman was not carrying duplicate trestles with his army and, consequently, believed that the destruction of bridges and track would slow Sherman's movements to a crawl for weeks. Johnston miscalculated. Standing near the Etowah River near Cartersville, Georgia, Hosea Rood of the 12th Wisconsin observed: "The bridge was burned by the Rebels, but nothing is too much for Yankee enterprise and perseverance; the bridge is being quickly rebuilt, and the road will soon be in running order clear to the front." J. W. Gaskill of the 104th Ohio wrote: "Rebel prisoners are surprised at the ingenuity of Yankees and the rapidity shown in repairing railroads and bridges. They declare that 'old Bill Sherman carries a supply of ready made bridges,' adding that it was useless to blow up Tunnel Hill for ' "old Bill" even carries a supply of ready made tunnels.' "[28]

Sherman's greatest advantage during the campaign, however, was the use of the railroads. The USMRR maintenance shops in Nashville repaired a hundred locomotives and a thousand cars each month. McCallum's construction teams built a rolling mill in Chattanooga and rerolled rails for one-third the cost of new rails. His men rebuilt eleven bridges and laid seventy-five miles of track during the campaign.[29]

Retreating Confederates had burned the bridge across the Oostanaula at Resaca, but W. W. Wright's Construction Corps included one of Herman Haupt's master bridge builders, E. C. Smeed. Smeed was able to repair the bridge in three days, although both Wright and McCallum got the credit. It was the same with the construction of the railroad bridge over the Etowah River. Smeed and his men arrived at the bridge on June 5 and started work on the following day. For three days, gangs removed the old structure and cut and hauled timber from the woods to the bridge site. The Construction Corps built a 600-foot trestle bridge made up of five trestles, 67 feet high. It was a magnificent piece of work, yet the most remarkable performance by Smeed and the Construction Corps was still to come.[30]

The railroad bridge over the Chattahoochee had to be 780 feet long and 90 feet high, almost twice as long as the Potomac Creek bridge that Haupt had built with Smeed's assistance in 1862. Again, using timber cut from area woods, Smeed used the stone piers left standing by the Confederates to form a sturdy foundation. In four and a half days the bridge was finished. Haupt called the building of the Chattahoochee Bridge "the greatest feat of the kind that the world has ever seen." The teacher's effusive praise for his student aside, word of Northern ingenuity and proficiency in bridge building soon traveled overseas, and in 1868 the British Association for the Advancement of Science invited Haupt to attend its meeting to describe how these remarkable feats of engineering were achieved.[31]

McCallum also deserved credit for the management system he put in place during the campaign. With some modifications, the system mirrored the one he had designed for the Erie Railroad back in 1857. Supplies such as rails, spikes, crossties, and iron were placed at collection points along the line, and detachments of men from the Construction Corps, with an ample supply of tools, were stationed at intervals. "Each detachment was under the command of a competent engineer or supervisor, who had orders to move in either direction, within certain limits, as soon as a break occurred, and make the necessary repairs without delay, working day and night when necessary. Under this arrangement small breaks were repaired at once, at any point on the line, even when the telegraph wires were cut and special orders could not be communicated to the working parties."[32] When larger breaks occurred, one or more divisions of the Construction Corps were moved there as quickly as possible.

After his army marched into Atlanta on September 2, 1864, General Sherman eventually ordered Poe "to take charge of the destruction with engineer troops [of] all railroads and property belonging thereto; all storehouses, machine shops, mills, factories, &c, within the lines of the enemy's defenses of Atlanta."[33] Private homes, hospitals, and churches were excluded from or at least not mentioned in that order.

Sherman had taken Atlanta, and on November 8, 1864, Abraham Lincoln recaptured the White House. As Northern church bells rang with the news of Lincoln's victory, in Atlanta, smokestacks toppled and buildings collapsed under the weight of sledgehammers wielded by Union engineers and pioneers. The engineers tore up and twisted rails and sometimes demonstrated a macabre sense of humor by bending the rails to form the letters U and S.[34] The objective was to deny the Confederates use of the railroad, so it was considered essential to damage the tracks beyond repair.

Now, as Sherman's juggernaut started for Savannah and as Lincoln, with four more years, prepared to prosecute the war to a successful conclusion, the Confederate Army of Tennessee and Army of Northern Virginia would each make one more desperate attempt to escape the grasp of the Union army and perhaps link up to form a larger army. Northern engineers had already proved there was no place the Confederates could safely hide. No swamp, river, mountain, unmarked roadway, or wilderness would block the Yankees from getting at their enemy. The end of the war was near at hand.

As 1865 opened, President Lincoln was preparing for his second term and, with congressional Republicans, planned to convince several Democratic congressmen to vote to adopt the Thirteenth Amendment to the US Constitution. Lincoln wanted to make emancipation permanent, and he wanted to end the war. His reelection had signaled that the Northern public was unwilling to negotiate a settlement with the Confederate government to allow the Confederacy to coexist peacefully with the Union. Lincoln's reelection was a mandate from the people—stay the course. Everyone felt victory was at hand, and the nation was euphoric. Sherman's army was on the move through the Carolinas, and Grant's forces continued to tighten their grip on Lee's army at Petersburg.

The war was not over, however, and Lincoln understood well that Northern hopes could turn sour if Union armies in the field were not successful or if Confederate forces under Lee and Joseph Johnston somehow managed to link up and operate against Union supply centers and garrisons in the South. Lincoln needed his generals, especially Sherman, to pursue Johnston tenaciously to prevent the Confederate commander from coming to Lee's rescue. The swamps and tangled forests of central South Carolina and North Carolina, however, were major obstacles to Sherman's movements to trap Johnston.

The operation through South Carolina and North Carolina required a massive road- and bridge-building project. It was an amazing effort and was performed by pioneers and engineers in what became known as the Carolina Campaign during the winter of 1865. Sherman had two important goals: first, to damage or destroy any manufacturing operations, agricultural surplus not consumed by the Union army, and railroad capacity essential to sustain the Confederate war-making effort; second, to occupy Columbia, South Carolina, and Goldsboro, North Carolina, and thus block rail traffic and shipment of supplies from the vital coastal ports of Charleston and Wilmington to Johnston's and Lee's armies. Occupying Goldsboro would give Sherman control of the Wilmington & Weldon Railroad, the critical supply link with the Army of Northern Virginia. Sherman's campaign began on January 30, 1865, from Pocotaligo, South Carolina, midway between Savannah and Charleston, and his forces arrived in Goldsboro on March 24.

The engineers had carefully chosen the line of march. During the march from Atlanta to Savannah, the army had followed a line parallel to the large watercourses. For the Carolina Campaign the same concept was applied. The march line was chosen near the junction between the clay of the uplands and the sand of the lower country, which, according to Colonel Orlando Poe, Sherman's chief engineer, "may be tolerably well defined by tracing [the] line through the lower rapids on each stream we crossed." This way, Poe hoped, the best roads would be used and the minimum amount of mud and swamp would need to be crossed.[35] (Interstate Highway 95 follows the same route today.)

Poe took great pains to organize the engineers and pioneers before the campaign began, and he readied the pontoon trains and inventoried the tool chests and other equipment, including a significant number of axes that would be carried by the infantry brigade wagons. The left wing of the army consisted of the XIV and XX corps, and the 58th Indiana Volunteer Infantry, under the command of Lieutenant Colonel Joseph Moore, served as the engineer troops. They hauled a pontoon train of eighty-five wagons, enough to construct a bridge 1,000 feet long.[36]

The right wing of the army consisted of the XV and XVII corps and was accompanied by the 1st Missouri Volunteer Engineers under the command of Lieutenant Colonel William Tweeddale. The 1st Michigan Engineers and Mechanics, led by Colonel J. B. Yates, were unassigned and used as Colonel Poe saw fit.[37] The role Poe assigned to the Michigan engineers was the destruction of Southern railroads.

The area Sherman's army needed to move through presented unique problems for his engineers and pioneers. For example, locals regarded the Big Salkehatchie River and its adjoining swamplands near the tiny communities of Allendale, South Carolina, to the south, and Barnwell, South Carolina, to the west, as impassable for troops. For the men of Captain George L. Searle's Pioneer Corps attached to Union General O. O. Howard's XVII Corps, the swamps were just places of cold, wet, misery. The pioneers and infantry detachments had bridged the Savannah River and the Coosawhatchie and Whippy swamps, and now, on February 3, 1865, their task was to get the XVII Corps over another river and through a swamp that their commanding officer described as "indescribably ugly."[38]

General Sherman wanted to cut the South Carolina Railroad that linked Charleston to Augusta, and the Big Salkehatchie was in the way of accomplishing this objective. Joe Johnston's Confederates understood the importance of blocking the determined Yankees and, consequently, had formed a battle line on the east bank of the river. The lead elements of Major General Joseph Mower's First Division had been directed to cross on Rivers' Bridge, but the enemy, dug in along a narrow causeway on higher ground, was waiting to greet Mower's men with an array of lead and iron projectiles that the Southern soldiers hoped would prove deadly.

The environment was as much an enemy to the bluecoats as were the Confederate soldiers. The winter weather was brutal. Water froze in the men's tin cups and on their clothing, especially on socks. Ears and hands were always exposed to the elements. Even when the temperature crept above thirty-two degrees, the frequent rains and chronic marshlands kept the men's hands arthritic and their feet wet, unsanitary, and cold. Soldiers would notice their feet blister and develop open sores. If they were unlucky, gangrene would set in. A warm and dry environment was the solution to some of the men's foot problems, but there was none of that around the Big Salkehatchie.

The swamp itself was eerie, dank, and wild. Shooting up from the mucky soil was a mixture of tall bald cypress, water tupelo, and green ash trees. Cypress knees surrounded the base of the bald cypress. These knees were stumps or woody projections that looked like stalagmites. Vegetation in the swamp included sawgrass, cattails, and pickerelweed. Alligators and cottonmouth snakes, normally active only in warmer weather, were in a state of torpor among the tangle of tree roots and shrubs. The cold weather had slowed their metabolism, but they would strike if disturbed.

The swamp extended approximately a mile on the western side of the river and another half mile on the eastern bank. The Confederates had built rifle pits and two redoubts for artillery on elevated ground just beyond the swamp on the eastern side. Since this artillery commanded the bridge and the narrow causeway leading out of the swamp, a Union frontal assault was out of the question. The river leaked a number of tiny streams, and because of the heavy freezing rains and occasional snow, the ground was roofed with one to eight inches of icy-cold water. Mower's aide-de-camp, Lieutenant Charles Christensen, had swum the river on the night of February 2 to determine the best place to build a road and bridge to bypass the Confederate defenses and exit the dreary swamp. He was almost captured by Southern pickets, but, even as he was suffering from hypothermia, he dove back into the river and swam the fifty yards back to Union lines.[39]

On the morning of February 3, under harassment from Confederate artillery, the pioneers began constructing a series of raised roadways through the swamp, both above and below Rivers' Bridge. Men from the 25th Wisconsin and 63rd Ohio cut trees and gathered planks from nearby houses and barns to use for both corduroying roads and bridging the Big Salkehatchie in three places, to move the entire brigade across the river as quickly as possible. As a diversion, Mower ordered the 43rd Ohio over Rivers' Bridge and up the causeway, while two other brigades moved over roadways and bridges built by the pioneers. By late in the day, a lodgment was gained on the eastern bank of the river, and with the Union threat of flanking the Confederates' entrenched positions, the Southern soldiers left their line and disappeared northeast. Mower wrote in his official report that "Captain Searle and his pioneers were for two days and nights in the water constructing roads, and at the conclusion of their labor were well-nigh exhausted."[40]

The following day at Buford's Bridge, just five miles north of Rivers' Bridge, pioneers from the XV Corps built twenty-two bridges scattered over a mile of swampland and, by early evening, had also built a corduroyed road through the swamp. Corduroy roads had been built for centuries and frontiersmen had found them helpful in aiding the movement of wagon trains and settlers over difficult terrain, but this knowledge of road building did not make the job any easier. For fourteen hours men felled trees, hauled lumber, hammered log piers into the ground, and in some cases, dug drainage ditches under the roads. All of this was done while standing in water sometimes as high as three feet. The pioneers had to remain in the frosty swamp while the corps passed so that they could repair shifting logs, which would make the roadway hazardous to wagons and dangerous to horses.

Not all roads were level with the ground and none were on a firm and stable surface. Pioneers had to build roads that could limit sliding logs and withstand heavy traffic, and these roads required considerable time to construct. The men laid longitudinal support timbers on either side of the road and notched the timbers so that the transverse logs could be fitted into the supports. The logs were tied into place, and a second longitudinal support was then laid and tied on either side across the top of the transverse logs. These log mats had to be anchored when the road was under water.[41] It was an amazing effort, and the work done at both Rivers' Bridge and Buford's Bridge was only a part of the extraordinary performance of pioneers and engineers during the campaign.

After the action around the Big Salkehatchie River, the army continued north and the engineers and pioneers continued to labor in the swamps and thickets, exceedingly cold, wet, tired, sick, and sore. Poe, in his official report, made the operations appear routine as he described the army's movement between February 9 and 18: "During the night of the 9th a pontoon bridge was thrown at Binnaker's, and the enemy driven away from the position he had taken to dispute crossing. Another pontoon bridge was thrown at Holman's, and all our force was across by the evening of the 11th . . . The enemy opposed the crossing of the North Fork of the Edisto River, but, as usual, he was driven away and three pontoon bridges built . . . A pontoon bridge was built at the Saluda River bridge [February 16], near a factory, and a portion of the Fifteenth Corps crossed during the night. The Left Wing pontoon bridge was built over the Saluda at Zion Church, nine and one-half miles above Columbia, and some force crossed. On the 17th a pontoon bridge was built just above the ruins of the former bridge over Broad River, three miles above Columbia . . . On the 18th the Left Wing crossed the Broad River on a pontoon bridge thrown at the mouth of Wateree Creek, near Freshly's Mills."[42]

Although Poe made the work of the pioneers and engineers appear routine, there was nothing routine about it. From Colonel Moore's description, the 58th Indiana's crossing of the Catawba River beginning on February 22 was anything but commonplace. His men had constructed a 660-foot pontoon bridge across the river during the night. The following day the skies opened up with a cold, torrential rain, and over the next two days as XX Corps and cavalrymen crossed the bridge, several problems made the march over the Catawba increasingly dangerous. The relentless rain turned the steep hill on the opposite bank

Using lateral thinking, Federal soldiers from the Second Brigade, Second Division, XV Corps, built and lashed together a string of makeshift rafts and crossed the North Edisto River on February 12, 1865, as part of the Union army's efforts to slice its way through South Carolina. William Waud, Library of Congress, Prints & Photographs Division, Civil War Photographs, LC-DIG-ppmsca-17671.

into an almost impassably muddy egress, and as the river rose its current became dangerously rapid. Moore's men placed heavy timbers on the lower ends of the pontoons to prevent them from sinking or filling with water, but at midnight on February 26, a 400-foot section of the span broke loose and washed away.[43]

Moore's men quickly placed a cable across the river, tied to trees on each side, and using the pontoon as a raft, moved over the rapidly running water, holding tightly to the cable. They collected the remaining pontoons, floated them to the western bank of the river, and hauled them out of the water. It was still raining hard. With the pontoons from the original bridge and those that remained on the wagons, the volunteer engineers then moved downriver about 500 yards and prepared to assemble another bridge. This operation was suspended until they received further orders.

The following day the weather improved, so moving once again to a spot on the river where the current was less swift, the 58th constructed a pontoon bridge 680 feet in length, and by February 28 the rest of the right wing could cross. Moore then ordered his exhausted men to take up the bridge and join the march to Haile's Ferry on the Great Pedee River, reaching there five days later.[44]

It was close to midnight when Moore's men arrived on the banks of the Great Pedee River, and as soon as the soldiers had finished unloading the pontoons, balk, and chess, several men rowed to the opposite bank to gather a measurement. The river was 920 feet wide, and the engineers had only 820 feet of pontoons and 460 feet of balk and chess. Colonel Moore was now incapacitated with rheumatism, and the regimental commander, Colonel Buell, was in charge of the operation.[45] Buell ordered the men to begin construction on a trestle bridge for the last 100 feet. The pioneers worked the entire day and night of March 6, cutting trees and trimming them to the proper length, framing a trestle section of the bridge and then placing it in the water. Anchoring the first trestle was not difficult, but driving in the legs of the other trestles became a huge problem because of the river's current. The pioneers had to determine the proper length of the legs by using a pole to measure the depth of the water, then a pontoon was brought alongside the first trestle, two balks were laid from the bridge, and the next trestle was laid on the balks. Pioneers moved the pontoon to the next spot for placing the trestle and it was righted into the water.[46] The legs were driven into the river floor—and the process was repeated.

The current was fastest in the middle of the river where the engineers hoped to use the pontoons. It took all morning of March 7 to get the upstream anchors to hold. Finally, at 3 p.m. the bridge was completed and the men of the XIV and XX corps started to cross. On March 8, a detachment of the 58th Indiana took up the bridge and then marched twenty miles toward Fayetteville. From February 23, when the 400-foot span of bridge had broken loose in the Catawaba River, to their arrival in Fayetteville, the 58th Indiana engineers had been constantly at work or on the march.

Hundreds of privates from infantry regiments had now joined the pioneers and engineers in corduroying roads and building bridges. A private noted in his diary: "Our whole division was put to work with engineers . . . It was not a pleasant job in our wet clothes with water up to our knees but we had the work done."[47] Sixteen days, four pontoon bridges, and endless feet of corduroyed roads later, all of Sherman's army walked into Goldsboro, North Carolina.

The army was exhausted and needed all kinds of supplies. In addition to essential items such as clean, dry clothes, new shoes, and good food, the engineers required new canvas pontoon covers. The Cumberland pontoons had held up well, but the covers, which had been in water almost every day for sixty days, were torn, mildewed, and rotting.[48]

Colonel Poe estimated that during the Carolina Campaign, engineers, pioneers, and infantry built 400 miles of corduroyed roads, many in the absence of fence rails, which, when found, made road construction less tedious. For the right wing of the army, the 1st Missouri Engineers built fifteen pontoon bridges, estimated to have an aggregate length of 3,720 feet. The 58th Indiana constructed about 4,000 feet of bridging. The XVII Corps pioneers kept their own records of the immense work they performed, which augmented Poe's report and pointed to the mechanical ability many Union soldiers possessed. General Mower's First Division built 39,405 feet of corduroyed roads, 399 feet of bridges, and six artillery batteries. Major General Mortimer Dormer Leggett's Third Division constructed 74,259 feet of roadway and 909 feet of bridges, while Major General Giles Alexander Smith's Fourth Division built 98,925 feet of roads and 1,317 feet of bridges.[49]

The Carolina Campaign broke the last major supply chain of the Confederacy, and the speed with which Sherman's army moved through the mud and swamps of central South Carolina and eastern North Carolina prevented General Joseph Johnston from coming to the assistance of Robert E. Lee. Within one month, on April 26, 1865, Johnston would surrender his army to Sherman near Durham, North Carolina. Johnston would remark: "When I learned that Sherman's army was marching through the Salk swamps, making its own corduroy roads at the rate of a dozen miles a day and more, I made up my mind that there had been no such army in existence since the days of Julius Caesar."[50]

The campaign was a remarkable achievement. Poe's staff, which included Amos Stickney, William Ludlow, and Chauncey Reese, were West Point–trained engineers, and they deserved credit for the proper execution of orders. The building of corduroy roads in the swamps, however, required improvisation, skill with an axe, saw, and hammer, and the ability to work out problems in brutal conditions. The 1st Missouri, the 1st Michigan Engineers and Mechanics, the 58th Indiana Infantry, and about 6.000 men who served as pioneers in the army did this work. These men were not trained at West Point but had learned their skills before the war in the mechanics' shops, mills, railroad yards, boatyards, farms, sawmills, and timbering operations in the North. By 1865, Union engineers could take the army anywhere their generals wanted to go over the vast and varied terrain of the Confederacy.

In the eastern theater, during the winter and early spring of 1865, the engineers were also hard at work against Robert E. Lee's Army of Northern Virginia. Their hope was to bring the American Civil War in Virginia to a close.

Grant had every intention of stretching Lee's lines as thin as possible around Richmond and Petersburg. This tactic not only included an extensive flanking movement southwest of Petersburg but also required maintaining pressure on the northeastern defenses near Dutch Gap Canal in the area of the Bermuda Hundred. Union supply distribution in this northeastern sector had used a pontoon bridge built in September, but by December, freshets, floating ice, and driftwood in the river made passage over the pontoon bridge dangerous.[51]

To continue to bring quartermaster stores and ordnance into this area, planning began for a permanent pile bridge. Grant's strategy was to extend Lee's lines southwest of Petersburg, but to do so effectively, he needed to prevent Lee from acquiring reinforcements by shifting Confederate soldiers from the northeastern to the southwestern sector. The northeastern sector had to remain active, and this required significant amounts of supplies. The pile bridge was essential, and troops from the 1st New York Volunteer Engineers began milling timber cut from the surrounding forests. On January 5, Lieutenant William R. King, Corps of Engineers, and pioneers under the command of Captain James W. Lyon, 4th Rhode Island Infantry, began work on the bridge.[52]

The weather was erratic, and this caused delays. Private John H. Westervelt recorded in his diary entry for January 10: "This is decidedly the wettest day I ever saw. All that I can see of Virginia is afloat." January 16 he described as beautiful and warm, and then the next day it snowed. Five days later he wrote: "This morning it rains and the earth here is covered with a complete glaze of ice. The day was one continual deluge but the earth was so completely coated that the water ran off as fast as it fell."[53]

These were difficult conditions in which to build a 1,368-foot-long bridge with piles averaging 40 feet high. Furthermore, the main channel was 25 feet deep, so the piles there had to be 150 feet. Each pier consisted of three piles driven into the riverbed and connected by a cap piece, and the piers were joined together to form bays 15 feet wide. To form icebreakers, an inclined brace was attached to the piles at one end, and the other end was chained to a new pile ready to be driven into the riverbed. The new pile was sawed nearly through just above the chain so that once it was driven into the bed, the pile could be broken off beneath the surface.[54]

On the left flank of Grant's army, the engineers were called upon to build bridges and roads intended to aid in the westward extension of the Federal line, to force General Lee to stretch his already thinly held defensive positions even further. Major M. Van Brocklin had taken four companies of the 50th New York engineers along the Vaughan Road to Hatcher's Run, just ten miles east of Five Forks, to build a pontoon bridge and log bridge for the passage of army supply trains. The engineers then began to construct a permanent corduroy road from Hatcher's Run to Fort Siebert, with the assistance of 2,000 men from the

II Corps and 2,000 men from the V Corps. By the middle of February, the work was completed and Van Brocklin was making a reconnaissance for an extension of the US Military Railroad to Hatcher's Run.[55]

Raw, damp, miserable weather throughout February and March continued to beleaguer the army's flanking movement west, and one result of heavy rains was that Van Brocklin's log bridge over Hatcher's Run was washed away. So, on March 15 the engineers completed a new bridge on Hatcher's Run, 285 feet long and supported by eleven cribs each 16 feet long, 6 feet wide, and 2 to 6 feet high. The new roadway of the bridge consisted of two tracks, each 8 feet wide, separated by a median strip nailed to the flooring. The rain had also damaged parts of the Vaughan Road, so time was spent making the necessary repairs there.[56]

Ten days later, 100 miles to the north, near the point where the Confederate left flank rested on the Appomattox River, Confederate General John Brown Gordon launched a surprise attack on the Union line at Fort Stedman, Petersburg. After the initial breakthrough, a Federal counterattack drove the Southerners back behind their own lines and signaled to an observant General Grant that this was Lee's last effort to take back the initiative before evacuating Petersburg.

In response, Grant wanted to move as quickly as possible to collapse the Confederate right flank and roll up Lee's army before it could escape and begin the race to link up with General Joseph Johnston's men, at that time in North Carolina. Accordingly, the Union engineers' work along the Vaughan Road and at Hatcher's Run allowed General Warren's V Corps to advance north, occupying ground along an east-west axis on Confederate General A. P. Hill's front. Major General Edward O. C. Ord's Army of the James then filled in between Warren on his left and the entrenched VI Corps of the Army of the Potomac on his right.[57]

Between March 30 and April 3, the engineers of the 1st New York (attached to Ord's army), the 50th New York, and the Engineer Battalion skillfully and swiftly repaired roads and built bridges that enabled Union infantry and cavalry to deliver a coordinated strike against Southern forces, bringing on the final chase to Appomattox Court House and Lee's surrender. Colonel Spaulding arrived at Hatcher's Run late in the afternoon of March 30 and found that persistent rain had raised the water in the stream to overflowing, making the approach to the log bridge unserviceable. Working until midnight and resuming at 4:00 a.m., some companies labored to repair washed-out roads, while other companies raised the abutments of the pontoon bridge about four feet and built a corduroy bridge, 100 yards long, over the flooded road to the small rise that ran to the south end of the bridge crossing the stream.[58]

With incessant rain turning the soil into sludge, the 50th New York continued carving out a path for soldiers and the troop trains. "During the 1st and 2nd of April," Spaulding wrote, "my whole command was . . . engaged in building a double corduroy track on the Vaughan road from the old stage road to Hatcher's Run. During the whole of this time Major Van Brocklin had a pontoon bridge [different from the one next to the log bridge] over Hatcher's Run near W. Perkins house, and also one over Gravelly Run, near the Friends' Meeting House. He was ordered to keep these bridges in use until the whole of the trains on the route of the old stage road had passed."[59]

The Engineer Battalion maintained the roads beyond Hatcher's Run to ensure that General Sheridan's sweeping flank movement toward Five Forks was successful. It was. On April 1, Sheridan and Warren combined forces to overwhelm George Pickett (of Gettysburg fame), and immediately orders were issued along the entire western sector of the Union army to move forward. Spaulding moved his command and pontoon trains along the Boydton Plank Road, River Road, and Cox's Road toward Burkeville. They repaired old roads and cut new roads for the movement of the V Corps.[60]

Meanwhile, the 1st New York engineers marched ahead of the Army of the James parallel to the South Side Railroad, also repairing roads and bridges. Other engineer troops were kept just as busy. In the city of Petersburg, General Benham ordered the 15th New York Volunteer Engineers to repair and reopen three bridges damaged by retreating Confederates. Under the command of Colonel Brainerd, the men entirely rebuilt one bridge, repaired the railroad bridge, and threw a pontoon bridge across the Appomattox, eliciting high praise from Benham. In the city of Richmond, a detachment of the 1st New York constructed a 2,400-foot pontoon bridge across the James River to connect the city with Manchester.[61]

During the momentous days of April 4 to April 9, 1865, it was General Ord's 1st New York engineers, the vanguard of his Army of the James, who first arrived at the Appomattox River at Farmville, Virginia, on April 7. The engineers had left the vicinity of Five Forks on April 3 and hauled their pontoon train for sixty miles in four days, at one point passing through the village of Blacks & Whites.[62] Just the day before, the Army of the Potomac's II, V, and VI corps, pressing toward Sailor's Creek—which meandered about two miles to the east of High Bridge and four miles east of Farmville—managed to cut off one-quarter of

Lee's army from its escape route over High Bridge. Sheridan's cavalry blocked the Confederates' access to the bridge, and Union infantry and artillery working in conjunction with cavalry delivered a devastating defeat to an already depleted Army of Northern Virginia. Almost 8,000 Southerners were captured, including nine generals.[63]

Now the 1st New York was ordered to throw a bridge over the Appomattox River at Farmville. Colonel Peter S. Michie, chief engineer of the Army of the James, reported that "the pontoon train of our army having been well kept up to the front, notwithstanding its overloaded condition, was fortunately able to pass over the artillery and train of the Sixth and Second Army Corps and enable them to follow in rapid pursuit of the enemy that night."[64] General Sheridan telegraphed Grant to say that with the VI and VII corps on Lee's heels, if Sheridan and the lead elements of the Army of the James could press ahead to the vicinity of Appomattox Court House, Union troops might surround the Army of Northern Virginia and force Lee to surrender. President Lincoln had seen Sheridan's correspondence with Grant. Now Lincoln telegraphed Grant on April 7 at 11:00 a.m.: "General Sheridan says, 'If the thing is pressed I think that Lee will surrender.' Let the *thing* be pressed."[65]

The final months of fighting for the Army of Northern Virginia, in March and April of 1865, were indicative of the difficulties Southern armies had in engineering operations. On April 2 and 3, the Army of Northern Virginia evacuated from Petersburg and started marching west toward Danville and the railroad there, hoping to find supplies for Lee's starving men and horses. Colonel Talcott, in advance of Lee's escape from Petersburg, had ordered Captain G. W. Robertson of the 1st Engineer Regiment to examine the crossing of the Staunton River above the railroad, at Moseley's Ferry to Russell's Ferry (eight miles northwest of Clarksville), and at South Boston on the Dan River.[66] These places would need to be bridged for the army to get to their waiting supplies.

Robertson managed to throw a pontoon bridge across the Staunton River near the railroad bridge on March 30. The following day, however, freshets from the recent downpours exploded down the river and tore away six bateaux, one-third of the chesses, half of the balks, and a quarter of the anchors. Robertson acted quickly. He ordered his men to salvage what they could of the bridge, and the captain sent off a message to Major Grandy, quartermaster in charge of river transportation, asking for additional pontoons and lumber.[67]

The response to Robertson's plea came six days later when Grandy's assistant informed the captain that the major was not available and, furthermore, he had no orders to allow the engineer troops to use the pontoons. This response was not the only questionable occurrence involving the work of the engineers. Because Lee found his southwestern line of march blocked by Union cavalry and the high water around Bevil's Bridge, he immediately redirected his columns to march north to Goode's Bridge where his entire Petersburg force would cross the Appomattox River. Lee inquired into the whereabouts of a pontoon bridge he had ordered to be shipped to Genito, fifteen miles north of Goode's Bridge. It turned out that some bureaucrat had appropriated the pontoons to float himself and his personal property up the James River. Therefore, with one bridge inaccessible due to high water and one bridge purloined, Lee ordered the engineers to plank over the railroad bridge at Mattoax. Once across, both the railroad bridge and, for some unexplained reason, the pontoon bridge were destroyed.[68]

Now the Army of Northern Virginia moved west toward Farmville with the hope of crossing the Appomattox over High Bridge. With a head start and perhaps some luck, Lee hoped to beat the Union army to the South Side Railroad somewhere west of Appomattox Court House, gather supplies, move on to Lynchburg, and then move south to link up with Johnston's army. This was the only possible route left. Taking the roads north, away from Grant's line of march, was an option, but eventually, by moving north, Lee's army would need to cross the James River. He had no pontoons with him to make the crossing.[69]

When the army reached High Bridge on April 6, it immediately began to cross the 2,400-foot span. The river at the bridge was only 100 feet wide, but a wide and steep valley on either side of the river required the structure to be built to such a length. The height of the bridge increased from 60 feet at the abutment to 125 feet near the river. Twenty-one brick piers on stone bases supported the railroad bridge. Just below the railroad tracks was a wagon bridge, which made it easy for horses and artillery pieces to cross.[70]

On the following morning, April 7, after Lee's army had crossed, Confederate soldiers prepared to burn the bridge. Then, from almost nowhere, the 19th Maine Infantry appeared and rushed the wagon bridge. As some Mainers used their bayonets to toss the lighted hay bales into the water, others crossed and established a skirmish line with both flanks anchored on the river. The upper bridge caught fire more readily, and Confederate engineers from Company G fought furiously to keep the Yankees off the bridge. Some portion of High Bridge was damaged, but the lower bridge was unharmed. Consequently, the slight distance Lee believed he had placed between his army and that of the pursuing Federals was gone because his men had failed to destroy both bridges. That afternoon Grant wrote a brief note to Lee: "The results of the last week must convince you of the hopelessness of further resistance on the part of the Army of Northern Virginia in this struggle. I feel that it is so, and regard it as my duty to shift from myself the responsibility of any further effusion of blood by asking of you the surrender of that portion of the Confederate State's army known as the Army of Northern Virginia."[71]

Why had the bridge not been burned sooner? The reason was a lack of flexibility and initiative. Colonel Talcott and his engineer troops were prepared to fire the bridge as soon as they received the order from their infantry commander, General William Mahone. The general, a railroad engineer before the war, most likely rode off with his rearguard without giving the order or assumed the engineers would proceed without his express permission. Either way, it was a costly miscommunication and demonstrated little initiative on Talcott's part. When Lee learned what had happened he lost his temper. As a staff member observed, "He spoke of the blunder with a warmth and impatience which served to show how great a repression he ordinarily exercised over his feelings."[72]

Lee surrendered his Army of Northern Virginia to General Grant on April 9, 1865. It was only fitting that in one of the final acts of the war, an engineer played a central role. After Grant had written out the surrender document in pencil and Lee had read and approved it, Grant turned the paper over to Assistant Adjutant General Theodore Bowers, requesting that he make a fair copy in ink. Bowers, feeling the solemnity of the proceedings and overcome by the weight of the historical moment, made several flawed attempts to copy the surrender document. Frustrated and nervous, Bowers turned the responsibility over to another member of Grant's staff, General Ely Samuel Parker.[73]

Parker was a Seneca Indian chief. After graduating from Rensselaer Polytechnic Institute as a civil engineer, he had worked on the Erie Canal and engineering projects in Illinois, where he met Grant. When the war broke out he asked to join the army as a Union engineer, but Secretary of War Simon Cameron told him he could not do so because he was Indian.[74] Parker contacted Grant, who needed engineers, and he assigned Parker to Brigadier General John Eugene Smith. Parker served as an engineer throughout the Vicksburg Campaign before joining Grant's staff.

Now the engineer made a handsome clean copy of the surrender document. Both Grant and Lee signed. The war in the east was over.

Know-How Triumphant

What is the trouble here? My people have been waiting for hours to cross
the stream. The bridge should have been finished long ago.

General Robert E. Lee, after ordering a bridge
to be built over the Rappahannock

Honor and shame from no Condition rise;
Act well your part, there all the honor lies.

Alexander Pope, An Essay on Man

The day after Confederate General Robert E. Lee surrendered to Union General Ulysses S. Grant, Lee wrote his Farewell Address, opening with: "After four years of arduous service, marked by unsurpassed courage and fortitude, the Army of Northern Virginia has been compelled to yield to overwhelming numbers and resources."[1] Beginning in the winter of 1862–63, when Lee's army went hungry and replacing soldiers lost to capture, wounds, or death became increasingly difficult, the Southern army fought a losing battle with declining resources, which gnawed away at its ability to sustain the fight. The lack of food, clothing, footwear, medical supplies, and manpower made conditions brutal. In the final days of the war, Lee believed and wanted everyone else to believe that overwhelming resources had crushed the Confederate army.

On paper, the North did have overwhelming resources. At the start of the war, 22 million Americans lived in the North and only 9 million lived in the South, including 3.5 to 4 million African American slaves, who would never support the Confederacy. The North was the manufacturing center of the country and, as such, had a monopoly on industry, iron, textiles, machine shops, railroad yards, and shipyards. The Northern railroad system was three times the size of the South's, which directly contributed to the flow of supplies from farmers in Minnesota and Wisconsin and factories in Pennsylvania and Connecticut to army depots in Philadelphia, Boston, New York, Louisville, and Cincinnati. These resource advantages were especially dominant when it came to manpower on the battlefield. In reviewing the ten costliest battles of the Civil War, in all but one—Chickamauga—the North outmanned the South.[2] During the Peninsula, Vicksburg, Atlanta, and Carolina campaigns, the South was again outnumbered. As historian Richard N. Current posited in the 1960s, the most common explanation of why the South lost the Civil War was the North's overwhelming resources.[3]

Yet, the overwhelming resources argument alone cannot account for Northern victory. First, delivering these abundant supplies of men and material to forward areas of advancing Union armies presented a unique challenge. It was one made particularly problematic by the South's great advantage in fighting on home soil, in number of civilians acting as scouts and spies, and in the operation of guerilla bands, which wreaked havoc on Union supply lines. Two further Confederate advantages neutralized the North's considerable resources: tactical warfare in the mid-nineteenth century and terrain.

Soldiers carried muzzle-loading rifles into combat, with an effective range of 150 to 200 yards. Generally, a soldier could fire three shots in a minute, discounting the panic and terror that would come from being shot at by an enemy intent on killing you. Therefore, as regiments attacked and attempted to close in on the opponent, the attackers' firepower was diminished as they ran forward toward the enemy's lines. For those on the defensive, the advantage was significant. Standing or kneeling behind barricades provided the soldier with a greater chance of firing and reloading quickly, getting off several shots before the attackers were close enough to use their bayonets. Artillery canister shot and grapeshot also played havoc with the attackers. As a result, to attack successfully required overwhelming numbers of men at the point of attack—at least a three-to-one ratio to drive the enemy from a well-defended position. In addition, defenders had the advantage of hastily moving men from a place in the defensive line to the point under attack. Operating on internal lines of communication, this gave the defender another distinct advantage over the attacker. Southern forces had fewer soldiers than their Northern enemy, but they had the tactical and strategic benefit of defending ground of their own choosing and moving men and material on more direct and shorter internal lines.

In addition to the South's strategic and tactical advantage, the terrain that had to be overcome and tamed by Northern troops as they extended into Southern territory was vast. The eleven seceding states formed an area of approximately 750,000 square miles. There was no central road system, and Southern railroads, built hastily and engineered as cheaply as possible, offered no help to the invading armies.[4] The Appalachian Mountains, the lowlands and swamps of the Carolinas, the Mississippi floodplain, the Cumberland, Tennessee, and Mississippi rivers, and the myriad lakes, streams, small rivers, and dense forests all offered fierce natural obstacles to the North's strategic planning.

With strategic, tactical, and terrain advantages in mind, many Southerners after the war rejected the overwhelming resources theory (yet accepted the noble experiment theory) in an attempt to explain why the South lost. General P. G. T. Beauregard claimed that "no people ever warred for independence with more relative advantages than the Confederates; and if, as a military question they must have failed, then no country must aim at freedom by means of war . . . The South would be open to discredit as a people if its failure could not be explained otherwise than by mere material conquest." Beauregard, no admirer of Jefferson Davis, blamed flawed strategy and the president's lack of leadership for Confederate defeat.[5]

Flawed strategy arguments remain difficult to prove. Beauregard, Joseph Johnston, and Braxton Bragg represented a western bloc of generals who, throughout the war, had insisted on focusing manpower resources in the western theater, much to the disapproval of Lee in the eastern theater. Davis was in the middle. He did not want to alienate either side and, for the most part, supported General Lee's suggestions. Davis's favoritism toward the eastern theater certainly had an impact on military operations in the west, but it did not result in the Confederacy's military failure.

Some Southerners blamed President Jefferson Davis for defeat. The argument went that the Confederacy collapsed because Davis was unable to see beyond the South's states' rights dogma, and he failed to manage the Confederate Congress and state governors in a more courageous fashion. For example, as historian David M. Potter pointed out, the Confederate Congress adopted an economic policy, which Davis supported, that lacked the central controls necessary to sustain the war effort financially. The embargo on cotton, the failure to pass a national tax to curb inflation, and the refusal to impress goods and services placed the South on financial quicksand. These policies drained Southern bank accounts and brought suffering and starvation to millions of Confederate civilians. Defeats on the battlefields and hardships at home eventually weakened the South's will to fight.[6]

Yet, even though Davis may have been trapped by states' rights dogma, limited by inadequate political or diplomatic skills, and hobbled by bickering generals, flawed economic policies, and declining morale, and even though Southern military and civilian leadership was unable to create a grand military strategy, the South still had an opportunity to win its independence. Certainly, in 1863, after Gettysburg and Vicksburg, it seemed unlikely that the Confederates would enjoy their own Cannae, Agincourt, or Rorke's Drift, but there was still a possibility that the South could force the Union to the bargaining table and, with a war-weary Northern citizenry and pressure from the Democratic Party, gain its independence. Therefore, the question to ask is: Up to what point in the war did Southern independence remain possible? Why, after this point, was Southern independence impossible? How should historians assess the years of fighting leading up to this critical breaking point?

In November 1864, when Abraham Lincoln was elected president for a second term—defeating the Democratic candidate George McClellan, who ran on a platform that left open the possibility of a negotiated peace—the hopes of the Confederacy vanished. Lincoln had promised to prosecute the war until the Confederacy collapsed and the Union was restored. Now Confederate morale fell precipitously, desertions in the army soared, and the economy, with three hundred percent inflation, was on the verge of collapse.

The Confederacy's hopes had rested on the outcome of the North's November election, and for some time it had looked as if McClellan and the Peace Democrats would win. The war had dragged on for four long, brutal, heartbreaking years, and although Union armies had achieved some success, the Northern public and press perceived the Wilderness, Cold Harbor, and Spotsylvania as accomplishing nothing but more dying and suffering. Was it all worth the cost? In early August, Grant was stalled outside Petersburg; Sherman was struggling around Atlanta; and Wilmington, North Carolina, and Mobile, Alabama, remained in Confederate hands. If Southern armies could prolong the war by maintaining control of Atlanta, Petersburg, and Richmond until at least the election, disillusioned Northern voters might put a different man in the White House.

Of course, they did not. Southern forces ran out of real estate between Chattanooga and Atlanta, and they had to evacuate Atlanta on September 1 to escape being completely surrounded. They were unable to delay Sherman's advance any longer. Confederate partisans, cavalry, and guerrillas had tried, with little success, to block Union soldiers from advancing farther south through Georgia in the summer of 1864. Union engineers were also increasingly skillful at circumventing those roadblocks when they did occur.

This is how things were during the four years of war. On numerous occasions, Southern forces had tried to cut off or delay Union supply trains attempting to reach combat areas, as Confederates damaged railroads, blocked roads with felled trees, and destroyed bridges. The natural geographical barriers such as cypress swamps, flooded rivers, and deep ravines also piled onto the obstacles the Union army had to overcome. If delays of just weeks or a month had stopped Union armies from controlling the Cumberland and Tennessee rivers or prevented them from getting to Vicksburg when they did, perhaps Sherman would not have arrived in Atlanta until November (if at all), too late to help swing the election in Lincoln's favor. So in looking back at the crucial campaigns and battles in 1862 at forts Henry and Donelson and Island No. 10, or in 1863 at Vicksburg and Chattanooga, if these events had turned out differently, or more importantly, if they had delayed the advance of the Union army longer than they did, could the Confederacy's chances for independence have been greater? It is difficult to speculate, but the evidence strongly suggests that the real difference maker in the war was the North's ability to sustain military operations deep inside the Confederacy and the South's problem in fighting a war of logistics.

The North was able to sustain military operations in the South over vulnerable and lengthy supply lines, subject to frequent sabotage and disruptions, not only because they had a greater number of soldiers and cavalry to defend the supply lines, but also because they had a large pool of trained civil engineers, mechanics, toolmakers, and managers to draw upon for building new roads and bridges and repairing damaged ones. Conversely, the South's pool of men with mechanical ability and managerial skills was too small, primarily because the South was blinded by the constraining factor of slavery, which contributed to a culture that denied ordinary Southern white men the benefits of common school education, prevented the dissemination of ideas out of fear, blocked attempts to expand industrial development, and promoted a top-down "planters know best" mentality.

Whereas Yankee logistical adroitness improved as the war went on, Southerners stumbled and opportunities were lost. Had the South been better able to manage a war of logistics, its generals would have achieved greater flexibility in both strategic and tactical planning. If Lee's army was better fed and supplied in the winter of 1862–63, would a Gettysburg campaign have been necessary? If better railroad management and engineering had been present in the winter of 1862, would Sidney Johnston have been able to hold on to upper Tennessee and the Cumberland and Tennessee rivers for more than several months, preventing Yankees from taking over Nashville so early in the war?

Confederate engineer Channing M. Bolton studied engineering at the University of Virginia for one year before joining the State of Virginia's Engineers in 1861, then served with the Army of Northern Virginia. He later recalled a time when he had traveled with General Lee. The engineers had been ordered to build a bridge over the Rappahannock River, but the work had been delayed far too long to suit the general. After learning who was in charge, Lee spoke to the lieutenant sternly about the lack of progress on the bridge The lieutenant responded that he did not have the necessary guy ropes to complete the construction. Dismounting, Lee walked toward the bridge and noticed the end of a rope under the seat in one of the boats. "He [Lee] called to a soldier to pull it out," Bolton recalled, "and this proved to be a coil of just such a rope as the lieutenant needed. The general told the soldier to drop it in front of the lieutenant, and without saying a word more, he mounted his horse and rode off."[7] The limited amount of competent Southern engineering, a lack of Yankee "can do" attitude among Southerners, and poor logistical management damaged the Confederate war effort.

Southern cultural attitudes and institutions contrived to limit the pool of skilled men who, before the war, were trained as professional engineers and managers. Foremost among these social realities was the constraining factor of slavery, which exerted its effect on many levels. Besides the horrific toll of slavery on the men, women, and children enslaved, it also was responsible for choking off intellectual growth among Southern citizens. The upper-class South carefully protected its status in society, and opportunities for financial and intellectual growth were consciously limited. Information and ideas were contained. When talk of bringing manufacturing to western Virginia reached the halls of the state's legislature, the controlling planter elite blocked efforts to pass laws that might assist in supporting industry because manufacturing was perceived as a threat to the establishment's power. When the few manufacturing businesses in the South did require training to adjust to changing technological developments, owners were careful to select white men who would not spread ideas about liberty and freedom to the predominately black labor force. Furthermore, ideas in the South flowed only one way: top down. Labor, white or black, did not dare suggest ideas to management.

This relationship between management and labor mirrored the relationship between owner and slave. The parochial worldview established in the years before the war was directly linked to how the Confederacy managed, not the fighting—for commanding men came naturally to Southerners—but the most critical business of *this* war: conducting massive logistical operations. In this critical ability, the South's vulnerability was created by the flaws in educational and transportation systems established during the antebellum period.

Before the war, the South paid considerable attention to higher education and the development and growth of private academies to instruct the sons of plantation owners and the small middle class. Following in the tradition of Thomas Jefferson, whose concept of the university was of an institution maintained by the state and free to young men who were intellectually qualified, the South pioneered the state university and before 1830 had established five of the six institutions of this type in the United States.[8]

The academies were represented by a variety of institutions, some private, some aided by the state, and many under religious denominational control. Yet, they all had one thing in common: these schools were designed to prepare the sons of the wealthy for college and, consequently, focused almost exclusively on providing a classical education, which included the study of Greek and Latin. Discipline was Spartan-like, so it was not uncommon for the famous headmasters of these schools, such as Moses Waddell and Robert L. Armstrong, to "ruin many a heavy pair of winter pantaloons at a single whipping."[9]

For the majority of Southern citizens, however, there was little opportunity to master the basics of learning. The common school movement in the South, with the exception of North Carolina, was virtually nonexistent before the war. The teachers in the few schools that did exist were often incompetent. State legislatures did not supervise or care about these schools, there were no district libraries, and the schoolhouses were in many cases without furniture and books. Also, many farmers wanted their sons to handle an axe and plow and believed formal learning was required only of the upper classes.

This lack of attention to the development of common schools was a choice. Comfortable in maintaining the social hierarchy, the men who held the power to change the system had no interest in doing so. This did not mean, of course, that illiterate Southerners did not possess remarkable good sense and intelligence. It did mean, however, that during the Civil War when men were needed to repair railroads or build bridges, those with more than good sense were hard to find. Bridge building and locomotive repair required mechanical knowledge and basic mathematics and science. Therefore, it should not be surprising that when the Confederate 1st Regiment Engineers was formed in 1863, only half the companies could be designated as pontoon bridge builders.

The Confederacy's inability to systematically manage the logistics of large-scale warfare also had its roots in the antebellum period. Management of railroads was a critical failure. Southern railroads, with the exception of half a dozen lines, were designed to transport the wealth of crops grown on the plantations to market. Many of these lines did not connect with others, creating serious gaps in the regional system. For example, the West Feliciana, the Clinton & Port Hudson, and the Baton Rouge, Grosse Tete & Opelousas railroads all ran from the countryside to the Mississippi River, without connecting spurs. Moreover, the managers of these short lines closely guarded their turf, running them as cheaply as possible because the low volume of passenger traffic generated little revenue. Since they were small lines, management could be highly centralized, supervising every aspect of the business. Relying on others who might have to manage a station on a longer line, say, 150 miles away, was not something Southern railroad management, for the most part, had to consider.

Repair work on the tracks, bridges, water stations, and locomotives was done locally, which impaired the manufacturing of railroad supplies and locomotives. There were only a small number of locomotive manufacturers and rolling mills in the South, and this kept demand low. There were also not enough skilled mechanics to support a booming manufacturing sector of the economy. Limited demand, few skilled workers, a transportation system with significant gaps, and a dominant planter class all worked in conjunction to create a business culture that was not prepared for the logistical challenges of the Civil War. In the final days of the Army of Northern Virginia, the Richmond & Danville Railroad had managed to rescue all of its engines and cars from the burning city. The president of the company, Lewis E. Harvie, wanted Jefferson Davis to order the small labor force working to build fortifications at Danville to work instead at changing the width of the Piedmont Railroad so it matched the gauge of Harvie's trains. The disagreement between Davis and Harvie was, by then, moot. It does reveal, however, that right until the end, the Confederacy's management issues haunted them.[10]

During the first two years of the war there was enough Confederate firepower to block Union advances into Tennessee, Mississippi, and Alabama. But poor planning and a limited number of skilled engineers and mechanics resulted in careless engineering efforts at forts Henry and Donelson, confusion and conflict between private railroad companies and the Railroad Bureau, the loss of Pemberton's army at Vicksburg, Bragg's failure to capture Grant's army at Chattanooga, and the Confederate army's unwillingness to form engineering regiments until the spring of 1863. Confederate difficulties during the first half of the war were due not to limited manpower but to the particular skills (or lack thereof) that men in both armies possessed. Engineering constituted the "tipping point," in author Malcolm Gladwell's phrase, of the war.[11] Both sides had skilled warriors and generals; only one side had an overwhelming number of skilled mechanics, craftsmen, and engineers.

A substantial number of skilled craftsmen in the South were African American slaves. These men were forced to construct forts and earthworks throughout the Confederacy during the war, but many slave owners were obstinate in their refusal to lend their slaves to the military, and the states' rights doctrine upheld their obstinacy. In December 1864, the War Department's Bureau of Conscription announced a plan, approved by General Lee, to impress African American slaves into Confederate labor gangs based on a state quota system. Governor A. G. Magrath of South Carolina asked his state agent, R. B. Johnson, to investigate how his citizens would respond to the new War Department directive. On January 10, 1865, Johnson reported to the governor: "I am fully assured, from my knowledge of the difficulties and embarrassments which attend the levying of slave labor, that the conscript authorities cannot successfully proceed with such impressment . . . The aid and the authority of the master is indispensable, and as the Confederate authorities possess no control over him, can impose upon him no pain or penalties, they must in the present condition of affairs be powerless to act effectually. I am satisfied that slaves can be impressed only through the agency of the State authorities in conformity with State law."[12]

With no money and a limited labor force, Southern engineer officers and their men would perform as well as possible, considering the circumstances. An example of this is provided by the 3rd Regiment Engineers, which accompanied the Army of Tennessee from the late spring of 1863 to the end of the war. Under the command of Lieutenant Colonel Stephen W. Presstman and Major John W. Green, the regiment spent the Atlanta Campaign building field fortifications and throwing pontoon bridges. On some occasions, as in the lines at "East Point" (Atlanta), Presstman's men served as infantry.[13]

The 3rd Regiment officially listed nine companies, but only eight were raised, and of those only five were effective bridge builders.[14] In December 1864, when General Hardee sought to leave Savannah one step ahead of Sherman's approaching army, engineers from companies B and F used three bridges to aid in the Southern army's escape. The first connected the wharves of Savannah with Hutchinson's Island and was 1,000 feet long. A second bridge was thrown over the Middle River, and a third spanned Back River to the rice dams of South Carolina. The bridges were made up of barges because of the shortage of pontoons. Once the army had crossed, the bridges were destroyed.[15]

Unfortunately, the successful extraction of Hardee's army from Savannah was not enough to prevent the eventual surrender of Joseph Johnston's army. In mid-January 1865, Company E, Hart's Engineers, and Company A, under the command of Captain Robert C. McCalla, were ordered to report to General Jubal Early to form the Engineer Battalion of the Department of Western Virginia and East Tennessee. They were assigned the impossible task of repairing thirty-three destroyed railroad bridges on the Virginia & Tennessee. It was a forlorn hope.[16]

It took days to get a pontoon bridge in place, and once men had crossed, engineers often failed to take up the bridge or destroy it. On February 17, 1865, a pontoon bridge was ordered across the Savannah River on the road between Abbeville, South Carolina, and Washington, Georgia. A Major McCrady was charged with the duty. Two days later the bridge was still not in place, and Captain A. H. Buchanan, Army of Tennessee engineers, reported that he would assume responsibility for its construction.[17]

The three other Confederate engineer regiments also experienced difficulties in the final months of the war. The 4th Regiment Engineers had formed in the summer of 1864 at Shreveport, and most of the men who signed on with the unit were from Louisiana, Arkansas, and Texas. They worked on the Mobile defenses, but no more than three companies were formed and the regiment was broken up in early 1865.

In the eastern theater, General Lee's pioneers and the 1st and 2nd Engineers did all in their power to maintain the defenses around Petersburg. According to engineer Lieutenant Henry Herbert Harris, the engineers spent their time countermining, working on fascines, and repairing bridges. He noted in his diary: "September 24, Saturday . . . Again . . . worked on the bridge . . . a showery day and I got quite wet . . . In the afternoon prepared a sermon for tomorrow's exercises." Four days later: "Got our bridge passable for footmen." Then finally, on September 30: "We finished our bridge."[18] Harris did not mention the type of bridge or its length, but six days was a long time to finish the structure.

In addition to suffering from a lack of skilled mechanics and engineers, the Confederate military had a thorny time trying to solve its persistent railroad problems. The Railroad Bureau in Richmond had tried valiantly to repair damaged lines, acquire and operate engines, cars, and repair facilities, manage a train schedule, and negotiate with owners of private companies on issues of military traffic and freight rates. The bureau could not, however, overcome poorly laid track, few repairmen, egotistical generals who commandeered trains for personal use, a dysfunctional command structure, and a government committed to ideological purity rather than military necessity.

For example, after the fall of Atlanta, the Army of Tennessee, led by General Hood, requested that the Mobile & Ohio and Memphis & Charleston be repaired as far as Decatur, Alabama. The request was sent to General Bragg, military adviser to Jefferson Davis. Bragg passed the message on to General Beauregard, recently appointed commander of the Military Division of the West. Beauregard finally approved the request, but no one communicated the decision to Major George Whitefield, the railroad engineer in charge of repairs. When Whitefield was notified of the need for his services, he quickly organized a workforce of impressed slaves, and they managed to mend the tracks so that supplies could be forwarded to Hood's army.[19]

In Virginia, General Lee had hoped to use the Piedmont, the South Side, and the Richmond & Danville railroads as his supply line as the army moved west after evacuating Petersburg. Depots were filled with food and ammunition at Greensboro, Danville, and Lynchburg, but the railroads themselves were in considerable disrepair. Since June 1862, Captain Edmund T. D. Myers of the Confederate Corps of Engineers had supervised construction of the forty-eight-mile Piedmont Railroad between Danville, Virginia, and Greensboro, North Carolina. It was completed in May 1864, and yet in the winter of 1865 the railroad still had no water stations, supplies of wood, or sufficient sidings.[20] Furthermore, throughout the month of January 1865, Colonel Sims, chief of the Railroad Bureau, had to negotiate with the management of both the South Side and the Danville & Richmond because neither was willing to interchange rolling stock. The two railroads finally compromised. They agreed to share their cars for one month until "neutral" cars could be brought in from the Virginia & Tennessee Railroad, which had been destroyed by Union General George Stoneman's cavalry.[21]

Government control of the railroads was needed, it had always been needed, and with the collapse of the Confederacy looming, this was finally addressed. On February 19, 1865, the Confederate Congress overwhelmingly passed "an act to provide for the more efficient transportation of troops, supplies and munitions of war upon the railroads, steamboats and canals in the Confederate States, and to control telegraph lines employed by the Government."[22] It was, of course, too late to change the outcome of the war.

For historians like Bruce Catton and Shelby Foote, the problems with the Confederacy's railroad system during the war only highlighted the accomplishments of Southern railroads and made manifest how remarkable it was that the South achieved any success at all—such as the transfer of Longstreet's men to Chickamauga—while working with so little.[23] The limited accomplishments were impressive, but these historians failed to address why the South was working with so little.

Historian James Huston remarked that the Confederacy did not have a Lincoln or Stanton, or a Tom Scott, Herman Haupt, or David McCallum, or the skillful mechanics to work in support of these great minds.[24] Huston's argument implies that these men just happened to live in the North before the Civil War. In reality, during the antebellum period, the South chose to invest in plantation farming in general and cotton in particular and not in manufacturing.

Conversely, the efforts of the railroad men of the North should not be taken for granted, nor should the sophisticated system of management and distribution oversight developed by the North in the prewar years. For example, Daniel McCallum, in his final report in 1866 for the United States Military Railroad (USMRR) department, wrote about the experimental nature of managing military railroads: "The fact should be understood that the management of railroads is just as much a distinct profession as is that of the art of war" and that "it was extremely difficult to induce those who were really valuable to leave the secure positions and enter upon a new and untried field of action." Moreover, "the attempt to supply the army of General Sherman in the field [1864], construct and reconstruct the railroad in the rear, and keep pace with its march" was regarded by many railroad men as a dangerous experiment that could result in disaster and defeat.[25]

Yet, the damage done by Southern guerillas and cavalry was so quickly repaired by the Construction Corps that Sherman never worried about delay, and he continued to press forward, confident that the railroad would catch up to him. Similarly, around Petersburg, Virginia, General Grant's campaign to capture the Army of Northern Virginia or drive Lee's army into the open was supported by logistical arrangements made possible by the USMRR. From City Point, Virginia, at the confluence of the James and Appomattox rivers, where supplies were unloaded, to beyond the left flank of the Union army, the Construction Corps restored 9 miles of the Petersburg & City Point Railroad and, over ten months, built an additional 21 miles of track.

A detachment of the Virginia Construction Corps, ordered to North Carolina in January 1865 to assist with logistical operations during Sherman's Carolina Campaign, worked until Construction Corps members could be sent from the military division of the Mississippi. Before Sherman reached Goldsboro, North Carolina, in late March, both the 95 miles of the Atlantic & North Carolina from Morehead City to Goldsboro and the 85-mile section of the Wilmington & Weldon from Wilmington to Goldsboro were repaired. For these roads, in addition to two others, 25 miles of main track were rebuilt and 5 miles of sidetrack laid. On the same roads, 3,263 lineal feet of bridges were built. At Morehead City, the Construction Corps built a wharf covering an area of 53,682 square feet, using 700,000 board-feet of timber.[26]

When the war ended, McCallum reported to the War Department a series of staggering numbers quantifying the work of the railroad organization during the war. The USMRR department obtained 419 locomotives during the conflict, and of that number, 312 were built in Northern shops. The department operated 2,105 miles of track, used or built 6,330 cars, constructed or rebuilt 137,418 feet (more than 26 miles) of bridges, and laid or relaid 641 miles of track. It employed a total of 24,964 men, with an average monthly rate of 2,378 men working in the Virginia, Pennsylvania, and Maryland sectors in 1864 and 11,580 men working monthly in the Division of the Mississippi, which covered Tennessee, Georgia, Kentucky, Alabama, Mississippi, and Arkansas.[27]

The Northern railroad crews could accomplish so much so rapidly in large measure because of the practical training men had received in running railroads in the years before the war. With many lines more than 150 miles long, new management systems had to be invented to operate a decentralized business, and men such as Daniel McCallum and Herman Haupt designed new organizational charts that were essential for large railroad companies. First Haupt then McCallum took these organizational models and transferred them to the military railroads. It was easy to look at the USMRR in 1865 and marvel at its efficiency and sophistication. It was easy to forget that in April 1861 the USMRR did not exist, and there was no previous military model.

The military railroads were only as good as the men who worked on them. Whether it was machinists in repair shops, carpenters building trestle bridges, masons working on foundations, laborers laying track or cutting timber, or mechanics and blacksmiths working on locomotives and cars, these men— educated in common schools, mechanics' institutes, or on the job in shipyards, clock factories, or textile mills—had learned how to work with their hands, repair their own broken tools, and innovate when necessary, which was often. For example, in October 1864, Private James Woodward of the 24th Wisconsin, Private William E. Ott of the 46th Pennsylvania, and Private James Nichols of the 13th New Jersey were ordered to report to the USMRR, Division of the Mississippi, for special assignment.[28] Ott and Nichols had been part of the Army of the Potomac until the XI and XII corps were transferred by rail to Sherman's army, and now these infantry soldiers were assigned railroad work. No doubt when asked, the regimental colonels had selected these men because of their mechanical skills.

Finally, the USMRR succeeded because President Lincoln and Secretary of War Edwin Stanton had the foresight to nationalize the railroads and the temerity and toughness to trust and support USMRR operations. Stanton's Special Order No. 337 summed up the secretary's conviction. In part it read: "No officer, whatever be his rank, will interfere with the running of the cars as directed by the superintendent of the road. Any one who so interferes will be dismissed from the service for disobedience of orders."[29]

Of course, the railroads were only one means of moving the Union army's men and supplies inside enemy territory over harsh terrain. Getting at the enemies' forces or, in one case, escaping from their clutches required the critical skills of the men who served as combat engineers and pioneers.

The North's better engineering efforts and superior management made the difference. Despite its vast material resources, the Union army still had to execute strategic, tactical, and logistical operations to destroy Southern forces. Delivering the Federal army to those key points of contact and providing the essential support to keep the army in ammunition and food were the job of the army engineers and the USMRR. The manpower resource to draw from could not come from a small cadre of railroad managers or West Point–trained engineers; the war was on much too grand a scale for that. Instead, the manpower came from the ranks of the citizen soldiers, many of whom had developed mechanical skills before the war. Even when volunteer engineers were unavailable, men from infantry regiments could be called upon to execute repairs to roads or build bridges, often with a pontoon train. It was remarkable.

For example, Sherman's Atlanta Campaign could have turned out differently had Union engineers and the railroad Construction Corps been unable to keep transportation lines repaired and open. The Construction Corps, engineers, and pioneers built roads and bridges that allowed Union generals to move their armies over difficult and problematic terrain. Grant's ability to move his army through the floodplain along the western side of the Mississippi River in eastern Louisiana in 1863 was the result of inventive engineering, not overwhelming resources. The rescues of Banks's army along the Red River, Grant's army in Chattanooga, and even McClellan's army on the Peninsula were the result of inventive thinking, the exchange of ideas, sometimes from the bottom up, and the mechanical skill of the men required to execute the plan.

The statistics alone are striking. For example, Colonel Ira Spaulding of the 50th New York wrote in June 1865 that his regiment had built eighty-six bridges

Throughout the war, engineers, pioneers, and infantrymen built and maintained corduroy roadways, such as this road from Belle Plain to Fredericksburg, Virginia. Dense woods and lowlands often made construction of these log roads difficult, and building them required constant attention and innovation. Arthur Lumley, Library of Congress, Prints & Photographs Division, Civil War Photographs, LC-DIG-ppmsca-20775.

from September 1862 to May 1865, totaling 21,248 feet or just over four miles. This number did not include the many trestle, timber, and corduroy bridges built by the 50th New York, nor the bridgework done by the 1st and 15th New York or the regular Engineer Battalion. In addition, over the course of the war, the Union adopted better management techniques. By 1864, each engineer company was furnished with a company wagon, commissary wagon, forage wagon, tool wagon, and carpenters' tool chest. "By this means," Spaulding wrote, "the whole or any portion of the regiment was prepared to move at any time of the day or night, with fifteen days' supplies and a complete outfit for the performance of all kind of engineer duty."[30]

The greatest feat of military engineering during the war, and perhaps the greatest in American military history, came at Vicksburg. Grant began the campaign with only the company of 1st Kentucky Engineers and Mechanics and three regular army engineers attached to his headquarters' staff. Recognizing

Displaying their ingenuity and skill, men from the 50th New York Volunteer Engineers built this remarkable gothic-style pine log church near the Petersburg siege lines at Poplar Grove, Virginia, in November 1864. Note the engineer insignia proudly carved above the door. Timothy O'Sullivan took this photograph in March 1865. Today, 6,718 Union remains, 4,600 of which are unknown soldiers, rest in Poplar Grove National Cemetery where the church once stood. Library of Congress, Prints & Photographs Division, Civil War Photographs, LC-DIG-cwpb-03885.

and utilizing the skills he had within his ranks, Grant tapped officers with engineering experience in civilian life and infantry units with men of mechanical ability. The results were incredible: two attempted canal projects, a series of floating and trestle bridges, corduroy roads through wetlands, wooden towers for sharpshooters, fabricated sap rollers, fascine from pork barrels, and gum tree mortars.

What made this work possible was the attention paid to industrial development, railroad management, and common school reform in the North in the years before the Civil War. No doubt the density of the Northern population helped ideas to bubble up during the antebellum period, but there was also a fundamental belief that ideas were important. Information was exchanged all the time, whether at agricultural fairs, lyceum meetings, mechanics' institutes, or universities. State governments recognized the need to reform common schools because an educated citizenry was both good for individuals, including those in the working class, and good for business.

Lincoln spoke often of providing opportunities for the laboring classes to improve their station in life. He believed that there was an important positive relationship between labor and capital, and the evidence suggests that many people in the North felt the same way. On March 6, 1860, in New Haven, Connecticut, Lincoln said: "So while we do not propose any war upon capital, we do wish to allow the humblest man an equal chance to get rich with everybody else. When one starts poor, as most do in the race of life, free society is such that he knows he can better his condition; he knows that there is no fixed condition of labor, for his whole life."[31]

Northern laborers recognized opportunities to improve their station in life. Risk, effort, skill, and reward were a part of the culture in the industrialized North. Risk and reward, business acumen, and entrepreneurship applied to the South as well, but generally only for those who ran the plantations and sold their crops for enormous profits. Labor, and especially slave labor, was viewed as the necessary "mudsill" of society. Education and ideas were reserved for the elite. Slavery represented the economic backbone of the Confederacy, and slavery contributed to the Confederacy's doom. This was not just because the war came as a result of the peculiar institution, but also because the institution of slavery permeated Southern culture and limited creativity, innovation, openness, and adaptation to change—all essential elements in a war that required enormous logistical considerations. Battles at Island No. 10, Vicksburg, Chattanooga, and Atlanta required ideas and innovations that had never been attempted before on such a scale. Generating new ideas was often the difference between victory and defeat. Chances of success increased when those ideas came from as many people as possible. It was not just the generals who were relied upon to offer a new approach to a new problem. In the North, a volunteer engineer captain from New York who had worked as a civil engineer before the war, an infantry sergeant from Pennsylvania who had been a miner, or a private from Illinois who had worked as a machinist—all generated ideas. This was the advantage the Federal army had over the Confederacy. It is why the North won the Civil War.

Between 1861 and 1865, the United States fought a Civil War to both maintain "the last best hope on earth" and to deliver "a new birth of freedom." The reason for the North's victory was that Union engineers, the majority of them volunteers, were able to apply ingenuity and innovation to the complex problems facing them. In the spring of 1861, both sides prepared for the one great, decisive battle that would determine the outcome. It never happened. Battles continued to unfold and men continued to die. These historic battles have captured the imagination and attention of Americans ever since. What ended the war, however, was not the battles per se but the Union army's ability to make roads and bridges through the forbidding terrain of the South and to quickly repair damage done to the railroads, bridges, and roads as retreating, maneuvering Confederates relentlessly tried to cut supply lines and stop the Yankees from pressing any advantage. The adroitness, proficiency, and versatility of the Union's citizen soldiers were the sine qua non of their army's success. They enabled the North to engineer victory in the Civil War.

Notes

INTRODUCTION. MASTERS AND MECHANICS

1. In December 1862, the Union army, under Grant's command, attempted to break through Vicksburg's outer defenses, striking at Chickasaw Bayou and Arkansas Post. Both actions failed to bring the desired results, although General McClernand was able to capture Arkansas Post. The other five failed operations were Grant's Canal, the Lake Providence Expedition, the Yazoo Pass Expedition, the Steele's Bayou Expedition, and the Duckport Canal. *The War of the Rebellion: A Compilation of the Official Records of the Union and Confederate Armies* (Washington, DC: Government Printing Office, 1880–1901), ser. 1, vol. 24, pt. 3, 230. The Official Records (hereafter, *OR*) was published in 73 volumes and 128 parts.

2. Regarding the number of mules, horses, and wagons, these are approximations based on the War Department's ratio of 20 wagons per 1,000 men. Artillery pieces usually required a six-mule team, and there were two horses per wagon. There were also 300 wagons to haul forage for the animals; this number is based on a ratio of 4 army wagons to 1 horse/mule wagon. Headquarters and cavalry horses were also factored into the number of animals.

3. *New York Times*, April 16, 1863.

4. *OR*, ser. 1, vol. 24, pt. 1, 491–495.

5. John Perkins owned and developed 17,500 acres of property known as Somerset Plantation. Some sources refer to the property as Perkins's Plantation, but I have chosen to identify it by its correct name.

6. *OR*, ser. 1, vol. 24, pt. 1, 187, 571–572. Colonel Keigwin and Lieutenant Tunica filed separate reports, differing in some details, especially regarding Lieutenant Tunica's role in the operation.

7. Ibid., 571–573.

8. Ibid.

9. Ibid., 126–127. See also "Vicksburg Diary of Henry Clay Warmoth," Henry Clay Warmoth Papers, 1798–1953, collection 00752, folder 123, Southern Historical Collection, Louis Round Wilson Special Collections Library, University of North Carolina, Chapel Hill, North Carolina.

10. *Harper's Weekly*, October 26, 1861.

11. Quoted in Martin Van Creveld, *Supplying War: Logistics from Wallenstein to Patton* (Cambridge: Cambridge University Press, 1977), 231–232.

12. Ulysses S. Grant, *Personal Memoirs of U. S. Grant* (New York: Penguin, 1999), 254.

13. See James H. Hammond, *Selections from the Letters and Speeches of the Honorable James H. Hammond of South Carolina* (New York: John F. Trow, 1866), 311–322.

14. Shelby Foote, *The Civil War, A Narrative: Red River to Appomattox* (New York: Random House, 1974), 3:955.

15. Richard N. Current, "God and the Strongest Battalions," in *Why the North Won the Civil War*, ed. David Herbert Donald (Baton Rouge: Louisiana State University Press, 1960), 21–22. See also James McPherson, "American Victory, American Defeat," in *Why the Confederacy Lost*, ed. Gabor S. Boritt (Oxford: Oxford University Press, 1992), 20; James M. McPherson, *Drawn with the Sword: Reflections on the American Civil War* (New York: Oxford University Press, 1996), 133.

16. McPherson, *Drawn with the Sword*, 116–117.

17. John Solomon Otto, *Southern Agriculture during the Civil War Era, 1860–1880* (Westport, CT: Greenwood Press, 1994). For a discussion of Confederate policies on taxation and cotton, among other things, see David M. Potter, "Jefferson Davis and the Political Factors in Confederate Defeat," in Donald, 13. For Confederate commercial and fiscal policy, see Stanley Lebbergott, "Why the South Lost: Commercial Purpose in the Confederacy, 1861–1865," *Journal of American History* 70 (June 1983): 58–74; Douglas B. Ball, *Financial Failure and Confederate Defeat* (Urbana: University of Illinois Press, 1991).

18. Gary W. Gallagher, "'Upon Their Success Hang Momentous Interests': Generals," in Boritt, 84.

19. Ibid. See also Emory M. Thomas, "Rebellion and Conventional Warfare: Confederate Strategy and Military Policy," in *Writing the Civil War: The Quest to Understand*, ed. James M. McPherson and William J. Cooper Jr. (Columbia: University of South Carolina Press, 1998), 54.

20. Thomas, 54.

21. Gallagher, 85. See also Richard E. Beringer, Herman Hattaway, Archer Jones, and William N. Still Jr., *Why the South Lost the Civil War* (Athens: University of Georgia Press, 1986; T. Harry Williams, "The Military Leadership of the North and South," in Donald, 23–47.

22. Brooks D. Simpson, "Facilitating Defeat: The Union High Command and the Collapse of the Confederacy," in *The Collapse of the Confederacy*, ed. Mark Grimsley and Brooks D. Simpson (Lincoln: University of Nebraska Press, 2001), 98. For an excellent introduction to Jomini and his principles, see Crane Brinton, Gordon A. Craig, and Felix Gilbert, "Jomini," in *Makers of Modern Strategy: Military Thought from Machiavelli to Hitler*, ed. Edward Mead Earle (Princeton: Princeton University Press, 1971), 77–92. For Russell F. Weigley's comparison of Grant's annihilation strategy and Lee's Napoleonic strategy, see his *The American Way of War: A History of United States Military Strategy and Policy* (Bloomington: Indiana University Press, 1977), 92–152.

23. Albert E. Castel, *Victors in Blue: How Union Generals Fought the Confederates, Battled Each Other, and Won the Civil War* (Lawrence: University Press of Kansas, 2011).

24. Thomas Lawrence Connelly and Archer Jones, *The Politics of Command: Factions and Ideas in Confederate Strategy* (Baton Rouge: Louisiana State University Press, 1973), 193. See also Grady McWhiney and Perry D. Jamieson, *Attack and Die: Civil War Tactics and Southern Heritage* (Tuscaloosa: University of Alabama Press, 1984).

25. Alan T. Nolan, *Lee Considered: General Robert E. Lee and Civil War History* (Chapel Hill: University of North Carolina Press, 1991). See also Emory M. Thomas, "Rebellion and Conventional Warfare: Confederate Strategy and Military Policy," in McPherson and Cooper, 55.

26. Merritt Roe Smith, ed., *Military Enterprise and Technological Change: Perspectives on the American Experience* (Cambridge, MA: MIT Press, 1985).

27. Dirk J. Struik, *Yankee Science in the Making: Science and Engineering in New England from Colonial Times to the Civil War* (New York: Dover, 1991).

28. Earl J. Hess, *In the Trenches at Petersburg: Field Fortifications & Confederate Defeat* (Chapel Hill: University of North Carolina Press, 2009). Other books by Earl Hess include *Field Armies and Fortifications in the Civil War: The Eastern Campaigns, 1861–1864* (Chapel Hill: University of North Carolina Press, 2005); *Trench Warfare under Grant & Lee: Field Fortifications in the Overland Campaign* (Chapel Hill: University of North Carolina Press, 2007); and *Into the Crater: The Mine Attack at Petersburg* (Columbia: University of South Carolina Press, 2010).

29. Phillip M. Thienel, *Mr. Lincoln's Bridge Builders: The Right Hand of American Genius* (Shippensburg, PA: White Mane, 200); Mark Hoffman, *"My Brave Mechanics": The First Michigan Engineers and Their Civil War* (Detroit: Wayne State University Press, 2007). For works on other engineering regiments or specific actions, see David F. Bastian, *Grant's Canal: The Union's Attempt to Bypass Vicksburg* (Shippensburg, PA: White Mane, 1995); Philip Katcher, *Building the Victory: The Order Book of the Volunteer Engineer Brigade Army of the Potomac* (Shippensburg, PA: White Mane, 1998); Phillip M. Thienel, *Seven Story Mountain: The Union Campaign at Vicksburg* (Jefferson, NC: McFarland, 1998); Philip L. Shiman, "Engineering and Command: The Case of General William S. Rosecrans, 1862–1863," in *The Art of Command*, ed. Steven E. Woodworth (Lincoln: University of Nebraska Press, 1998); Colonel Wesley Brainerd, *Bridge Building in Wartime: Colonel Wesley Brainerd's Memoir of the 50th New York Volunteer Engineers*, ed. Ed Malles (Knoxville: University of Tennessee Press, 1997); Paul Taylor, *Orlando M. Poe: Civil War General and Great Lakes Engineer* (Kent, OH: Kent State University Press, 2009); Hess, *Into the Crater*.

30. James L. Nichols, *Confederate Centennial Studies Number Five: Confederate Engineers* (Tuscaloosa, AL: Confederate Publishing, 1957); Harry L. Jackson, *First Regiment Engineer Troops P. A. C. S.: Robert E. Lee's Combat Engineers* (Louisa, VA: R. A. E. Design, 1998). See also J. Boone Bartholomees, *Buff Facings and Gilt Buttons: Staff and Headquarters Operations in the Army of Northern Virginia, 1861–1865* (Columbia: University of South Carolina Press, 1998); Kent Masterson Brown, *Retreat from Gettysburg: Lee, Logistics & the Pennsylvania Campaign* (Chapel Hill: University of North Carolina Press, 2005).

31. Grant described his theory of war in a letter to John Russell Young: "If the Vicksburg campaign meant anything, in a military point of view, it was that there are no fixed laws of war which are not subject to the conditions of the country, the climate, and the habits of the people . . . I don't underrate the value of military knowledge, but if men make war in slavish observances of rules, they will fail." Quoted in Donald, v–vi.

CHAPTER 1. COMMON SCHOOL REFORM AND SCIENCE EDUCATION

Epigraph. Quoted in Diana Ross McCain, *It Happened in Connecticut* (Guilford, CT: Guilford Press), 137.

1. Mary Bobbitt Townsend, *Yankee Warhorse: Biography of General Peter Osterhaus* (Columbia: University of Missouri Press, 2010), 3.

2. William Watts Folwell, *William Watts Folwell: The Autobiography and Letters of a Pioneer of Culture*, ed. Solon J. Buck (Minneapolis: University of Minnesota Press, 1933), 13–14.

3. United States Bureau of the Census, 1840.

4. Catherine Fennelly, *Town Schooling in Early New England, 1790–1840* (Sturbridge, MA: Old Sturbridge Village, 1962), 39, 40.

5. Folwell, 35–36.

6. Ibid., 44–45.

7. Rush Welter, *Popular Education and Democratic Thought in America* (New York: Columbia University Press, 1962), 48.

8. *Report of a Committee of Philadelphia Workingmen* (1830), in *American Writings on Popular Education: The Nineteenth Century*, ed. Rush Welter (Indianapolis: Bobbs-Merrill, 1971), 38, 39.

9. Thomas Frederick Woodley, *Great Leveler: The Life of Thaddeus Stevens* (Freeport, NY: Books for Libraries Press, 1969 [1937]), 110.

10. Ibid., 111.

11. Christopher Collier, *Connecticut's Public Schools: A History, 1650–2000* (Orange, CT: Clearwater Press, 2009), 74.

12. Henry Barnard, 2nd, *Second Annual Report of the Board of Commissioners of Common Schools in Connecticut together with the Second Annual Report of the Secretary of the Board, May 1840* (Hartford, CT: Case, Tiffany and Burnham, 1840), 42. Barnard reported that in 1839 the average cost to each child attending school was $3.20. The state's "school fund" contributed $1.25 to the cost; the "Town Deposite Fund," consisting of funds from the 1836 federal surplus, contributed 30 cents; and the rest was raised through a Town Tax, District Tax, and tax on parents of children attending school.

13. Collier, 80.

14. Barnard, 28.

15. Collier, 103.

16. Fairfield North School District Journal, 1851–1861, Connecticut Historical Society, Hartford, Connecticut. The winter term was seventy days long; it opened on December 21, 1857, and closed on March 27, 1858. The data collected for the winter term beginning on October 8, 1860, and ending on January 14, 1861, were more substantial than for previous years. During this winter term, seventeen students were enrolled, ten boys and seven girls. The oldest girl was thirteen years old and attended school for forty-four days. Five girls between seven and eleven years old attended for, on average, sixty-one days. The one six-year-old girl attended for fifteen days. The oldest boy was sixteen years old and was in school for thirty-one days. Four boys between fourteen and eleven years old attended for, on average, thirteen days. The nine- and ten-year-old boys were in school for fifty-nine days, and the six-year-old boy for sixty-one days. The seven- and eight-year-old boys were present for twenty-six days.

17. Collier, 150. Two boys under fourteen years old, John B. Stow and George Barnett, were reported by an eyewitness to have calculated the following in forty minutes: Multiply 253,412,003,520,155,102,350 by 521,342,125,145,534,142,125. The product is 132,114,352,452,585, 239,925,224,746,717,418,821,493,750.

18. William J. Anderson and William A. Anderson, eds., *The Wisconsin Blue Book, 1929* (Madison, WI: Democrat Printing Company, State Printer, 1929), 155.

19. The Board of Regents supervised all educational institutions within the State of New York.

20. Enos T. Throop, Governor, 1831 January 4 Legislature, 54th Session, in *State of New York Messages from the Governors Comprising Executive Communications to the Legislature and Other Papers Relating to Legislation from the Organization of the First Colonel Assembly in 1683 to and Including the Year 1906. With Notes, vol. 3, 1823–1842*, ed. Charles Z. Lincoln (Albany, NY: J. B. Lyon Company, 1909), 336.

21. Ibid., 337.

22. Document of the Senate, *Second Report of the Special Committee for Promoting the Intro-*

duction of Agricultural Books in Schools and Libraries, no. 85, 79th sess., vol. 1, no. 1 to 40, 1856, from the *69th Annual Report of the Regents of the University of the State of New York Made to the Legislature January 15, 1856*, 381.

23. *Abstract of Reports in 1854 from the following Counties: Chemung, Elmira, Kings, Monroe, Odgen, Westchester, Otsego, Albany, Broome, Maine, Erie, Buffalo, Oswego, and Oneida*, New York State Library, Albany, New York. Most students attended school for six months, and the number of children in school as a percentage of all children in each county varied. In Albany County it was forty-four percent, but in Oneida County it was eighty percent. Monroe boasted seventy percent attendance; Kings County, sixty-nine percent; and Oswego County, sixty-seven percent. Some counties operated "Colored Schools." Erie and Buffalo counties had 194 black children in school; Kings County, 130; and Albany County, 104. Many, however, reported zero, and no one made an account of Native American children.

24. E. Powell to V. M. Rice, December 1, 1856, State of New York Archives, Albany, New York.

25. *Speeches of Henry Lord Brougham: Upon Questions Relating to Public Rights, Duties, and Interests; with Historical Introductions*, vol. 2 (Philadelphia: Lea and Blanchard, 1841), 145–149.

26. Holbrook entered Yale in 1806 and graduated in 1810. Silliman, as Yale's first chemistry professor, delivered the first science lectures ever given at Yale (in 1804). In 1818 he founded the *American Journal of Science*.

27. Josiah Holbrook, "American Lyceum for Science and the Arts," in *The Massachusetts Lyceum during the American Renaissance: Materials for the Study of the Oral Tradition in American Letters: Emerson, Thoreau, Hawthorne, and Other New England Lectures*, ed. Kenneth Walter Cameron (Hartford, CT: Transcendental, 1969), 46.

28. Ibid., 46–47.

29. Kenneth Walter Cameron, ed., *Historical Sketch of the Salem Lyceum with a List of the Officers and Lecturers since Its Formation in 1830 in Massachusetts Lyceum during the American Renaissance* (Hartford, CT: Transcendental, 1969), 16.

30. Ibid., 16–17.

31. *Lyceum Records of Lincoln, Massachusetts*, in Cameron, *Historical Sketch*, 213.

32. United States Bureau of the Census, Eighth Census of the United States, 1860.

33. Ibid.

34. Carl F. Kaestle, *Pillars of the Republic: Common Schools and American Society, 1780–1860* (New York: Hill and Wang, 1983), 194.

35. George G. Kundahl, *Confederate Engineer: Training and Campaigning with John Morris Wampler* (Knoxville: University of Tennessee Press, 2000), 1–2.

36. Ibid., 4.

37. Ibid., 23, 68.

38. Kaestle, 205.

39. Welter, *Popular Education*, 133–134.

40. "Instruction in Schools and Colleges," *Southern Quarterly Review*, October 1860.

41. Henry A. Wise, "Address to His Constituents, 1844," in Welter, *American Writings*, 122, 128.

42. Ibid., 131.

43. Kaestle, 210.

44. "A State of Convenience: The Creation of West Virginia," Proceedings of the First Wheeling Convention, May 15, 1861, West Virginia Archives and History, at www.wvculture .org/history/statehood/wheelingconvention10515.html.

45. C. H. Wiley, *First Annual Report of the General Superintendent of Common Schools* (Raleigh, NC, 1854), in *Documenting the American South* (Chapel Hill: University of North Carolina, 2003), at docsouth.unc.edu/nc/schools1854/schools1854.html.

46. Charles Lee Smith, *The History of Education in North Carolina* (Washington, DC: US Government Printing Office, 1888), 170.

47. Ibid.

48. *Report of the Committee on Education, Louisiana Constitutional Convention, 1845*, in Welter, *American Writings*, 139.

49. Edward Alfred Pollard, "Hints on Southern Civilization," *Southern Literary Messenger* 32 (April 1861): 310.

50. "Editor's Table," *Southern Literary Messenger* 33 (July 1861): 75.

51. Jennifer R. Green, "Networks of Military Educators: Middle-Class Stability and Professionalism in the Late Antebellum South," *Journal of Southern History* 73 (February 2007): 41.

52. Ibid., 42, 45–46. See also Bruce Allardice, "West Points of the Confederacy: Southern Military Schools and the Confederate Army," *Civil War History* 43 (December 1997): 314–315.

53. Green, 45.

54. Bode, 75.

55. Ibid., 79–84.

56. John Majewski, *Modernizing a Slave Economy: The Economic Vision of the Confederate Nation* (Chapel Hill: University of North Carolina Press, 2009), 44.

57. Thomas R. Dew, in *Review of the Debates of the Virginia Legislature*, at https://archive .org/details/reviewofdebateon00dewt. See also "Thomas R. Dew Defends Slavery (1852)," at wwnorton.com/college/history/archive/resources/documents/ch15_03.htm; Thomas R. Dew, *Digest of the Laws, Customs, Manners, and Institutions of Ancient and Modern Nations* (New York: D. Appleton & Company, 1853).

58. A. J. Angulo, *William Barton Rogers and the Idea of MIT* (Baltimore: Johns Hopkins University Press, 2009), 22–25.

59. W. B. Rogers to Judge J. F. May, University of Virginia, March 13, 1841, in *Life and Letters of William Barton Rogers*, ed. Emma Savage Rogers (Boston: Houghton, Mifflin, 1896), 1:182–183.

60. Angulo, 26.

61. W. B. Rogers to Robert Rogers, University of Virginia, September 11, 1841, in Emma Savage Rogers, 1:191–193.

62. Emma Savage Rogers, 1:239.

63. W. B. Rogers to Henry Rogers, University of Virginia, March 13, 1846, in Emma Savage Rogers, 1:259. At UVA Rogers included with his lectures some instruction in laboratory demonstrations. This was considered radical in the 1830s and 1840s. The standard university classroom required students to study textbooks and recite memorized lessons learned in Greek and Latin. Expansion of the natural sciences raised questions about classical education, and these questions were put to rest in what became known as the Yale Report of 1828. Yale President Jeremiah Day wrote: "Its [university education's] object is to lay the foundation of a superior education." Collegiate education was to distinguish a man from those receiving mere practical studies and training. Rogers wanted to inspire "inventive thought." In his report to the Virginia legislature in 1845, he wrote: "But along with this modest deference to the oracles of knowledge, he [the student] cherishes that manly self-dependence of

thought which springs from the conscious vigour due to the free training of his facilities." See Emma Savage Rogers, 1:399. See also Angulo, 76.

64. Angulo, 87–88.

65. In the years before Folwell attended Hobart, the school was known as Geneva College. See Folwell, 53.

66. Ibid., 56.

67. Ibid., 77.

68. Kundahl, 73. The graphodometer was a piece of surveying equipment used to measure angles.

69. Dirk J. Struik, *Yankee Science in the Making: Science and Engineering in New England from Colonial Times to the Civil War*, rev. ed. (New York: Dover, 1991), 425.

70. "Fair of the American Institute," *Mechanics' Magazine and Journal of the Mechanics' Institute* 8, no. 6 (December 1836): 289, at lhldigital.lindahall.org/cdm/ref/collection/rrjournal/id/4597.

CHAPTER 2. MECHANICS' INSTITUTES AND AGRICULTURAL FAIRS

1. Library of Virginia, "A Guide to the David Ross Papers, 1813, 1822," at http://ead.lib.vir ginia.edu/vivaead/published/lva/vi00317.document. See also Ronald L. Lewis, *Coal, Iron, and Slaves: Industrial Slavery in Maryland and Virginia, 1715–1865* (Westport, CT: Greenwood Press, 1979), 27.

2. Lewis, 28, 30, 33, 35.

3. David Ross to Robert Richardson, n.d. [probably December 1812], David Ross Papers, accession 37815, Library of Virginia, Richmond, Virginia.

4. Ibid.

5. Lewis, 190.

6. Charles B. Dew, *Bond of Iron: Master and Slave at Buffalo Forge* (New York: W. W. Norton, 1994), 108–115.

7. Ibid., 107–108.

8. Ibid., 333. For an outstanding discussion on how and why Southern industrialization during the antebellum period was built by African American miners, mechanics, engineers, artisans, and craftsmen, see James E. Newton and Ronald L. Lewis, eds., *The Other Slaves: Mechanics, Artisans, and Craftsmen* (Boston: G. K. Hall, 1978).

9. See Charles B. Dew, *Ironmaker to the Confederacy: Joseph R. Anderson and the Tredegar Iron Works* (New Haven: Yale University Press, 1966), 26–32. Dew pointed out that slave labor at Tredegar Iron Works in Richmond centered on the rolling mill operations. The owner of the iron works, Joseph R. Anderson, believed slaves helped to control white labor strikes and reduced labor costs. Yet, in other areas of Anderson's business, Northern or foreign-born white workers made up the vast majority of the labor force. The machine shops and locomotive works were staffed almost exclusively with white labor.

10. Edmund Fuller, *Tinkers and Genius: The Story of the Yankee Inventors* (New York: Hastings House, 1955), 246–248.

11. Ibid., 248.

12. Ibid.

13. In Chauncey Jerome's *History of the American Clock Business*, the author estimates his profit margin was approximately forty to fifty percent of his costs.

14. National Clock Repair, "Jerome Clock History," at www.nationalclockrepair.com/Jerome_Clock_History.php.

15. Barnum was a cousin to my great grandmother, Mabel Watson Whaley, who was as kind and compassionate as Barnum was devious and cunning.

16. Lewis, 231.

17. David Jaffee, *A New Nation of Goods: The Material Culture of Early America* (Philadelphia: University of Pennsylvania Press, 2010), 186–187.

18. Hugo A. Meier, "The Ideology of Technology," in *Technology and Social Change in America*, ed. Edwin T. Layton Jr. (New York: Harper & Row, 1973), 81.

19. Ibid., 81, 85.

20. US Department of the Interior, *Manufactures of the United States of America in 1860; Compiled from the Original Returns of the Eighth Census under the Direction of the Secretary of the Interior* (Washington, DC: US Government Printing Office, 1865), v.

21. Ibid., xv.

22. Clement Eaton, *The Growth of Southern Civilization, 1790–1860* (New York: Harper & Row, 1961), 166.

23. Ibid., 223–224.

24. Ibid., 225.

25. US Department of State, "Aggregate Value and Produce, and Number of Persons Employed in Mines, Agriculture, Commerce, Manufactures, etc., Exhibiting a Full View of the Pursuits, Industry, and Resources of the United States of America Including the District of Columbia, and the Territories of Wiskonsan, Iowa, and Florida," in *Statistics of the United States of America as Collected and Returned by the Marshals of the Several Judicial Districts under the 13th Section of the Act for Taking the Sixth Census Corrected at the Department of State June 1, 1840* (Washington, DC: Blair and Rives, 1841), 4:16–22. These numbers are not entirely helpful because they do not differentiate between white men and women, free black men and women, slave men and women, and both white and black boys and girls. The following is a breakdown of the total invested in manufacturing for these states in 1840: eastern Virginia, $7,443,024—of which $2,999,108 was invested in cotton textile mills and $241,840 in machinery manufactured; western Virginia, $11,360,861—$5,184,669 in cotton textile mills and $429,858 in machinery manufactured; North Carolina, $3,838,900—$1,670,228 in cotton mills and $43,285 in machinery manufactured; South Carolina, $3,216,970—$1,668,804 in cotton textile mills and $65,561 in machinery manufactured; Georgia, $2,899,565—$1,491,973 in cotton textile mills and $131,238 in machinery manufactured.

26. *American Industry and Manufactures in the 19th Century: A Basic Source Collection Compiled from U.S. Government Documents* (Elmsford, NY: Maxwell Reprint Company, 1970), 251. Middlesex County reported capital investments in manufacturing of $26,946,527 (larger than Virginia); Worcester County, $13,934,769 (larger than Georgia); and Essex County, $20,885,580 (larger than North Carolina and South Carolina combined).

27. *American Industry and Manufactures in the 19th Century*, 252. See also *Statistics of the United States of America as Collected and Returned by the Marshals*, 56–57.

28. Douglass C. North, *The Economic Growth of the United States, 1790–1860* (Englewood Cliffs, NJ: Prentice-Hall, 1961), 162.

29. Paul E. Rivard, *A New Order of Things: How the Textile Industry Transformed New England* (Hanover, NH: University Press of New England, 2002), 51, 52.

30. United States Bureau of the Census, Eighth Census of the United States, 1860.

31. Bruce Sinclair, "The Direction of Technology," in *Technology and Social Change in America*, ed. Edwin T. Layton Jr. (New York: Harper & Row, 1973), 65–66.

32. Nathan Rosenberg, ed., *The American System of Manufactures* (Edinburgh: Edinburgh University Press, 1969), 20.

33. Simon North had partially accomplished interchangeability in the firearms industry when John Hall's new machinery produced a uniformity of parts, described in December 1837 by John Tipton of Indiana, chairman of the Senate committee on military affairs, as "the greatest improvement in the mechanical arts ever made by one man." Hall also produced a set of gauges used to measure parts and detect deformities in machine work. See Merritt Roe Smith, *Harpers Ferry Armory and the New Technology: The Challenge of Change* (Ithaca, NY: Cornell University Press, 1977), 223–225.

34. Sinclair, "Direction of Technology," 71. Philadelphia was considered the machine-tool building center in the country. Sellers began his career as an apprentice in his uncle's machine shop in Wilmington, Delaware. He later became superintendent of Fairbanks, Bancroft and Company, a family-connected machine shop in Providence, Rhode Island. He returned to Philadelphia in 1848 and started his own business with Bancroft. They manufactured machine tools and mill gearing. See Bruce Sinclair, "At the Turn of a Screw: William Sellers, the Franklin Institute, and a Standard American Thread," *Technology and Culture* 10 (January 1969): 20–34.

35. Joseph Whitworth, "New York Industrial Exhibition. Special Report of Mr. Joseph Whitworth. Presented to the House of Commons by Command of Her Majesty, in Pursuance of Their Address of February 6, 1854," in Rosenberg, 357, 358.

36. Ibid., 389.

37. George Wallis, "New York Industrial Exhibition. Special Report of Mr. George Wallis. Presented to the House of Commons by Command of Her Majesty, in Pursuance of Their Address of February 6, 1854," in Rosenberg, 205.

38. Smith, 129.

39. Ibid., 139.

40. Ibid., 145, 146, 151.

41. Ibid., 272–273.

42. The new journal was entitled *Journal of the Franklin Institute and American Repertory of Mechanical and Physical Science, Civil Engineering, the Arts and Manufactures, and of American and Other Patented Inventions.*

43. Bruce Sinclair, *Philadelphia's Philosopher Mechanics: The History of the Franklin Institute* (Baltimore: Johns Hopkins University Press, 1974), 285. See also *Journal of the Franklin Institute* index for Trautwine's "Rough Notes of an Exploration for an Interoceanic Canal Route by Way of the Rivers Atrato and San Juan, in New Granada, South America"; the first installment is in vol. 57 (April 1854): 217–231.

44. Sinclair, *Philadelphia's Philosopher Mechanics*, 285–286, 294. Trautwine's books included *The Field Practice of Laying out Circular Curves for Railroads* (1851) and *A New Method of Calculating the Cubic Content of Excavations and Embankments* (1852). The future chief of the Union's Engineer Bureau, Joseph Gilbert Totten, wrote *Essay on Hydraulic and Common Mortars, and on Lime-Burning, Translated from the French of G. Treussart, M. Petot, and M. Courtois, and an Account of Some Experiments Made Therewith at Fort Adams, Newport, Rhode Island, from 1825 to 1838* (Philadelphia: Franklin Institute, 1838).

45. Sinclair, *Philadelphia's Philosopher Mechanics*, 295, 298.

46. *The Sixth Exhibition of the Massachusetts Charitable Mechanics Association: Faneuil and Quincy Halls, in the City of Boston, September 1850* (Boston: Eastburn's Press, 1850), v, x, at books .google.com/books?id=8QwAAAAAMAAJ&pg=RA3-PA144&lpg=RA3-PA144&dq="georgiana +ball+Hughes"&source.

47. Victor S. Clark, *History of Manufactures in the United States, Volume I, 1607–1860* (New York: Peter Smith, 1949 [1929]), 465.

48. Ibid., 467. See also *Niles' Register* 50, no. 378 (August 6, 1836).

49. Clark, 467, 471. See also *American Annual Register* 2 (1826–1827): 460; Buffalo, *Annual Statement of Trade and Commerce* (Buffalo, NY, 1856), 26.

50. Morgan Bibliography of Ohio Imprints, 1796–1850, at ocl7.ohiolink.edu/morgan/titles .cgi?chunk=title-15.

51. Eighth Census of the United States.

52. Joseph Whitworth, "New York Exhibition. Special Report of Mr. Joseph Whitworth. Presented to the House of Commons by Command of Her Majesty, in Pursuance of Their Address of February 6, 1854," in Rosenberg, 387, 388–389.

53. Paul W. Gates, *The Farmer's Age: Agriculture, 1815–1860*, vol. 3 of *Economic History of the United States* (New York: Holt, Rinehart, and Winston, 1962), 135.

54. "Second Report of the Special Committee for Promoting the Introduction of Agricultural Books in Schools and Libraries, with Letters from Deputy Superintendent, Common School Department of New York," *Transactions of the New York State Agricultural Society* 4, no. 85 (1844): 372–391, reel 1 microfilm, New York State Library, Albany, New York.

55. Adelaide R. Hasse, *Index of Economic Material in Documents of the States of the United States: New York, 1789–1904* (Washington, DC: Carnegie Institution, 1907); "Transactions of the American Institute of the City of New York for the Year 1850," document 149, pp. 31, 53–62, 133–139, 139–140, in *Documents of the Assembly of the State of New York*, 74th sess., 1851, vol. 5, nos. 132–149, Albany, New York. A snath is the shaft or handle of a scythe.

56. Hasse, 53, 180, 317.

57. Gates, 314.

58. "Agricultural Education," *Western Prairie Farmer* 7 (September 1847): 266–267.

59. Margaret W. Rossiter, *The Emergence of Agricultural Science: Justus Liebig and the Americans, 1840–1880* (New Haven: Yale University Press, 1975), 69.

60. Growing a Nation, "Historical Timeline—1840," at www.agclassroom.org/gan/time line/1840.htm.

61. Washington Hunt, January 7, 1857, Legislature, 74th Session Annual Message, in *Messages from the Governors of the State of New York* (1843–1856), 4:553.

62. Abraham Lincoln, "Address to the Wisconsin State Agricultural Society, Milwaukee, Wisconsin," in *Abraham Lincoln: Speeches and Writings, 1859–1865*, ed. Don E. Fehrenbacher (New York: Library of America, 1989), 90–101.

63. "Education in Rural Districts," *Southern Cultivator* 15 (July 1857): 218–219.

64. John Majewski, *Modernizing a Slave Economy: The Economic Vision of the Confederate Nation* (Chapel Hill: University of North Carolina Press, 2009), 35–36.

65. Majewski, 35, 44. See also Lee Soltow and Edward Stevens, *The Rise of Literacy and the Common Schools in the United States: A Socioeconomic Analysis to 1870* (Chicago: University of Chicago Press, 1981), 166–176.

66. Leading Southern nationalists included William Yancey (AL), Edmund Ruffin (VA), Joseph Brown (GA), Henry Wise (VA), Robert Barnwell Rhett (SC), and Robert Toombs (GA).

See John McCardell, *The Idea of a Southern Nation: Southern Nationalists and Southern Nationalism, 1830–1860* (New York: W. W. Norton, 1981).

67. Steven G. Collins, "System, Organization, and Agricultural Reform in the Antebellum South, 1840–1860," *Agricultural History* 75 (Winter 2001): 12–13.

68. *De Bow's Review* 16 (March 1854): 331, Hathi Trust Digital Library, at catalog.hathitrust .org/Record/008923645.

69. Majewski, 60–63. See "Address of Mr. Ruffin," *Journal of Transaction of the Virginia State Agricultural Society* (Richmond: P. D. Bernard, 1853), 16. For a full reprint of Ruffin's address, see Edmund Ruffin, "Address," *Southern Planter* 12 (February 1852): A8.

70. Carl L. Becker, "Everyman His Own Historian" (annual address of the president of the American Historical Association, delivered at Minneapolis, December 29, 1931), *American Historical Review* 37, no. 2 (1932): 221–236.

71. William Watts Folwell, *William Watts Folwell: The Autobiography and Letters of a Pioneer of Culture*, ed. Solon J. Buck (Minneapolis: University of Minnesota Press, 1933), 26.

CHAPTER 3. BUILDING THE RAILROADS

Epigraph. American Railroad Journal, September 2, 1854.

1. Victor S. Clark, *History of Manufactures in the United States* (New York: Peter Smith, 1949 [1929]), 1:235.

2. Dirk J. Struik, *Yankee Science in the Making: Science and Engineering in New England from Colonial Times to the Civil War* (Mineola, NY: Dover, 1991 [1948]), 310.

3. Ibid., 314.

4. George W. Cullum, *Biographical Register of the Officers and Graduates of the United States Military Academy at West Point, New York since Its Establishment in 1802*, 1:161–166, at Penelope .uchicago.edu/Thayer/E/Gazetter/Places/America/United_States/Army/USMA/Cullums_Reg ister/1075*.html#no. See also Struik, 314.

5. Struik, 315

6. Ibid., 321.

7. *American Industry and Manufactures in the 19th Century: A Basic Source Collection Compiled from U.S. Government Documents* (Elmsford, NY: Maxwell Reprint Co., 1970), clxxxviii– clxxxix.

8. Ibid., clxxxix. The table "Statistics of locomotive engines produced in the United States during the year ending June 1, 1860" indicated that the shop in Baltimore produced six locomotives with sixty workers, and Tredegar produced nineteen locomotives with thirty workers. The Census data are not specific, but most workers in the Baltimore and Richmond facilities were likely to be slaves, not counted by the Department of the Interior.

9. Robert C. Black III, *The Railroads of the Confederacy* (Chapel Hill: University of North Carolina Press, 1998), 23–25. Factories producing iron railings and car wheels were usually separate from those fabricating engines, although one Taunton, Massachusetts, engine manufacturer made its own wheels. Worcester, Massachusetts, made 7,000 car-wheels, which were "somewhat celebrated, being made of cold-blast charcoal iron chilled in sand-pits." Some wheels were manufactured in Cincinnati and Chicago, but most were made in Troy, Albany, Rochester, Buffalo, Jersey City, Philadelphia, and Wilmington, Delaware. The Wilmington foundry also built "chilled tires" used primarily as the drive wheel on locomotives. Chilled tires were harder than wrought iron tires and wore better. They were not as round as wrought iron or steel tires, which were turned on a lathe, and were very heavy, but they were cheaper

to manufacture. H. W. Moore at a company in Jersey City created a process of casting the tire hollow, thus reducing the tire weight from 1,000 to 720 pounds and preserving the strength. See "Scientific Notes; The Manufacture and Defects of Cast-Iron Car-Wheels—An Improved Wheel," *New York Times*, March 2, 1860, at www.nytimes.com/1860/03/02/news/scientific -notes-manufacture-defects-cast-iron-car-wheels-improved-wheel.html; *American Industry and Manufactures in the 19th Century*, clxxxv.

10. Black, 20. Types of cars included passenger, mail, baggage, and freight cars, the latter occasionally referred to as merchandise cars.

11. Henry V. Poor, *History of the Railroads and Canals of the United States* (New York: John H. Schultz, 1860), 1:104–105, 210. See *American Railroad Journal*, January 19, 1861, January 26, 1861, March 23, 1861, and August 9, 1862. The South Carolina Railroad operated on 130 miles of track between Charleston and Augusta, and the Georgia Central Railroad ran 166 miles between Savannah and Macon, in addition to operating a trunk line (unusual for a Southern railroad) of 38 miles between Eatonton and Gordon, Georgia. The New York & New Haven had 53.56 miles of main line and 63.82 miles of secondary track. The Boston & Worcester ran 68.4 miles of main track and 59.13 miles of secondary rail.

12. John E. Clark Jr., *Railroads in the Civil War: The Impact of Management on Victory and Defeat* (Baton Rouge: Louisiana State University Press, 2001), 14.

13. Charles F. O'Connell Jr., "The Corps of Engineers and the Rise of Modern Management, 1827–1856," in *Military Enterprise and Technological Change: Perspectives on the American Experience*, ed. Merritt Roe Smith (Cambridge, MA: MIT Press, 1985), 87, 91–94.

14. Ibid., 100–101. See also Alfred D. Chandler Jr., *The Visible Hand* (Cambridge, MA: Belknap Press of Harvard University Press, 1977).

15. O'Connell, 111–112. The four departments under the General Transportation Office were responsible for the day-to-day operation of the railroad and concentrated on "problems of cost determination, competitive rate making, and strategic expansion." See James A. Ward, *That Man Haupt: A Biography of Herman Haupt* (Baton Rouge: Louisiana State University Press, 1973), 28.

16. Alfred D. Chandler Jr., *Henry Varnum Poor: Business Editor, Analyst, and Reformer* (Cambridge, MA: Harvard University Press, 1956), 11–12.

17. Ibid., 38–42, 45.

18. *American Railroad Journal* 28 (September 8, 1855): 568. See also *American Railroad Journal* 28 (March 3, 1855): 129–130.

19. *American Railroad Journal* 28 (September 1, 1855): 555–556. See also *Reports of the President and the Superintendent of the New York and Erie Railroad to the Stockholders for the Year Ending September 30, 1855* (New York: Press of the New York and Erie Railroad Company, n.d.), 40. Although there is some debate over the origins of modern industrial management in the United States—and, certainly, the development of the railroads and McCallum's system had much to do with this—the Army Corps of Engineers was one of the first organizations to produce a regular reporting mechanism. In 1825, Brigadier General Winfield Scott wrote the *General Regulations for the Army*, which established guidelines throughout the service. In 1847, J. Edgar Thomson hired West Point graduate Herman Haupt as "principle [*sic*] assistant."

20. Ward, 108.

21. Nelson Morehouse Blake, *William Mahone of Virginia: Soldier and Political Insurgent* (Richmond, VA: Garrett & Massie, 1935), 6, 10–11.

22. Ibid., 10, 12–13.

23. Ibid., 23–24. William Gregg of South Carolina wrote: "The plank road is capable of meeting all the wants of our country, and superior to the railroad in every particular but that of indulging our fancy in rapidly passing from one point to another; it is so simple and cheap in its construction and management, that there is scarcely a village or an agricultural section of our country that cannot afford to build and maintain one." See Mitchell Broadus, *William Gregg, Factory Master of the Old South* (Chapel Hill: University of North Carolina Press, 1928), 152–156.

24. "William Mahone (1826–1895)," at www.thelatinlibrary.com/chron/civilwarnotes/mahone.html.

25. See the Florida Internal Improvement Act of 1855; Biographical Directory of the United States Congress, "Yulee, David Levy (1810–1886)," at http://bioguide.congress.gov/scripts/biodisplay.pl?index=Y000061. See also Leon Hühner, "David L. Yulee, Florida's First Senator," in *Jews in the South*, ed. Leonard Dinnerstein and Mary Dale Palsson (Baton Rouge: Louisiana State University Press, 1973), 68.

26. Maury Wiseman, "David Levy Yulee: Conflict and Continuity in Social Memory" (University of Florida, Gainesville, Florida), at fch.ju.edu/fch-2006/wiseman-davidlevyyulee.htm.

27. Ibid., 1.

28. Allen W. Trelease, *The North Carolina Railroad, 1849–1871, and the Modernization of North Carolina* (Chapel Hill: University of North Carolina Press, 1991), 59.

29. Ibid., 111. Trelease points out that two-thirds of the railroad's labor force consisted of African American slaves. Most, like white men, worked as unskilled laborers, but some were employed as machinists. Blacks also worked as boilermakers, patternmakers, and coppersmiths. There were few white machinists in the South. Most of them were recruited from the North. The master machinist was from Massachusetts. Of the fifty-two men listed as machinists in 1857, twenty-seven were from outside North Carolina, and twenty of these were from the North. Firemen and brakemen were also considered skilled positions, and in 1857, the NCRR employed thirty; twenty-seven of these were black, of which twenty-one were slaves. See ibid., 62–69 and 391, n. 50.

30. Ibid., 118.

31. Ibid., 118–119.

32. George W. Cullum, *Biographical Register of the Officers and Graduates of the United States Military Academy at West Point, New York since Its Establishment in 1802*, 2:81, at Penelope.uchicago.edu/Thayer/E/Gazetter/Places/America/United_States/Army/USMA/Cullums_Register/1075*.html#no. According to Cullum, when Fremont graduated, his name was recorded as Sewall L. Fish.

33. Confederate Railroads, "Biography of William M. Wadley," at www.csa-railroads.com/Essays/Biography_of_William_M._Wadley.htm.

34. Alden Partridge, West Point Class of 1806, founded the American Literary, Scientific, and Military Academy (now Norwich University) in 1819 in Norwich, Vermont. Partridge served as a first lieutenant of engineers and assistant professor of mathematics before establishing his own school. He moved his school to Middletown, Connecticut, in 1823 and relocated it back to Norwich in 1829. See Cullum, *Biographical Register*; Lester A. Webb, *Captain Alden Partridge and the United States Military Academy, 1806–1833* (Northport, AL: American Southern, 1965).

35. Black, 28. The president of the New Orleans, Jackson & Great Northern Railroad faced a number of challenges. Ranney wrote in his March 1, 1861, annual report: "It was a mistaken

policy of the engineers, who had charge of the construction of the road [from New Orleans to Arberdeen, Mississippi], who, to save a small amount on the first cost of construction, fixed the levels on these portions of the road below high water mark, from which the road has suffered in loss of business, reputation and extraordinary repairs, not less than One Million of Dollars since its construction." See "Annual Report of the New Orleans, Jackson & Great Northern Railroad as of March 1, 1861, President's Report," Essays & Documents: Index of Confederate Railroad Presidents, at www.csa-railroads.com.

36. Black, 28–29.

37. David Brooks, "Psst! 'Human Capital,'" (editorial), *New York Times,* November 2005. Brooks defines "cognitive capital" as the ability to evaluate situations, solve problems, share information, and develop brainpower. His "aspirational capital involved the ability to achieve something society believed you could not accomplish." This is "the fire-in-the-belly ambition to achieve."

38. Black, 30.

39. *Richmond Dispatch*, March 22, 1861, at www.csa-railroads.com/Essays/Originial Docs/ NP/RD/NP, RD 3-22A-61.htm.

40. *Daily Richmond Examiner*, December 16, 1861.

41. *American Railroad Journal*, August 9, 1862.

42. The junction farthest north was the Memphis & Ohio meeting at Humboldt, Tennessee. Next was the Memphis & Charleston intersecting at Corinth, Mississippi. Freight and passengers traveling from Memphis to Charleston, however, would be required to take four different railroads—the Memphis & Charleston, the Western & Atlantic, the Georgia Railroad, and the South Carolina Railroad. The final junction was at Meridian, Mississippi, along the Southern Railroad of Mississippi.

43. OR, ser. 4, vol. 1, 394, 485–486.

44. Thomas W. Chadwick, ed., "Diary of Samuel Edward Burges, 1860–1862," *South Carolina Historical and Genealogical Magazine* 48, no. 3 (July 1947): 157. See Colton's *Map of Savannah* (1855) and G. W. Colton, *Colton's Atlas of the World, Illustrating Physical and Political Geography*, vol. 1 (New York: J. H. Colton, 1856).

45. An examination of the Wilmington & Weldon Railroad in 1860 illustrates this point. Since 1839 the company had built two engines in company shops and purchased the remaining twenty-eight from locomotive manufacturers, twenty-five of which were located in the North. The Baldwin Locomotive Works produced thirteen of the engines, and the most dominant producer of locomotives in the mid-nineteenth century, the Norris Locomotive Works, manufactured eight. Both works were located in Philadelphia. The Manchester Locomotive Works, in Manchester, New Hampshire, built three engines, and the Hinkley Company, officially known as the Boston Locomotive Works, built one. Before 1855, a small Richmond company, Burr, Pea & Simpson (known before it closed as Burr & Ettinger), manufactured two engines, and the Tredegar Iron Works built one. See North Carolina Business History, "Railroads—Wilmington & Raleigh (later Weldon) Locomotives," at www.historync.org/rail road-WWRRLocomotives.htm.

CHAPTER 4. WANTED: VOLUNTEER ENGINEERS

1. Wesley Brainerd, *Bridge Building in Wartime: Colonel Wesley Brainerd's Memoir of the 50th New York Volunteer Engineers*, ed. Ed Malles (Knoxville: University of Tennessee Press, 1997), 269.

2. Ibid., 6.

3. Ibid., 269.

4. Before the war, army engineers William Rosecrans, George Meade, Robert E. Lee, and P. G. T. Beauregard, for example, all worked on harbor improvement and fortification projects.

5. George T. Ness Jr., "Engineers of the Civil War," *Military Engineer* 44 (May–June 1952): 187.

6. These army groups operated independently from each other; the most famous became known as the Army of the Potomac, Army of the Tennessee, Army of the Ohio, Army of the Mississippi, Army of the Cumberland, and Army of the James.

7. Mark M. Boatner III, *The Civil War Dictionary* (New York: David McKay, 1959), 25, 26, 169, 858. See also "Size of the Union and Confederate Armies: Comparative Strength," at www.civilwarhome.com/armysize.htm; Edward Hagerman, *The American Civil War and the Origins of Modern Warfare: Ideas, Organization, and Field Command* (Bloomington: Indiana University Press, 1988), 61. It is difficult to determine in Hagerman's numbers whether he includes horses for artillery. For an excellent overview of artillery operations at Gettysburg, including the number of horses used, see Philip M. Cole, *Civil War Artillery at Gettysburg: Organization, Equipment, Ammunition, and Tactics* (Cambridge, MA: DaCapo Press, 2002).

8. Two other sets of numbers draw attention to the fact that more students and graduates of West Point who entered the engineers were from the North than from the South by a factor of 3 to 1. Between 1802 and 1861, 137 men served in the Corps of Engineers but did not necessarily begin their army careers in the engineers; 109 came from Union states and 28 from Confederate states. Between 1802 and 1861, of those men who did begin their army careers in the engineers, 87 were from the North and 24 from the South.

9. George W. Cullum, *Biographical Register of the Officers and Graduates of the United States Military Academy at West Point, New York since Its Establishment in 1802,* at http://penelope.u chicago.edu/Thayer/E/Gazetteer/Places/America/United_States/Army/USMA/Cullums_Reg ister/home.html. Pierre G. T. Beauregard, William Henry Whiting, Edward Porter Alexander, and George Washington Custis Lee all served as field commanders for the Confederacy. Robert E. Lee was not on the list of corps members because at the time Virginia seceded he was in command of cavalry.

10. Massachusetts produced nineteen engineers, Pennsylvania thirteen, Ohio nine, and Vermont six. Eight cadets born in Virginia served in the Corps of Engineers, four from Tennessee, four from North Carolina, three from Louisiana, and two each from South Carolina, Georgia, Alabama, and Mississippi.

11. The battle was deemed devastating for two reasons. First, the Lincoln government fully expected to defeat the upstart Confederates in one crushing Napoleonic victory. Second, the battle turned into a full-blown panic, with Union soldiers and civilians (who had come out from Washington to witness the battle) fleeing back to Washington. Morale in the days that followed was torturous.

12. George B. McClellan to Mary Ellen McClellan, July 27, 1861, in *The Civil War Papers of George B. McClellan: Selected Correspondence, 1860–1865,* ed. Stephen W. Sears (New York: Ticknor & Fields, 1989), 70.

13. Hagerman, 65.

14. Captain Alexander J. Swift was given command of the new company of engineer soldiers. Swift had spent two years as a student at the School of Application for Artillery and

Engineers in France, before returning to the United States in early 1846, just before Congress authorized the new unit.

15. Russell F. Weigley, *History of the United States Army* (Bloomington: Indiana University Press, 1984), 182–187.

16. The Crimean War was fought between the Russian Empire and the Ottoman Empire, Great Britain, and France.

17. Phillip Thienel, "Engineers in the Union Army, 1861–1865: Part I. Engineer Organization," *Military Engineer* 47, no. 315 (January–February 1955): 36.

18. Captain J. C. Duane, *Manual for Engineer Troops* (New York: D. Van Nostrand, 1862).

19. McClellan to Mary Ellen McClellan, October 11, 1861, in Sears, 106.

20. Ness, 180–181.

21. Earl J. Hess, *Field Armies and Fortifications in the Civil War: The Eastern Campaigns, 1861–1864* (Chapel Hill: University of North Carolina Press, 2005), 139.

22. Ron Field, *American Civil War Fortifications (2): Land and Field Fortifications* (Oxford: Osprey, 2005), 26.

23. Ibid., 31. A lunette was a detached fieldwork shaped like an irregular pentagon with the base open and the vertex facing in the direction of the enemy. Lunettes were used as advanced works in front of the main line, covering the gaps between major fortifications.

24. Barnard was promoted from a major in the regular army to brigadier general of United States Volunteers. His general's rank remained, as well as the pay, for the duration of the war. In 1865, when the volunteers were mustered out of service, his rank reverted to its prewar grade, pay, and pension of a major.

25. India-rubber pontoons were made from three cylinders, each twenty feet long and twenty inches wide, bound together by rubber straps. Each cylinder had a brass air nozzle, and the pontoon was inflated using a large bellows that fit over the nozzle. This type of bridge was easy to transport and easy to repair, but if damaged while supporting a large number of men and horses it could prove disastrous.

26. OR, ser. 1, vol. 5, chap. 14, 616–617.

27. Ibid., 617.

28. Hagerman, 234

29. Bruce Catton, *Reflections on the Civil War*, ed. John Leekley (Garden City, NY: Doubleday, 1981), 181–182. The engineer insignia was a metal replica of an armory with battlemented towers.

30. OR, ser. 1, vol. 5, 617–619.

31. Ibid., 619, 620.

32. Ibid., 621.

33. Brainerd, 44. See also Brainerd, 348, n. 5: "The Washington Engineer Depot was located one-half mile north of the Navy Yard, near the foot of East 14th and/or East 15th Streets."

34. "Soldiers in the Civil War from Broome County Clerk's Office, Binghamton, NY: Miscellaneous Records," at freepages.genealogy.rootsweb.ancestry.com/~marcri/civilwar.html.

35. Frederick Phisterer, *New York in the War of Rebellion, 1861 to 1865* (Albany, NY: Weed, Parsons, 1890), 273.

36. "Abstract of New York State Muster Rolls, Volunteers," 13775-83 MUI, reel 959, New York State Archives, Albany, New York. Sappers were specialists in digging crisscrossing trenches, called saps, that shielded foot soldiers from angled rifle fire.

37. OR, ser. 1, vol. 6, 245–247.

38. A corduroy road was built from felled trees or logs to aid travelers over impassable

muddy roads. First, logs were laid parallel along each side of the roadway, then logs were placed perpendicular to these and fastened at each end to the parallel logs, thus providing an elevated surface and making it easier for traffic to pass over.

39. The boats were 31 feet long, 3½ feet deep, and approximately 3 feet wide at the bow and 4½ feet wide at the stern. The thirty-four boats that made up the bridge train weighed about half a ton each. When building a bridge, the pontoons were placed 20 feet apart, and every alternate one was anchored upstream and downstream. Crosstie wood spacers were laid across each pontoon to serve as the underside of the bridge floor, or deck. The chess, or flooring, was laid over the balks. The chesses were about 1 foot wide and 15 feet long; they were not nailed to the balks but held in position by side rails laid over them, one at each side, and lashed to the balks underneath. This formed the bridge.

40. Fascines were tightly bound bundles of brushwood used in building the foundation for earthworks and batteries. Gabions were cylindrical wicker basket–like objects, open at both ends, 3 feet tall and 2 feet in diameter. The baskets were filled with earth to form a stationary defense.

41. Benjamin F. Cooling, *Symbol, Sword, and Shield: Defending Washington during the Civil War* (Hamden, CT: Archon, 1975), 79–80. See also *Congressional Globe*, 37th Congress, 2nd sess., January 13, 1862, 286.

42. Herman Hattaway and Archer Jones, *How the North Won: A Military History of the Civil War* (Urbana: University of Illinois Press, 1983), 56–57.

43. Ibid., 58.

44. Ibid.

45. The highest-ranking officer in the Confederate army was Samuel Cooper, the adjutant- and inspector-general.

46. Using the basic formula distance = rate $\times$ time, we can approximate the time it would take to move men by rail from Columbus, Kentucky, to Bowling Green, Kentucky, over the available railroads. In 1860, trains averaged between 15 and 20 mph. It would take almost eleven hours to travel the distance between the two cities by train and about fifty-four hours on foot.

47. John F. Marszalek, *Commander of All Lincoln's Armies: A Life of General Henry W. Halleck* (Cambridge, MA: Belknap Press of Harvard University Press, 2004), 108.

48. Three independent engineer companies were formed, two of which served for only ninety days: Balz's Company Sappers and Miners and Wolster's Independent Company Sappers and Miners. The latter company's known accomplishment was that, under General Nathaniel Lyon's orders, it repaired the road from Rolla to Springfield, Missouri, in May and June 1861. Gerster's Independent Company Pioneers served twelve months for the Department of Missouri in the southwest region of the state.

49. W. A. Neal, *An Illustrated History of the Missouri Engineer and the 25th Infantry Regiments* (Chicago: Donhue and Henneberry, 1889), 9. Neal suggested that Bissell was colonel of the 10th Missouri Volunteer Infantry before he approached Frémont about an engineering regiment, yet there is no corroborating evidence. *The Military Order of the Loyal Legion of the United States War Papers* (1887–1915; repr., Wilmington, NC: Broadfoot, 1992) (hereafter, MOLLUS), Missouri Commandery, listed Bissell as belonging only to the volunteer engineers, not the infantry. See "Index to the Officers of Missouri Volunteers and Missouri State Militia," at home .usmo.com/~momollus/MOofficersB1.htm.

50. Stephen Beszedits, "Hungarians with General John C. Frémont in the American Civil War," at www.skszeged.hu/statikus_html/vasvary/newsletter/03dec/beszedits.html.

51. Neal, 9–15.

52. "Record of Events for Josiah W. Bissell's Engineer Regiment of the West, July 1861–April 1864," in *Supplement to the Official Records of the Union and Confederate Armies Part II—Record of Events*, ed. Janet B. Hewett (Wilmington, NC: Broadfoot, 1996), 36 (no. 48): 214.

53. Ibid., 195–196.

54. *OR*, ser. 1, vol. 4, 263.

55. Whitelaw Reid, *Ohio in the War: Her Statesmen, Generals, and Soldiers* (Cincinnati, OH: Robert Clarke, 1895), 1:591. An impact crater on the planet Mars is named in Mitchell's honor. During the war, his men referred to him as "Old Stars."

56. *OR*, ser. 1, vol. 4, 276.

57. Ibid. Confederate Brigadier General Felix Kirk Zollicoffer was in command of one of Major General George B. Crittenden's brigades and was responsible for protecting the Cumberland Gap.

58. Frederick H. Dyer, *A Compendium of the War of the Rebellion* (Des Moines, IA: Dyer, 1908).

59. Mark Hoffman, *"My Brave Mechanics": The First Michigan Engineers and Their Civil War* (Detroit: Wayne State University Press, 2007), 4. See also "Howland's Engineers," at www.michiganinthewar.org/engineers/howland.htm. The Chicago regiment was never formed, and continuous attempts to organize the regiment were finally abandoned in February 1862.

60. Charles R. Sligh, *History of the Services of the First Regiment Michigan Engineers and Mechanics, during the Civil War, 1861–1865* (Grand Rapids, MI: White Printing, 1921), 7.

61. *OR*, ser. 3, vol. 1, 497, 509.

62. Hoffman, 9–10, 11.

63. Sligh, 7; Hoffman, 320.

64. Robert E. L. Krick, *Staff Officers in Gray: A Biographical Register of the Staff Officers in the Army of Northern Virginia* (Chapel Hill: University of North Carolina Press, 2003), 15–16. See also *OR*, ser. 1, vol. 5, 953.

65. James L. Nichols, *Confederate Centennial Studies Number Five: Confederate Engineers* (Tuscaloosa, AL: Confederate Publishing, 1957), 10–11. The two majors were William H. C. Whiting and Danville Leadbetter. The captains were Walter H. Stevens; William R. Boggs, who served as first lieutenant of ordnance in the US army; Edward P. Alexander; Samuel H. Lockett; and William H. Echols, a topographical engineer.

66. Ibid., 15, 24.

67. Ibid., 17–19.

68. List of Officers in the Engineer Service of the State of Virginia, October 28, 1861, Military Affairs: Engineer Department Index to Letters Received 1861, Index to Letters Sent 1861, accession number 36887, box 4, no. 73, Library of Virginia, Richmond, Virginia (hereafter, Letters Received and Sent).

69. Ibid., no. 437.

70. Pay Roll of Sundry Persons Employed by the Commonwealth of Virginia for Coast, Harbor, and River Defenses, and on the Defensive Works at Richmond, Index to Material Book of Engineer Department of Virginia 1861, Library of Virginia, Richmond, Virginia (hereafter, Pay Roll); Engineer Department, Slave Rolls, July, August, September 1861 for Richmond, accession number 36887, folder 9, box 8, Library of Virginia, Richmond, Virginia (hereafter, Slave Rolls). Slaves working on the Richmond fortifications had to pay for their

own shoes, clothing, and food. A slave's owner decided whether the slave could keep whatever money was left over.

71. Pay Roll; Slave Rolls.

72. Letters Received and Sent, no. 387.

73. Krick, 40.

74. Thomas Lawrence Connelly, *Army of the Heartland: The Army of Tennessee, 1861–1862* (Baton Rouge: Louisiana State University Press, 1967), 8.

75. Ibid., 9.

76. Ibid., 25.

77. Alexis de Tocqueville said of Southerners: "The citizen of the Southern states becomes a sort of domestic dictator from infancy; the first notion he acquires in life is that he was born to command, and the first habit he contracts is that of ruling without resistance." See David Donald, *Lincoln Reconsidered* (New York: Vintage, 1961), 223.

78. Forts Harris, Wright, and Pillow and defenses at Memphis were being built.

79. Bromfield L. Ridley, *Battles and Sketches, Army of Tennessee 1861–1865* (Mexico, 1906), 63–66.

80. Benjamin Franklin Cooling, *Forts Henry and Donelson: The Key to the Confederate Heartland* (Knoxville: University of Tennessee Press, 1987), 46.

81. Ibid., 47.

82. Captain Jesse Taylor, "Defense of Fort Henry," in *Battles and Leaders of the Civil War*, 2nd ed., ed. Robert U. Johnson and Clarence C. Buel (New York: Century Company, 1956), 1:368–369.

83. See ibid., 368; Connelly, 19.

84. Cooling, 56.

85. OR, ser. 1, vol. 4, 463, 488.

86. Ibid., 456, 463–469. See also Nichols, 43, 44.

87. Cooling, 54.

88. Nichols, 44.

89. OR, ser. 1, vol. 4, 496–497, 506, 513.

90. OR, ser. 1, vol. 7, 685, 692–700, 703–704, 709, 719, 723–724, 733–735. See also Connelly, 104–105. Pillow was scheming. By late November he started to report an increase in enemy forces in front of him: the numbers went from 25,000 to 100,000 Union soldiers. Pillow was attempting to get reinforced so that he could launch a winter offensive in the vicinity of Cairo, Illinois. Unfortunately, Pillow's alarming reports focused attention on him (part of his plan) and away from the twin forts.

91. Other engineer units were formed and served anywhere from three to thirteen months: Chadwick's Engineers (August–October 1861), Howland's Battle Creek Engineers (October–December 1861), Wolster's Independent Company of Sappers and Miners (May–September 1861), Balz's Company of Sappers and Miners (October 1861–February 1862), and Gerster's Independent Company of Pioneers (August 1861–September 1862).

CHAPTER 5. EARLY SUCCESSES AND FAILURES

1. The battle was also known as Logan's Cross Roads, Fishing Creek, Somerset, and Beech Grove.

2. Martin Van Creveld, *Supplying War: Logistics from Wallenstein to Patton* (Cambridge: Cambridge University Press, 1977), 231–232.

3. *OR*, ser. 1, vol. 7, 137.

4. Spencer C. Tucker, *Unconditional Surrender: The Capture of Forts Henry and Donelson* (Abilene, TX: McWhiney Foundation Press, 2001), 48.

5. *OR*, ser. 1, vol. 7, 131–132.

6. Ibid., 147.

7. Benjamin Franklin Cooling, *Fort Henry and Fort Donelson: The Key to the Confederate Heartland* (Knoxville: University of Tennessee Press, 1988), 88.

8. *OR*, ser. 1, vol. 7, 139.

9. Ibid., 149.

10. Richard D. Goff, *Confederate Supply* (Durham: Duke University Press, 1969), 56.

11. The Cumberland River acted as a thirty-one-mile highway between the train station at Clarksville and Fort Donelson. It was sixty-three miles between Paris and Clarksville and forty-nine miles between Clarksville and Nashville. These were short distances by train. Ten miles separated the two forts, connected by the Telegraph Road and, just south of this footpath, Bottom Road. There was no direct road from Paris to Fort Heiman. All the railroads were on five-foot gauges. From Bowling Green running due south was the Louisville & Nashville Railroad. From Bowling Green, moving southwest, was the Memphis, Clarksville & Louisville, which ended at Paris, Tennessee; this line then became the Memphis & Ohio, which connected Memphis, Humboldt, and Johnsonville, Tennessee, to Paris. The Mobile & Ohio Railroad, in its northern section, connected General Polk's army in Columbus, Kentucky, with Union City and Humboldt on a north-south axis

12. *OR*, ser. 1, vol. 7, 818–819.

13. Ibid., 819.

14. See A. S. Johnston's letter to Secretary of War J. P. Benjamin, January 5, 1862, in *OR*, ser. 1, vol. 7, 820–822.

15. Cooling, 84.

16. Robert C. Black III, *The Railroads of the Confederacy* (Chapel Hill: University of North Carolina Press, 1998).

17. *Journal of the Confederate States Congress* 1 (1862): 654, 720, 721; *OR*, ser. 4, vol. 1, 884–885.

18. *OR*, ser. 4, vol. 1, 884.

19. Ibid.

20. *OR*, ser. 1, vol. 7, 895.

21. Stanley F. Horn, *The Army of Tennessee* (Norman: University of Oklahoma Press, 1993), 99.

22. In an undated response to a congressional inquiry regarding the Nashville evacuation, Colonel Nathan B. Forrest reported that only a portion of the stores were removed before the city surrendered. Because the army could not remove all the supplies, "The quartermaster's stores were open, and the citizens were invited to come and help themselves, which they did in larger crowds, if possible, than at the other department [commissary]." See *OR*, ser. 1, vol. 7, 429–430.

23. *Journal of the Confederate States Congress* 5 (1862): 82.

24. Ibid., 251–253.

25. Ibid., 152, 188, 215, 253–254, 269.

26. Ibid., 269.

27. Message of President Davis on Railroads, *Wilmington Journal*, February 17, 1862, at www.csa-railroads.com/Essays/Confederate_Railroads_Original_Documents.htm.

28. At the time of the forts Henry and Donelson campaigns, the western command was divided between, first, Halleck (Department of the Missouri), headquartered in St. Louis, and then Brigadier General Don Carlos Buell (Department of the Ohio), headquartered in Louisville. On March 11, 1862, Halleck was promoted to command the newly created Department of the Mississippi, giving him unified command of western forces.

29. *OR*, ser. 1, vol. 3, 617–619, 654.

30. There were woods in the area between where the river turned north toward New Madrid and where it turned south. This parcel of land west of Island No. 10 was in Tennessee.

31. Larry J. Daniel and Lynn N. Bock, *Island No. 10: Struggle for the Mississippi Valley* (Tuscaloosa: University of Alabama Press, 1996), 43.

32. *OR*, ser. 1, vol. 8, 80. The swamp was approximately thirty miles wide.

33. *Cincinnati Commercial*, March 10 and March 17, 1862; *New York Times*, March 17, 1862.

34. *OR*, ser. 1, vol. 8, 80.

35. Daniel and Bock, 44.

36. Ibid., 41.

37. Janet B. Hewett, ed., *Supplement to the Official Records of the Union and Confederate Armies*, pt. II (Wilmington, NC: Broadfoot, 1996), 36:199, 220.

38. Ibid., 225, 226, 229, 233, 240.

39. James Buchanan Eads built Foote's gunboats in St. Louis, Missouri. Known as City-class ironclads, each vessel had 251 officers and men.

40. James M. Powles, "Steaming through the Trees," *American Civil War*, March 2005, 34.

41. J. W. Bissell, "Sawing out the Channel above Island Number Ten," in *Battles and Leaders of the Civil War: From Sumter to Shiloh*, ed. Robert C. Johnson and Clarence C. Buel (1887–1888; repr., New York: Thomas Yoseloff, 1956), 1:460.

42. Ibid.

43. W. A. Neal, *An Illustrated History of the Missouri Engineers and the Twenty-Fifth Infantry Regiment* (Chicago: Donohue and Henneperry, 1889), 37, 42–43. Pope wrote to Halleck on March 19, 1862: "Have had the country examined between here [New Madrid] and Islands 8 and 10. Had to be done in skiffs, as the whole region is under water . . . Am having an examination made, to see if by digging across one or two ridges I cannot connect Island No. 8 with the river below Island No. 10 by connecting two bayous." On the same day, Pope wrote to Bissell: "I desire, you . . . to make an examination of the peninsula opposite Island No. 10, to ascertain whether a short canal, not to exceed 2 miles in length, cannot be dug, so that boats can enter above Island No. 10 and come out into the river below it." See *OR*, ser. 1, vol. 8, 625.

44. *OR*, ser. 1, vol. 8, 104. See also "The Schuyler Hamilton Canal," *New York Herald*, April 13, 1862; "Comment by General Schuyler Hamilton, Major General, U. S. V.," in Johnson and Buel, 1:462.

45. *OR*, ser. 1, vol. 8, 79.

46. Bissell, 461.

47. Ibid. A snatch block is a pulley inside a casing that is used to assist with winching duties.

48. Ibid.

49. Powles, 36; Bissell, 461.

50. *Memphis Appeal*, April 4, 1862.

51. *OR*, ser. 1, vol. 8, 132.

52. "How Colonel Bissell's Engineers Fores [Force] Their Way to General Pope," *Philadelphia Inquirer*, April 11, 1862; "The Great Western Stump Cutter," *Philadelphia Inquirer*, April 16, 1862.

53. "Great Exploit of the Engineers," *Macon Telegraph*, April 24, 1862.

54. For an excellent discussion and explanation of logistical problems, see Edward Hagerman, *The American Civil War and the Origins of Modern Warfare* (Bloomington: Indiana University Press, 1992), 58–64.

55. Frank J. Welcher, *The Union Army 1861–1865: Organization and Operations* (Bloomington: Indiana University Press, 1993), 2:82. See also Mark Hoffman, *"My Brave Mechanics": The First Michigan Engineers and Their Civil War* (Detroit: Wayne State University Press, 2007), 58.

56. F. A. Mitchell, *Ormsby MacKnight Mitchell, Astronomer and General: A Biographical Narrative* (New York: Houghton, Mifflin, 1887), 267–272.

57. Hoffman, 58. Running southeast from Nashville was the Nashville & Chattanooga Railroad. Stops included Murfreesboro, Tullahoma, and finally Stevenson, Alabama. In Stevenson it met the Memphis & Charleston, which ran east to Bridgeport, Alabama, and beyond Bridgeport forked northeast to Chattanooga and continued to Richmond, Virginia. The southeast/south fork went through Dalton and Atlanta, Georgia, terminating in Charleston, South Carolina. Running southwest from Nashville, three railroads—the Tennessee & Alabama to Columbia, the Central Southern to the state line, and the Tennessee & Central Alabama to Decatur—completed the line to the Memphis & Charleston. West from Decatur the railroad ran to Corinth and ended at Memphis. These lines were vital to Confederate armies' internal transportation and supply network and equally vital to the Union strategy for taking the war into the Confederate heartland. Ambrose Bierce's "Occurrence at Owl Creek Bridge," first published in the San Francisco *Examiner* in 1890, is a vivid fictional measure of this reality.

58. Mitchell, 267–272. See also Henry J. Haynie, *The Nineteenth Illinois: A Memoir of a Regiment of Volunteer Infantry Famous in the Civil War of Fifty Years Ago for Its Drill, Bravery, and Distinguished Service* (Chicago: M. A. Donohue, 1912), 163–164.

59. Hoffman, 59.

60. *OR*, ser. 1, vol. 10, pt. 1, 329.

61. Ibid.

62. George Edgar Turner, *Victory Rode the Rails: The Strategic Place of Railroads in the Civil War* (Indianapolis: Bobbs-Merrill, 1953), 125.

63. Russell F. Weigley, *Quartermaster General of the Union Army: A Biography of M. C. Meigs* (New York: Columbia University Press, 1959), 5.

64. John E. Clark Jr., *Railroads in the Civil War: The Impact of Management on Victory and Defeat* (Baton Rouge: Louisiana State University Press, 2001), 22–23.

65. *Congressional Globe*, 37th Congress, 2nd sess., 1862, 32:1022–1030.

66. Hoffman, 351, n. 29. A sergeant in the regular army engineers received $24 a month, whereas infantry sergeants received $17 a month. Corporals were paid $20 in the engineers and $13 in the infantry.

CHAPTER 6. McClellan Tests His Engineers

1. Telegram from G. B. McClellan to E. M. Stanton, February 26, 1862, in *The Civil War Papers of George B. McClellan: Selected Correspondence, 1860–1865*, ed. Stephen W. Sears (New York: Ticknor & Fields, 1989), 191.

2. Captain J. C. Duane, *Manual for Engineer Troops* (New York: D. Van Nostrand, 1862), 7–48.

3. Communications to the Secretary of War and to Congress, No. 10, April 22, 1859 to January 12, 1863, RG 77, entry 8, Letters, Reports, and Statements Sent to the Secretary of War, etc., Engineer Department, February 25, 1862, Totten to Stanton, Letter 347, National Archives, Washington, DC.

4. Unofficial and Private letter from Geo. B. McClellan to Joseph G. Totten, March 28, 1862, in Sears, *Civil War Papers*, 218.

5. Frederick Phisterer, *New York in the War of the Rebellion: 1861 to 1865* (Albany: State of New York, 1912), 2:1650–1651, 1669–1670. See also "A Century of Lawmaking for a New Nation: U.S. Congressional Globe," Appendix to the *Congressional Globe*, July 17, 1862, Library of Congress / American Memory, at http://memory.loc.gov/ammen/amlaw/lwcg.html.

6. In a Civil War peculiarity, regular army engineers retained their regular army rank. When in command of volunteer troops, they would also carry a volunteer rank. Thus, McClellan listed Woodbury as "Brigadier General D. P. Woodbury, Major U.S. Engineers." See George B. McClellan, *Report of the Organization and Campaigns of the Army of the Potomac: To Which Is Added an Account of the Campaign in Western Virginia with Plans of Battle-Fields* (Freeport, NY: Books for Libraries Press, 1970 [1864]), 63.

7. Adrian G. Trass, *From the Golden Gate to Mexico City: The U.S. Army Topographical Engineers in the Mexican War, 1846–1848* (Washington, DC: Center of Military History, 1993), 222. The Smithsonian Institution, established in 1846, was first designated as a center for scientific research but soon became a repository for ethnographic artifacts and animal and plant specimens from around the world. The Naval Hydrographic Office made nautical charts, measured the physical features of oceans, lakes, and rivers, and made predictions about changes to those features in the future. The Pacific Wagon Road Office surveyed and built a wagon road from New Mexico to California in the 1850s.

8. Office of Coast Survey, National Oceanic and Atmospheric Administration, "History of Coast Survey," at www.nauticalcharts.noaa.gov/staff/hist.html.

9. Report of Brigadier General Andrew A. Humphreys, U.S. Army, Chief of Topographical Engineers, February 20, 1863, in *OR*, ser. 1, vol. 11, pt. 1, 152.

10. McClellan, 65.

11. Earl J. Hess, *Field Armies and Fortifications in the Civil War: The Eastern Campaigns, 1861–1864* (Chapel Hill: University of North Carolina Press, 2005), 13.

12. Thomas B. Buell, *The Warrior Generals: Combat Leadership in the Civil War* (New York: Three Rivers Press, 1997), 70.

13. Ibid., 70.

14. George T. Ness Jr., "Engineers of the Civil War," *Military Engineer*, May–June 1952, 181. See also Robert E. L. Krick, *Staff Officers in Gray: A Biographical Register of the Staff Officers in the Army of Northern Virginia* (Chapel Hill: University of North Carolina Press, 2003), 71–72, 101, 324–325, 335, 355.

15. Lee ranked second in his West Point class of 1829 and after graduation entered the Corps of Engineers. He had worked on harbor, fortification, and surveying projects by the time war with Mexico broke out in the spring of 1846. He served with distinction as an engineer on General Winfield Scott's staff, but after the war he longed for a combat command. His wish was granted in 1855 when he was transferred to the cavalry, and by 1860 he was promoted to colonel.

16. Krick, 90.

17. James L. Nichols, "Confederate Map Supply," *Military Engineer*, January–February 1954, 28.

18. Peter W. Roper, *Jedediah Hotchkiss: Rebel Mapmaker and Virginia Businessman* (Shippensburg, PA: White Mane, 1992), 193.

19. Earl B. McElfresh, *Maps and Mapmakers of the Civil War* (New York: Harry N. Abrams, 1999), 29.

20. Ibid., 17, 21.

21. *OR*, ser. 1, vol. 11, pt. 1, 320–321.

22. Ibid., 327.

23. Phillip M. Thienel, *Mr. Lincoln's Bridge Builders: The Right Hand of American Genius* (Shippensburg, PA: White Mane, 2000), 36.

24. Phillip M. Thienel, "Engineers in the Union Army, 1861–1865: Part I, Engineer Organization," *Military Engineer* 47, no. 315 (January–February 1955): 38. This floating wharf was a forerunner of the floating Mulberry Harbor used by the Allies during the Normandy invasion eighty-one years later.

25. *OR*, ser. 1, vol. 11, pt. 1, 142–145. Woodbury's report also included work done by the Engineer Brigade on railroad bridges and roads. A trestle is a rigid frame consisting of a series of short spans supported by a complex network of cross-braced posts or framed latticework. A crib is a frame of timbers placed horizontally on top of each other to form a wall or a square structure used to support a road. Cribs made effective approaches to bridges.

26. William J. Miller, "Scarcely Any Parallel in History: Logistics, Friction and McClellan's Strategy for the Peninsula Campaign," in *The Peninsula Campaign of 1862: Yorktown to the Seven Days*, ed. William J. Miller (Cambridge, MA: Da Capo Press, 1995), 2:137.

27. Charles B. Haydon, *For Country, Cause & Leader: The Civil War Journal of Charles B. Haydon*, ed. Stephen W. Sears (New York: Ticknor & Fields, 1993), 221.

28. *OR*, ser. 1, vol. 11, pt. 1, 161.

29. Miller, "Scarcely Any Parallel," 157.

30. Ibid., 150.

31. William J. Miller, "Weather Still Execrable: Climatological Notes on the Peninsula Campaign March through August 1862," in *The Peninsula Campaign of 1862: Yorktown to the Seven Days*, ed. William J. Miller (Campbell, CA: Savas, 1997), 3:191.

32. No. 2—Reports of Brigadier General John G. Barnard, U.S. Army, Chief Engineer of Operations from May 23, 1861 to August 15, 1862, in *OR*, ser. 1, vol. 11, pt. 1, 106.

33. Stephen W. Sears, *To the Gates of Richmond: The Peninsula Campaign* (New York: Ticknor & Fields, 1992), 136.

34. Oliver Otis Howard, *Autobiography of Oliver Otis Howard, Major General United States Army* (New York: Baker & Taylor, 1907), 1:237–239.

35. *OR*, ser. 1, vol. 11, pt. 1, 113–122.

36. Ibid., 148.

37. Ibid., 151.

38. Douglas Southall Freeman, *Manassas to Malvern Hill*, vol. 1 of *Lee's Lieutenants: A Study in Command* (New York: Charles Scribner's, 1942), 561. The Grapevine Bridge was known by several names, including the Upper Trestle Bridge and Sumner's Upper Bridge.

39. Ibid. See also Sears, *To the Gates of Richmond*, 269; R. L. Dabney to Jed Hotchkiss, April

22, 1896, Papers of Jedediah Hotchkiss, accession no. 2822 and 2907, Special Collections, University of Virginia, Charlottesville, Virginia.

40. *OR*, ser. 1, vol. 11, pt. 1, 146–147.

41. Ibid., 123.

42. Ibid., 124. Barnard's report commended the following volunteer engineer officers: from the 50th New York, Colonel W. H. Pettes, Major James A. Magruder, captains Hine, Beers, Ford, Brainerd, Perry, and Spaulding; from the 15th New York, captains Ketchum and Bowers, lieutenants Slosson, Farrell, and Hassler, and C. S. Webster and H. C. Yates, both of whom died of disease during the campaign.

43. Ibid., 130.

44. Ibid., 126–127.

45. Ibid. See also George T. Ness, Jr., "Engineers of the Civil War," *Military Engineer* 44, no. 299 (May–June 1952).

46. *OR*, ser. 1, vol. 11, pt. 1, 127.

47. Ibid., 128. The Birago trestle was shaped like an isosceles trapezoid but with no top or bottom line; the sides were held together by a 20-foot-long two-by-four piece of wood in the middle of the two sides. It looked like a sawhorse. Chains were attached to the top of the sides. The general principle was that the trestle would be positioned and the legs pushed through a cap that acted like a shoe to anchor the trestle to the floor. The chains were used to help anchor the base. The balk was then used to connect two trestles. The chess was laid at ninety-degree angles to the balk, and the floor of the bridge was built atop the chess. See Duane, 20, plates 2 and 4; George W. Cullum, *Systems of Military Bridges in Use by the United States Army: Those Adopted by the Great European Powers and Such as Are Employed in British India with Directions for the Preservation, Destruction, and Re-establishment of Bridges* (New York: D. Van Nostrand, 1863), 156–157.

48. *OR*, ser. 1, vol. 11, pt. 1, 128.

49. Ibid., 126.

50. William Rattle Plum, *The Military Telegraph during the Civil War in the United States*, ed. Christopher H. Sterling (New York: Arno Press, 1974), 1:63–64, 68. See also David Homer Bates, *Lincoln in the Telegraph Office: Recollections of the United States Military Telegraph Corps during the Civil War* (New York: D. Appleton-Century, 1907), 20.

CHAPTER 7. THE BIRTH OF THE UNITED STATES MILITARY RAILROAD

1. A city ordinance in Baltimore prevented railroad lines from running through the city. Consequently, trains coming from Wilmington and Philadelphia had to stop at President Street Station. The train was then coupled to a horse team and, on trolley lines, pulled ten blocks west along Pratt Street to Camden Station, where the Baltimore & Ohio began. After the riots in which four soldiers and twelve civilians died, Maryland Governor Thomas Hicks asked Lincoln to avoid sending troops through Baltimore. This was an untenable request. Hicks then asked the militia to block the railroad bridges entering the city.

2. Office of the Quartermaster General, Military Railroads, RG 92, Brief History of Military Railroads during the Civil War, box no. 10, entry 1525, National Archives, Washington, DC. See also "Correspondence between Governor Hicks and General Butler, *New York Times*, April 29, 1861, at www.nytimes.com/1861/04/29/news/gov-hicks-gen-butler-following-corre spondence-between-governor-maryland.htm.

3. William Rattle Plum, *The Military Telegraph during the Civil War in the United States*, ed. Christopher H. Sterling (New York: Arno Press, 1974), 1:63–64, 68. See also David Homer Bates, *Lincoln in the Telegraph Office: Recollections of the United States Military Telegraph Corps during the Civil War* (New York: D. Appleton-Century, 1907), 20.

4. Plum, 68.

5. Bates, 27. See also Rebecca Robins Raines, *Getting the Message Through: A Branch History of the U.S. Army Signal Corps* (Washington, DC: Center of Military History, 1996), 16–17. The original operators of the Telegraph Corps were David Strouse, Samuel M. Brown, Richard O'Brian, and David H. Bates. In November, Anson Stager was commissioned as a colonel and appointed chief superintendent of the corps.

6. Eric Ethier, "The Operators," *Civil War Times* 45, no. 10 (January 2007): 17–18. See also *The Civil War Papers of George B. McClellan: Selected Correspondence, 1860–1865*, ed. Stephen W. Sears (New York: Ticknor & Fields, 1989), 75–76.

7. Appendix to the *Congressional Globe*, 37th Congress, 2nd sess., January 31, 1862, at http://memory.loc.gov.html.

8. Samuel Richey Kamm, *The Civil War Career of Thomas A. Scott* (Philadelphia: University of Pennsylvania Press, 1940), 38, 68, 136.

9. Ibid., 89–90.

10. Alfred D. Chandler Jr., *The Railroads: The Nation's First Big Business* (New York: Harcourt, Brace & World, 1965), 97–98.

11. Daniel C. McCallum, "Superintendent's Report, March 25, 1856," in *Annual Report of the New York and Erie Railroad Company for 1855* (New York, 1856), 33–37, 39–41, 50–54, 57–59.

12. George Edgar Turner, *Victory Rode the Rails: The Strategic Place of the Railroads in the Civil War* (Lincoln: University of Nebraska Press, 1992), 149.

13. Eulogies of the Press, "Memorial Tributes to Daniel L. Harris, with Biography and Extracts from his Journal and Letters," Internet Archive, at www.archive.org/stream/memorial tributesooburtuoft/memorialtributesooburtuoft_djvu.txt.

14. After some time, Harris finally responded in the affirmative to Stanton's offer, but received no reply. Harris then traveled to Washington, where he was told that Haupt had taken the job. Stanton offered Harris a job as Haupt's assistant; Harris immediately declined.

15. James A. Ward, *That Man Haupt: A Biography of Herman Haupt* (Baton Rouge: Louisiana State University Press, 1973), 115; Herman Haupt, *Reminiscences of General Herman Haupt* (Milwaukee: Wright & Joys, 1901).

16. Ward, 116.

17. Ibid.; Phillip M. Thienel, *Mr. Lincoln's Bridge Builders: The Right Hand of American Genius* (Shippensburg, PA: White Mane, 2000), 65–66.

18. Haupt, *Reminiscences*, 48–49.

19. Ibid., 46.

20. Note by General McDowell in *OR*, ser. 1, vol. 12, pt. 1, 281.

21. Jackson had defeated General Nathaniel P. Banks at Front Royal on May 23 and at Winchester on May 25.

22. Ward, 118.

23. Turner, 155.

24. *OR*, ser. 1, vol. 12, pt. 3, 275.

25. Ibid., 334.

26. Haupt, *Reminiscences*, 64–66.

27. Ibid., 69.

28. Ibid., 70.

29. Records of the Office of the Quartermaster General, Transportation 1834–1917: Office of the United States Military Railroad, 1860–1867, Headquarters, Correspondence Letters Received by Colonel H. Haupt, Chief of Construction and Transportation, 1862, RG 92, box 1, National Archives, Washington, DC.

30. Halleck to Haupt, August 23, 1862, Military Telegraph, RG 92, National Archives, Washington, DC.

31. Haupt to Sturgis, August 23, 1862, Military Telegraph, RG 92, National Archives, Washington, DC. See also Ward, 127.

32. Ward, 127.

33. Ibid., 131, 168.

34. *OR*, ser. 3, vol. 2, 548–549.

35. D. C. McCallum, *United States Military Railroads: Report of Bvt. Brigadier General D. C. McCallum, Director and General Manager, from 1861 to 1866* (ULAN Press, 2013 [n.d.]), 10; see also *OR*, ser. 3, vol. 5, 974–1005. The numbers are extrapolated from McCallum's figures. He reported that the average number of workers per month in the Construction Corps for the year ending June 30, 1862, was 750. For the year ending June 30, 1863, the average number of workers increased to 1,974. John H. Devereux was superintendent of all railroads running from Alexandria. Erasmus L. Wentz was superintendent of the Richmond & York River Railroad and the Norfolk railroads. William W. Wright, former Gettysburg College student and assistant engineer on the Pennsylvania Railroad, was superintendent of the Aquia Creek Railroad. Adna Anderson and J. J. Moore served as construction supervisors.

36. Railroads and the Making of Modern America, "Letter from E. L. Wentz to Daniel McCallum, October 13, 1862," University of Nebraska, Lincoln, at http://railroads.unl.edu/documents. *Contraband* referred to the new status of slaves who had escaped or were moved behind Union lines and were not returned to their Confederate owners.

37. Francis A. Lord, *Lincoln's Railroad Man: Herman Haupt* (Teaneck, NJ: Fairleigh Dickinson University Press, 1969), 108–112. See also *Report of the Joint Committee on the Conduct of the War*, pt. 1 (1863), 682–687.

38. Lord, 108–110.

39. Robert C. Black III, *The Railroads of the Confederacy* (Chapel Hill: University of North Carolina Press, 1998), 102.

40. *OR*, ser. 1, vol. 4, 617, 634.

41. *Journal of the Confederate States Congress* 2 (1862): 267. See also Black, 102.

42. *OR*, ser. 1, vol. 18, 859.

43. Black, 148–153.

44. *OR*, ser. 1, vol. 18, 779.

45. John B. Jones, *A Rebel War Clerk's Diary at the Confederate States Capital*, ed. Howard Swiggett (New York: Old Hickory Bookshop, 1935), 1:183.

46. Black, 153.

47. The six railroads Bragg used were the Mobile & Ohio, Mobile & Great Northern, Alabama & Florida of Alabama, Montgomery & West Point, Atlanta & West Point, and Western & Atlantic.

CHAPTER 8. SUMMER–FALL 1862

1. Stephen W. Sears, *Landscape Turned Red: The Battle of Antietam* (Boston: Houghton Mifflin, 1983), 65.

2. *OR*, ser. 1, vol. 19, pt. 2, 590.

3. John G. Walker, "Jackson's Capture of Harper's Ferry," in *Battles and Leaders of the Civil War: North to Antietam* (New York: Castle, 1956), 605. See also Sears, 67.

4. *OR*, ser. 1, vol. 19, pt. 2, 593.

5. Ibid.

6. Douglas Southall Freeman, *R. E. Lee: A Biography* (New York: Charles Scribner's, 1937), 1:643.

7. James L. Nichols, *Confederate Centennial Studies Number Five: Confederate Engineers* (Tuscaloosa, AL: Confederate Publishing, 1957), 92–93.

8. William W. Blackford, *War Years with Jeb Stuart* (New York: Charles Scribner's, 1945), 63–65.

9. *OR*, ser. 1, vol. 25, pt. 2, 735.

10. Keith S. Bohannon, "Dirty, Ragged, and Ill-Provided For: Confederate Logistical Problems in the 1862 Maryland Campaign and Their Solutions," in *The Antietam Campaign*, ed. Gary W. Gallagher (Chapel Hill: University of North Carolina Press, 1999), 118–119. See also William Miller Owen, *In Camp and Battle with the Washington Artillery* (Boston: Ticknor, 1885), 158.

11. Bohannon, n. 57; Owen, 160.

12. *OR*, ser. 1, vol. 19, pt. 2, 957.

13. Walker, 606.

14. Robert E. L. Krick, *Staff Officers in Gray: A Biographical Register of the Staff Officers in the Army of Northern Virginia* (Chapel Hill: University of North Carolina Press, 2003), 270, 403.

15. "Gross Mismanagement," *Memphis Daily Appeal*, October 21, 1862; this includes the comment from the *Guardian*.

16. *OR*, ser. 1, vol. 19, pt. 1, 143.

17. John B. Jones, *A Rebel War Clerk's Diary at the Confederate States Capital*, ed. Howard Swiggett (New York: Old Hickory Bookshop, 1935), 1:183.

18. Herman Haupt, *Reminiscences of General Herman Haupt* (Milwaukee, WI: Wright & Joys, 1901), 159–160.

19. *OR*, ser. 1, vol. 21, 764.

20. Ibid., 856.

21. Gilbert Thompson, "The Engineer Battalion in the Civil War," *Occasional Papers* (Engineering School, US Army), no. 44 (1910): 148–149.

22. Roy P. Basler, ed., *The Collected Works of Abraham Lincoln* (New Brunswick, NJ: Rutgers University Press, 1953), 5:514. See also John G. Nicolay and John Hay, *Abraham Lincoln: A History* (New York: Century Company, 1890), 6:200.

23. Mathew Brady, "Pontoon Bridge on the Rappahannock," in *Mathew Brady's Illustrated History of the Civil War*, ed. Benson J. Lossing (New York, Portland House, 1996), 234; Timothy O'Sullivan, "Meade's Pontoon Bridge at Fredericksburg," in *Civil War: A Complete Photographic History*, ed. William C. Davis and Bell L. Wiley (New York: Tess Press, 2000), 326.

24. *OR*, ser. 1, vol. 21, 802–807, 808–819.

25. George C. Rable, *Fredericksburg! Fredericksburg!* (Chapel Hill: University of North Carolina Press, 2002), 136–139.

26. Ibid., 156–157.

27. The distance between the eastern bridges and western bridges was about two miles.

28. *OR*, ser. 1, vol. 21, 345–346.

29. Wesley Brainerd, *Bridge Building in Wartime: Colonel Wesley Brainerd's Memoir of the 50th New York Volunteer Engineers*, ed. Ed Malles (Knoxville: University of Tennessee Press, 1997), 113–114.

30. Phillip M. Thienel, *Mr. Lincoln's Bridge Builders: The Right Hand of American Genius* (Shippensburg, PA: White Mane, 2000), 87.

31. Mark Hoffman, *"My Brave Mechanics": The First Michigan Engineers and Their Civil War* (Detroit: Wayne State University Press, 2007), 93.

32. Ibid., 96–99.

33. James Lee McDonough, *War in Kentucky: From Shiloh to Perryville* (Knoxville: University of Tennessee Press, 1994), 150–151.

34. Richard McCormick, "Bridging North and South: The Pontoon Bridge from Cincinnati to Kentucky," *My Civil War Obsession* (blog), September 15, 2010, at www.civilwarobsession.com/2010/09bridging-north-and-south-pontoon-bridge.html. See also Chester Geaslen, *Our Moment of Glory in the Civil War: When Cincinnati Was Defended from the Hills of Northern Kentucky* (Fort Wright, KY: James A. Ramage Civil War Museum and City of Fort Wright, 2007 [1972]).

35. Larry J. Daniel, *Days of Glory: The Army of the Cumberland, 1861–1865* (Baton Rouge: Louisiana State University Press, 2004), 188–189.

36. Philip L. Shiman, "Engineering and Command: The Case of General William S. Rosecrans 1862–1863," in *The Art of Command in the Civil War*, ed. Steven E. Woodworth (Lincoln: University of Nebraska Press, 1998), 92.

37. *OR*, ser. 1, vol. 16, pt. 1, 297–298, 300.

38. *OR*, ser. 1, vol. 20, pt. 2, 65, 120, 215–216. The two officers from the Corps of Engineers were Lieutenant George Burroughs and Second Lieutenant H. C. Wharton. Both graduated from West Point in June 1862.

39. Ibid., 6–7.

40. Shiman, 91.

41. Daniel, 189.

42. John Fitch, *Annals of the Army of the Cumberland* (Philadelphia: J. B. Lippincott, 1863), 184–186.

43. Thomas E. Parson, "Shovels and Pickaxes," *Civil War Times*, August 2001, 80. Each squad carried the following tools:

Six Felling Axes	Six Hammers
Six Hatchets	Two Half-Inch Augurs
Two Cross-Cut Saws	Two Inch Augurs
Two Cross Cut Files	Two Two-Inch Augurs
Two Hand-Saws	Twenty lbs. Nails, Assorted
Two Hand-Saw Files	Forty lbs. Spikes, Assorted
Six Spades	One Coil Rope
Two Shovels	One Wagon, with four horses, or mules
Three Picks	

See General William Rosecrans's General Order No. 3, November 3, 1862, quoted in Geoffrey L. Blankenmeyer, "The Pioneer Brigade," at www.thecivilwargroup.com/pioneer.html.

44. Bobby L. Lovett, "Nashville's Fort Negley: A Symbol of Blacks' Involvement with the Union Army," *Tennessee Historical Quarterly* 41 (Spring 1982): 3–9. See also Fitch, 619–620, 632–633; Peter Maslowski, *Treason Must Be Made Odious: Military Occupation and Wartime Reconstruction in Nashville, Tennessee, 1862–1865* (Millwood, NY: KTO Press, 1978), 100.

45. Shiman, 93.

46. Hoffman, 114–120.

47. OR, ser. 1, vol. 20, pt. 2, 83, 94, 97, 98.

48. The Birago trestle was also invented in Europe under the direction of Karl Ritter von Birago, an Austrian military engineer in the 1830s and 1840s.

49. OR, ser. 1, vol. 20, pt. 2, 102, 120, 133–134.

50. Another option besides the wooden pontoon was the canvas pontoon. Instead of a pontoon boat, a wood-framed bateau was covered with common cotton canvas. The benefit of such a pontoon was that the frame was light and easy to transport; the problem was that the canvas was not dependable. Rosecrans suggested to Cullum a double-canvas paulin (tarpaulin) pulled over a light foldable frame, which was a prototype. The idea for the frame would not be fully developed until 1864.

51. Shiman, 95; OR, ser. 1, vol. 30, pt. 3, 56, 85–86.

52. Bruce Catton, *Reflections on the Civil War*, ed. John Leekley (Garden City, NY: Doubleday, 1981), 189–190.

CHAPTER 9. VICKSBURG

1. David F. Bastian, *Grant's Canal: The Union's Attempt to Bypass Vicksburg* (Shippensburg, PA: White Mane, 1995), 7.

2. Charles Ellet Jr., *The Mississippi and Ohio Rivers* (Philadelphia: Lippincott, Grambo, 1853), 191–192.

3. Bastian, 3.

4. Ibid., 14–17; *Official Records of the Union and Confederate Navies in the War of Rebellion* (Washington, DC: Government Printing Office, 1905), ser. 1, vol. 18, 625, 675 (hereafter, *ORN*).

5. The *Chicago Tribune* correspondent's description was published in *Harper's Weekly* 6, no. 292 (August 2, 1862): 482.

6. Quoted in Thomas B. Buell, *The Warrior Generals: Combat Leadership in the Civil War* (New York: Three Rivers Press, 1997), 238.

7. Ibid., 247.

8. Phillip M. Thienel, *Seven Story Mountain: The Union Campaign at Vicksburg* (Jefferson, NC: McFarland, 1998), 12.

9. Grant's staff engineer was Captain Frederick E. Prime and his staff topographical engineer was Lieutenant James H. Wilson. The third engineer was on Sherman's staff, Captain William L. B. Jenney.

10. *ORN*, ser. 1, vol. 24, 711–712.

11. OR, ser. 1, vol. 24, pt. 1, 373.

12. Ibid., 371–390.

13. Thienel, 63.

14. OR, ser. 1, vol. 24, pt. 1, 409–411.

15. Ibid., 371–390.

16. OR, ser. 1, vol. 24, pt. 3, 113–114.

17. OR, ser. 1, vol. 24, pt. 1, 432–433.

18. Ibid., 435.

19. James R. Arnold, *Grant Wins the War: Decision at Vicksburg* (New York: John Wiley, 1997), 50–51.

20. Ibid., 51. See also William L. B Jenny, "Personal Recollections of Vicksburg," MOL-LUS, Illinois Commandery, vol. 3, 256.

21. Jenny, 252.

22. Edwin Cole Bearss, *Vicksburg Is the Key*, vol. 1 of *The Campaign for Vicksburg* (Dayton, OH: Morningside House, 1985), 439. See also OR, ser. 1, vol. 24, pt. 3, 10; John A. Bering and Thomas Montgomery, *History of the Forty-eighth Ohio Veteran Volunteer Infantry, Giving a Complete Account of the Regiment, from Its Organization at Camp Dennison, Ohio in October, 1861, to the Close of the War, and Its Final Muster Out, May 10, 1866* (Hillsboro, OH, 1880), 70–72; OR, ser. 1, vol. 24, pt. 1, 437, 441, 445–447, 450–454; T. B. Marshall, *History of the Eighty-third Ohio Volunteer Infantry* (Cincinnati: Eighty-third Ohio Volunteer Infantry Association, 1912).

23. Henry G. Ankeny, *Kiss Josey for Me: Letters to My Dear Wife*, ed. Florence M. A. Cox (Santa Ana, CA: Friis Pioneer Press, 1974).

24. Bastian, 38. See also OR, ser. 1, vol. 24, pt. 1, 120.

25. Prime to Grant, February 16, 1863, RG 107, Records of the Office of the Secretary of War, M-504, reel 184, National Archives, Washington, DC.

26. Henri Lovie to Editor, *Frank Leslie's Illustrated Newspaper*, March 28, 1863.

27. OR, ser. 1, vol. 24, pt. 3, 621–622.

28. OR, ser. 1, vol. 24, pt. 1, 119–121. Captain Patterson's Kentucky Company of Mechanics and Engineers received orders to join General McClernand's corps, and Patterson was placed in charge of McClernand's pioneer corps. A revetment is a sloped wall that supports a parapet or breastworks. The wall is built of logs, fence rails, gabions (crude cylindrical wicker baskets, two feet in diameter and three feet tall, filled with dirt), or fascines (tightly bound bundles of small, straight, tree branches). With the sloped wall design, incoming enemy artillery shot and shell would, in theory, strike only a glancing blow.

29. Ibid., 122.

30. Civil War dredges were floating steam shovels. The length of the beam holding the shovel limited the depth of dredging. Dredges were not self-propelled. They were moved by winching cables that ran from the machine's platform to the shore.

31. OR, ser. 1, vol. 24, pt. 1, 122–123.

32. ORN, ser. 1, vol. 20, 8–9.

33. OR, ser. 1, vol. 24, pt. 3, 32–33.

34. Bruce Catton, *Grant Moves South* (Boston: Little, Brown, 1960), 378.

35. OR, ser. 1, vol. 24, pt. 3, 41–42, 76.

36. Ibid., 78–79.

37. William W. Belknap, *History of the Fifteenth Regiment, Iowa Veteran Volunteer Infantry, from October, 1861, to August, 1865, When Disbanded at the End of the War* (Keokuk, IA, 1887), 245–246.

38. OR, ser. 1, vol. 24, pt. 3, 86–87, 120, 134; Belknap, 243–246.

39. OR, ser. 1, vol. 24, pt. 1, 20.

40. Warren E. Grabau, *Ninety-Eight Days: A Geographer's View of the Vicksburg Campaign* (Knoxville: University of Tennessee Press, 2000), 60–62; OR, ser. 1, vol. 24, pt. 1, 46, 151–152.

41. Mary Bobbitt Townsend, *Yankee Warhorse: A Biography of Major General Peter J. Osterhaus* (Columbia: University of Missouri Press, 2012), 7.

42. "Thomas W. Bennett," Fayette, Franklin, Union and Wayne Counties, Indiana: Portrait and Biographical Record, Indiana County History Preservation Society, at www.county history.org/books/doc.fayet/513.htm.

43. Grabau, 62; OR, ser. 1, vol. 24, pt. 1, 495.

44. Grabau, 32-33.

45. OR, ser. 1, vol. 24, pt. 1, 171, 490-494. A scow is a flat-bottomed boat with a broad shallow hull.

46. Ibid., 571; OR, ser. 1, vol. 24, pt. 3, 230.

47. OR, ser. 1, vol. 24, pt. 3, 151-152.

48. Grabau, 82.

49. Ulysses S. Grant, *Personal Memoirs of U. S. Grant* (New York: Penguin, 1999), 253-254.

50. OR, ser. 1, vol. 24, pt. 1, 126-127.

51. Ibid.

52. Ibid., 601.

53. James H. Wilson, *Under the Old Flag* (New York: D. Appleton, 1912), 169.

54. Grant, 254.

55. Edwin Cole Bearss, *Grant Strikes a Fatal Blow*, vol. 2 of *The Campaign for Vicksburg* (Dayton, OH: Morningside House, 1986), 283.

56. OR, ser. 1, vol. 24, pt. 1, 127-128.

57. Elias Moore, Diary, 114th Ohio Infantry, April 1–July 4, 1863, files, Vicksburg National Military Park, Vicksburg, Mississippi.

58. OR, ser. 1, vol. 24, pt. 1, 571.

59. W. H. H. Terrell, *49th Indiana Infantry Officers Roster: Report of the Adjutant General of the State of Indiana* (Indiana, 1865), 2:483. Fullyard was from New Albany, Indiana, on the Ohio River. This was the wealthiest area of the state before the war, and the city's fortune was derived from steamboat manufacturing and ancillary industries such as machine shops, foundries, furniture makers, and silversmiths. The American Plate Glass Works was the second-largest business in the town. New Albany opened the first public high school in the state in 1853, and by 1859 it hosted the Indiana State Fair. Fullyard was exposed to the perfect storm of change sweeping the North before the war: educational reform, industrial development, and mechanized farming.

60. OR, ser. 1, vol. 24, pt. 1, 186; Bearss, *Grant Strikes a Fatal Blow*, 286.

61. Bearss, *Grant Strikes a Fatal Blow*, 287; OR, ser. 1, vol. 24, pt. 1, 571-572.

62. Terrell, 484. Peckinpaugh was from Leavenworth, Indiana, the second county east of New Albany, next to Jeffersonville, Colonel Keigwin's home. Although there is no record of Peckinpaugh's occupation before the war, Leavenworth was on the Ohio River, and the town made its reputation as an important boatbuilding and brick-manufacturing center. Around 1845 the town had opened a stagecoach line, primarily for students who wanted to attend the new state college.

63. OR, ser. 1, vol. 24, pt. 1, 573.

64. Henry Clay Warmoth, "The Vicksburg Diary of Henry Clay Warmoth, Part II (April 28, 1863–May 26, 1863)," ed. Paul H. Hass, *Journal of Mississippi History* 32 (1970): 63; ORN, ser. 1, vol. 24, 608-628.

65. Israel Ritter, Private, Company D, 24th Iowa Infantry, Diary, September 10, 1862–May 31, 1863, files, Vicksburg National Military Park, Vicksburg, Mississippi; *History of the Forty-*

sixth Regiment Indiana Volunteer Infantry, September, 1861–September, 1865, compiled by committee (Logansport, IN, 1888), 56.

66. *OR,* ser. 1, vol. 24, pt. 1, 128; Captain William F. Patterson, Kentucky Company of Mechanics and Engineers, Daily Activity Cards, May 3, 1863, microfilm M 397, roll 472, National Archives, Washington, DC.

67. Thomas B. Marshall, *History of the 83rd Ohio Volunteer Infantry: The Greyhound Regiment* (Cincinnati, 1912), 76; Grabau, 171–172.

68. *OR,* ser. 1, vol. 24, pt. 1, 129; Thienel, 181.

69. Brigadier General A. Hickenlooper, "Our Volunteer Engineers," MOLLUS, Ohio Commandery, vol. 3, 301–318; *OR,* ser. 1, vol. 24, pt. 1, 203–206.

70. Hickenlooper, "Volunteer Engineers," 304–305.

71. The battles were fought at Port Gibson, Raymond, Jackson, Champion Hill, and Big Black River.

72. Grabau, 340.

73. A Union artillery battery consisted of six guns and about 100 horses to pull the caissons, additional ammunition, a portable forge and blacksmith equipment, and forage for the animals.

74. *OR,* ser. 1, vol. 24, pt. 1, 125–126.

75. Captain Prime was with the men constructing the India-rubber pontoon bridge; Hickenlooper and Tresilian supervised the bridge at Coaker's Ferry; and lieutenants Francis Tunica and Peter Hains, along with the Kentucky engineers, worked at replacing the railroad bridge. The *Official Records* makes no mention of who supervised the bridge at Hooker's Ferry. Since Ransom was a civil engineer before the war, it is likely that he supervised its construction.

76. Bearss, *Grant Strikes a Fatal Blow,* 680.

77. *OR,* ser. 1, vol. 24, pt. 2, 203–206

78. Ibid., 27.

79. Wilson, 204–207.

80. Clarence C. Buell and Robert U. Johnson, eds., *Battles and Leaders of the Civil War* (New York: Castle, 1956), 3:482–484. A redoubt was a fortification laid out in the shape of a regular or irregular convex polygon. A redan was a fieldwork consisting of two faces joined to form an outward projecting angle. These works were frequently used to guard roads and bridges. A redoubt and redan were built inside or were part of a larger fortification such as a star fort.

81. The five roadways were, from north to south, the Graveyard Road, Jackson Road, Baldwin's Ferry Road, Alabama & Vicksburg Railroad, and Hall's Ferry Road. Ten forts were built, tied together by rifle pits; from north to south: Fort Hill, the 26th Louisiana Redoubt, the Stockade Redan (including the 27th Louisiana Lunette and Green's Redan) guarding Graveyard Road, the 3rd Louisiana Redan and 21st Louisiana Redan guarding Jackson Road, the 2nd Texas Lunette guarding Baldwin's Ferry Road, the Railroad Redoubt (forts Pettus and Beauregard), the Square Fort, the Salient work guarding Hall's Ferry Road, and South Fort.

82. Ron Field, *American Civil War Fortifications (3): The Mississippi and River Forts* (Oxford, UK: Osprey, 2007), 34. The palisade was a crude stockade or stick wall, and the glacis was a wide and gently sloped parapet that gave the defenders a clear field of fire.

83. Samuel H. Lockett, "The Defense of Vicksburg," in *Battles and Leaders of the Civil War: Retreat from Gettysburg* (New York: Castle, 1956), 484, 488; Mark M. Boatner III, *The Civil War*

Dictionary (New York: Vintage, 1976), 876. Boatner's numbers are from Thomas L. Livermore, *Numbers and Losses in the Civil War in America 1861–65* (Boston: Houghton Mifflin, 1901).

84. Lockett, "Defense of Vicksburg," 488.

85. Samuel H. Lockett, "May 19–July 4, 1863: Report of the Siege of Vicksburg, Mississippi," at www.civilwaralbum.com/vicksburg/lockett_or.htm.

86. Field, 33.

87. Lockett, "Defense of Vicksburg," 490.

88. The following members of the Corps of Engineers were with Grant at Vicksburg: Captain Frederick E. Prime, Captain Miles Daniel McAlester, Captain Cyrus Ballou Comstock, and First Lieutenant Peter C. Hains. Lieutenant Clemens C. Chaffee, a West Pointer, was an ordnance officer detailed on engineer duty with General Sherman, as was aide-de-camp Captain William Le Baron Jenney. Jenney had considerable engineering experience. An 1846 graduate of Phillips Academy in Andover, Massachusetts, he then studied at the Lawrence Scientific School at Harvard before transferring to Paris to study engineering and architecture at the École Centrale des Arts et Manufactures. After the war, Jenney became known as the Father of the American Skyscraper, designing the first one in Chicago—the ten-story Home Insurance Building—in 1884.

89. *OR*, ser. 1, vol. 24, pt. 1, 176–177; *OR*, ser., 1, vol. 24, pt. 2, 168–179.

90. *OR*, ser. 1, vol. 24, pt. 2, 168–206.

91. Ibid., 176.

92. Thienel, 216.

93. J. C. Duane, *Manual for Engineer Troops* (New York: D. Van Nostrand, 1862), 67, 78.

94. *OR*, ser. 1, vol. 24, pt. 2, 168–206, 171.

95. Ibid., 172. Thayer's Approach was named after Brigadier General John M. Thayer, Third Brigade, First Division, XV Army Corps.

96. Richard Billies, "Union Siege Operations at Vicksburg," *North against South: Understanding the American Civil War on Its 150th Anniversary* (blog), April 18, 2012, at http://north againstsouth.com/union-siege-operations-at-vicksburg. Logan's Approach was named after Major General John A. Logan, Third Division, XVII Army Corps.

97. Ibid.

98. Andrew Hickenlooper, "The Vicksburg Mine," in *Battle and Leaders*, 539.

99. Ibid., 540–542.

100. Ibid., 542.

101. Ibid.

102. *OR*, ser. 1, vol. 24, pt. 2, 173–176. The Union approaches were Thayer's, Ewing's, Buckland's, Lightburn's, Giles A. Smith's, Ransom's, Logan's, Andre J. Smith's, Carr's, Hovey's, Lauman's, Herron's, and Slack's.

103. Captain William Patterson, Kentucky Company of Mechanics and Engineers, Letters and Papers, Manuscript Division, MMC 1180, F79 1802, Library of Congress, Washington, DC.

104. Albert Castel, *Victors in Blue: How Union Generals Fought the Confederates, Battled Each Other, and Won the Civil War* (Lawrence: University Press of Kansas, 2011), 203–204.

105. Herman Hattaway and Archer Jones, *How the North Won: A Military History of the Civil War* (Urbana: University of Illinois Press, 1983), 421–423. See also Albert Castel, "Vicksburg: Myths and Realities," *North and South*, November 2004, 62–69; Thomas L. Connelly, "Vicksburg: Strategic Point or Propaganda Device?" *Military Affairs* 34 (April 1970): 49–53.

106. Lincoln to James C. Conkling, August 26, 1863, in *Abraham Lincoln: Speeches and Writings, 1859–1865*, ed. Don E. Fehrenbacher (New York: Library of America, 1989), 495–499.

107. Ibid., 498.

108. Ibid., 496.

109. *OR*, ser. 1, vol. 24, pt. 2, 178; *OR*, ser. 1, vol. 24, pt. 1, 62.

110. Lincoln to Conkling, 499.

111. James R. Arnold, *Grant Wins the War: Decision at Vicksburg* (New York: John Wiley, 1997), 304.

112. Bell Irvin Wiley, ed., *"This Infernal War": The Confederate Letters of Sergeant Edwin H. Fay* (Austin: University of Texas Press, 1958), 290.

113. Frank E. Vandiver, *Ploughshares into Swords: Josiah Gorgas and Confederate Ordnance* (Austin: University of Texas Press, 1952), 194–195; Frank E. Vandiver, ed., *The Civil War Diary of General Josiah Gorgas* (Tuscaloosa: University of Alabama Press, 1947), 55.

114. Arnold, 310.

115. William C. Everhart, *Vicksburg and the Opening of the Mississippi River: A History and Guide Prepared for Vicksburg National Military Park, Mississippi* (Washington, DC: National Park Service, 1954), 40.

116. *OR*, ser. 1, vol. 24, pt. 2, 177–178.

117. Justus Scheibert, *A Prussian Observes the American Civil War*, ed. and trans. Frederic Trautmann (Columbia: University of Missouri Press, 2001), 140.

CHAPTER 10. GETTYSBURG

1. *OR*, ser. 1, vol. 24, pt. 2, 171.

2. Edward Hagerman, *The American Civil War and the Origins of Modern Warfare: Ideas, Organization, and Field Command* (Bloomington: Indiana University Press, 1992), 238–239.

3. *Congressional Globe*, 37th Congress, 3rd sess., bill S. 528, in *A Century of Lawmaking for a New Nation: U.S. Congressional Documents and Debates, 1774–1875*, at http://lcweb2.loc.gov/cgi-bin/ampage?collId=llsb&fileName=037/llsb037.db&recNum=1987. The bill also called for thirty more lieutenants and ten more second lieutenants.

4. "An Act to Promote the Efficiency of the Corps of Engineers and of the Ordnance Department, and for Other Purposes," *Statutes at Large*, 37th Congress, 3rd sess., chap. 78, 743.

5. Lincoln to Major General Joseph Hooker, January 26, 1863, in Roy P. Basler, ed., *The Collected Works of Abraham Lincoln* (New Brunswick, NJ: Rutgers University Press, 1953), 6:78–79.

6. *OR*, ser. 1, vol. 25, pt. 1, 215.

7. *OR*, ser. 1, vol. 27, pt. 1, 226. The Engineer Battalion was the descendant of the regular army's First Engineer Battalion formed at the start of the Mexican-American War in 1846. A member of the army's Corps of Engineers commanded the battalion. The Engineer Brigade was made up of volunteer regiments (ten companies) and commanded by volunteer officers, but overall command of the brigade was given to a regular army engineer.

8. Warren was originally designated chief of topographical engineers for the Army of the Potomac, but after Senator Wilson's bill passed, Warren became chief of engineers.

9. *OR*, ser. 1, vol. 27, pt. 1, 226–227.

10. *OR*, ser. 1, vol. 21, 1110.

11. *OR*, ser. 1, vol. 18, 873.

12. *OR*, ser. 1, vol. 25, pt. 2, 610–611.

13. Ibid., 693, 730.

14. Armistead L. Long, *Memoirs of Robert E. Lee* (New York: J. M. Stoddart, 1886), 268; Sir Frederick Maurice, ed., *An Aide-de-Camp of Lee, Being the Papers of Colonel Charles Marshall, Sometimes Aide-de-Camp, Military Secretary, and Assistant Adjutant General on the Staff of Robert E. Lee, 1862–1865* (Boston: Little Brown, 1927), 185–186; Edwin B. Coddington, *The Gettysburg Campaign: A Study in Command* (New York: Simon & Schuster, 1968), 8; Thomas Lawrence Connelly and Archer Jones, *The Politics of Command: Factions and Ideas in Confederate Strategy* (Baton Rouge: Louisiana State University Press, 1973), 128.

15. Henery Reinish, editorial, *New York Times*, September 4, 1863. Some questions remain about the validity of Reinish's comments. It was also alleged that Vallandigham wrote to Robert Ould, a high-ranking member of the Confederate Secret Service, strongly suggesting that Lee not invade the North in the summer of 1863 for fear of uniting Northern public opinion against the Confederacy. This reasoning seems suspect since most Democrats vehemently opposed Lincoln's Emancipation Proclamation and many still clamored for a peace settlement leaving slavery intact.

16. Richard D. Goff, *Confederate Supply* (Durham, NC: Duke University Press, 1969), 195–196.

17. *OR*, ser. 1, vol. 20, pt. 2, 445.

18. Goff, 5.

19. *OR*, ser. 4, vol. 2, 881–883.

20. *OR*, ser. 1, vol. 28, pt. 2, 295–296, and vol. 30, pt. 4, 713–714.

21. The Battle of Gettysburg cost the Confederacy 3,903 killed, 18,735 wounded, and 5,425 missing, or thirty-seven percent of Confederate forces engaged. Lee's army could ill afford to lose so many men. He had already lost 12,821 killed, wounded, or missing at Chancellorsville— twenty-two percent of his force. See Mark M. Boatner III, *The Civil War Dictionary* (New York: David McKay, 1976), 140, 339.

22. Earl B. McElfresh, "Fighting on Strange Ground," *Civil War Times*, August 2013, 31–37.

23. Ibid., 37.

24. *Atlas to Accompany the Official Records of the Union and Confederate Armies* (New York: Barnes & Noble, 2003 [1891–1895]), plate 43, no. 2.

25. McElfresh, 37.

26. Robert E. L. Krick, *Staff Officers in Gray: A Biographical Register of the Staff Officers in the Army of Northern Virginia* (Chapel Hill: University of North Carolina Press), 366–406.

27. John D. Imboden, "The Confederate Retreat from Gettysburg," in *Battles and Leaders of the Civil War: Retreat From Gettysburg* (New York: Castle, 1956), 428.

28. "Summerfield Smith, MA, Captain 1st Regiment Engineer Troops," University Memorial Biographical Sketches of Alumni of the University of Virginia Who Fell in the Confederate War, *University Memorial*, vol. 4, 1864, at www.fsu.edu/~ewoodwar/uva_mem.html.

29. Kent Masterson Brown, *Retreat from Gettysburg: Lee, Logistics, & the Pennsylvania Campaign* (Chapel Hill: University of North Carolina Press, 2005), 321.

30. Ibid.

31. *OR*, ser. 1, vol. 27, pt. 2, 361.

32. G. Moxley Sorrel, *Recollections of a Confederate Staff Officer*, ed., Bell Irwin Wiley (Jackson, TN: McCowat-Mercer Press, 1958), 165.

33. Edward Porter Alexander, *Fighting for the Confederacy: The Personal Recollections of General Edward Porter Alexander*, ed. Gary W. Gallagher (Chapel Hill: University of North Carolina Press, 1998), 272.

34. "Summerfield Smith."

35. *OR*, ser. 4, vol. 2, 609–610.

36. J. Boone Bartholomees Jr., *Buff Facings and Gilt Buttons: Staff and Headquarters Operations in the Army of Northern Virginia, 1861–1865* (Columbia: University of South Carolina Press, 1998), 101–102; Confederate States of America War Department, *Regulations for the Army of the Confederate States, 1863* (Richmond: West & Johnson, 1863), 366, at https://archive .org/stream/regulationsfor00conf#page/366/mode/2up; Captain J. C. Duane, *Manual for Engineer Troops* (New York: D. Van Nostrand, 1862).

37. *OR*, ser. 1, vol. 27, pt. 3, 1017.

38. Used as a defensive obstacle, a cheval-de-frise was a log or beam embedded with many long, sharpened stakes or spears.

39. Bartholomees, 103–104.

40. Bernard H. Nelson, "Confederate Slave Impressment Legislation, 1861–1865," *Journal of Negro History* 31, no. 4 (October 1946): 398–400.

41. James L. Nichols, *Confederate Centennial Studies Number Five: Confederate Engineers* (Tuscaloosa, AL: Confederate Publishing, 1957), 38–39.

42. *OR*, ser. 1, vol. 27, pt. 3, 67.

43. Herman Haupt, *Reminiscences of General Herman Haupt* (Herman Haupt, 1901), 236.

44. *OR*, ser. 1, vol. 27, pt. 3, 523.

CHAPTER 11. CHATTANOOGA

1. Philip L. Shiman, "Engineering and Command: The Case of General William S. Rosecrans 1862–1863," in *The Art of Command in the Civil War*, ed. Steven E. Woodworth (Lincoln: University of Nebraska Press, 1998), 109.

2. *OR*, ser. 1, vol. 49, pt. 2, 502–503. See also James St. Clair Morton, *Memoir Explaining the Situation and Defense of Fortress Rosecrans* (Fortress Rosecrans, TN: Pioneer Press, 1863); David Russell Wright, "History of Fortress Rosecrans," in *Civil War Fortifications*, thesis, Middle Tennessee State University (Murfreesboro, TN: Rutherford County Historical Society, 1984), 1–6, at digital.mtsu.edu/cdm/ref/collection/rchs/id/22.

3. Shiman, 97.

4. William Merrill, "The Engineer Service in the Army of the Cumberland," in *History of the Army of the Cumberland: Its Organization, Campaigns, and Battles*, ed. Thomas B. Van Horne (Cincinnati: Robert Clarke, 1885), 2:456.

5. Ibid., 456–457.

6. Ibid.

7. William P. Loughlin, "Detached Service," in *History of the Ninety-sixth Regiment, Illinois Volunteer Infantry*, ed. Charles Addison Partridge (Chicago: Brown, Pettibone, 1887), 630; James R. Willett, "Rambling Recollections of a Military Engineer," MOLLUS, Illinois Commandery, 1888, 9–10.

8. Mark Hoffman, *"My Brave Mechanics": The First Michigan Engineers and Their Civil War* (Detroit: Wayne State University Press, 2007), 157. A bent is part of the framework of an overall structure, including rafters, joists, posts, and pilings. Bents are the building blocks that define the shape of a structure. Stringers are bridge girders laid longitudinally between bents to support both railroad ties and track.

9. Haupt's prefabricated trusses were designed so that the top frame, or chord, of the bridge was straight but the lower chord was shaped like the arc of a circle. The shad belly

truss got its name because the arches in the bridge made it look like a fish. See Herman Haupt, *Military Bridges: With Suggestions of New Expedients and Constructions for Crossing Streams and Chasms* (New York: D. Van Nostrand, 1864), at http://archive.org/details/militarybridge soohaupgoog. In September 1863, Rosecrans notified the president of the Louisville & Nashville Railroad, James Guthrie, that a contract had been made "with the McCollum [*sic*] Bridge Company" to build a bridge across the Tennessee River at Bridgeport. This company was no doubt USMRR superintendent D. C. McCallum's company, established in 1858. See OR, ser. 1, vol. 30, pt. 3, 297; ser. 1, vol. 31, pt. 3, 38; and ser. 3, vol. 5, 933–971.

10. OR, ser. 1, vol. 30, pt. 1, 34, 398, 439, 497.

11. Ibid., 52; OR, ser. 1, vol. 30, pt. 3, 187, 203, 235.

12. OR, ser. 1, vol. 30, pt. 3, 299, 304, 326.

13. William B. Hazen, *A Narrative of Military Service* (Boston: Ticknor, 1885), 407–408.

14. Thomas Lawrence Connelly and Archer Jones, *The Politics of Command: Factions and Ideas in Confederate Strategy* (Baton Rouge: Louisiana State University Press, 1973), 135. Members of the informal yet politically powerful Western Coalition included Joseph Johnston, Kirby Smith, Braxton Bragg, Pierre G. T. Beauregard, William Hardee, and Leonidas Polk.

15. E. P. Alexander, *Military Memoirs of a Confederate: A Critical Narrative* (New York: Charles Scribner's Sons, 1907), 449. See also Thomas Lawrence Connelly, *Autumn of Glory: The Army of Tennessee, 1862–1865* (Baton Rouge: Louisiana State University Press, 1971), 152.

16. G. Moxley Sorrel, *Recollections of a Confederate Staff Officer*, ed. Bell Irwin Wiley (Jackson, TN: McCowat-Mercer, 1958), 189.

17. Collected papers of Edwin M. Stanton, September 1863, Library of Congress, Washington, DC.

18. OR, ser. 1, vol. 29, pt. 1, 151.

19. Felton was president of the Philadelphia, Wilmington & Baltimore Railroad; Scott, vice president of the Pennsylvania Railroad; Garrett, president of the Baltimore & Ohio; and Smith, master of transportation for the B&O.

20. OR, ser. 1, vol. 29, pt. 1, 146, 147, 149, 150, 152, 153, 155.

21. Larry J. Daniel, *Days of Glory: The Army of the Cumberland, 1861–1865* (Baton Rouge: Louisiana State University Press, 2004), 344–346. Numbers can be misleading. Several historians put the number of men sent to Bridgeport at approximately 25,000. This estimate is accurate if you include Sherman's and Hurlburt's corps from Vicksburg. Otherwise, about 16,000 men from the XI and XII Corps were sent to rescue Rosecrans.

22. James L. McDonough, *Chattanooga: A Death Grip on the Confederacy* (Knoxville: University of Tennessee Press, 1984), 76.

23. Daniel, 346.

24. OR, ser. 1, vol. 30, pt. 4, 102, 207, 244, 307, 361, 415, 435; General William F. Smith to General Reynolds, November 5, 1863, Records of the Chief Engineer, vol. 106, Department of the Cumberland, Record Group 393, National Archives, Washington, DC.

25. OR, ser. 1, vol. 30, pt. 1, 219–220.

26. Crackers, or hardtack, were a staple food in both armies. Hardtack was made of flour, water, and salt. It was easy to carry and stayed edible for a very long time. During the siege of Chattanooga, to demonstrate their incessant hunger, Union soldiers on fatigue duty would shout to their officers: "Crackers!" Thus, when a final plan was adopted to open a new supply line it was given the sobriquet the "Cracker Line." See Ulysses S. Grant, *Personal Memoirs of U. S. Grant* (New York: Penguin, 1999), 329.

27. Albert Castel, *Victors in Blue: How Union Generals Fought the Confederates, Battled Each Other, and Won the Civil War* (Lawrence: University Press of Kansas, 2011), 236–237; Herbert M. Schiller, ed., *Autobiography of Major General William F. Smith, 1861–1864* (Dayton, OH: Morningside, 1990), 72.

28. Alfred L. Hough, *Soldier in the West: The Civil War Letters of Alfred Lacy Hough* (Philadelphia: University of Pennsylvania Press, 1957), 153.

29. Horace Porter, *Campaigning with Grant* (New York: Century Company, 1897), 8.

30. Ibid., 2, 5. See also Shelby Foote, *The Civil War: A Narrative* (New York: Random House, 1963), 2:807–808.

31. Schiller, 75–76; Grant, 411.

32. Grant, 339.

33. Ibid., 340.

34. William G. Le Duc, "The Little Steamboat That Opened the 'Cracker Line,'" in *Battles and Leaders of the Civil War: Retreat from Gettysburg* (New York: Castle, 1956), 676.

35. Ibid., 677.

36. Ibid., 678.

37. Charles E. Belknap, *History of the Michigan Organizations at Chickamauga, Chattanooga, and Missionary Ridge* (Lansing, MI: R. Smith Printing, 1897), 159.

38. *OR*, ser. 1, vol. 31, pt. 1, 77–78.

39. Ibid., 84–85. See also Edward S. Cooper, *William Babcock Hazen: The Best Hated Man* (Madison, NJ: Fairleigh Dickinson University Press, 2005), 90–95. The Minié bullet was a cylindrical lead projectile with a conical point that, when fired from a rifle, expanded and fit into the grooves of the rifle barrel, thus leaving it spinning tightly. This action on the bullet allowed it to travel farther and more accurately than a traditional musket ball. The bullet was named after one of its co-developers, Claude-Etienne Minié.

40. Hough, 162–163.

41. Merrill, 454; emphasis added.

42. Arnold Gates, ed., *The Rough Side of War: The Civil War Journal of Chesley A. Mosman, First Lieutenant, Company D, 59th Illinois Volunteer Infantry Regiment* (Garden City, NY: Basin, 1987), 140–156.

43. *OR*, ser. 1, vol. 32, pt. 2, 73, 365, 372. The telegram giving McCallum authority to make the necessary changes to his department was sent on January 12, 1864, and provides another example of efficient management. See *OR*, ser. 1, vol. 32, pt. 2, 73. Halleck wrote to Thomas: "Your telegram of yesterday was shown to the Secretary of War, who says that Colonel McCallum has full authority to immediately adopt any measures he may deem necessary to put the railroad in efficient running order. He has authority to make any changes he may deem proper in its management. It is not necessary that he should make any previous explanations to the War Department. Tell him to go right ahead, and he will be sustained by the Secretary."

44. Haupt had refused the appointment as brigadier general in September 1862. He preferred to serve without rank or pay because he did not want to limit his freedom to continue his private business interests. In September 1863, Secretary of War Stanton again offered him a generalship. Haupt replied that he would consider the promotion if Stanton agreed to establish the USMRR bureau (eventually Stanton did agree) and if the secretary promised to grant considerable authority to the bureau chief. Stanton did not want to negotiate with Haupt over a generalship and so refused his requests. Haupt then resigned. He continued to serve as chief

engineer or general manager for various railroads, and he invented a drilling machine. He died while traveling, appropriately enough, on a train in New Jersey on December 14, 1905.

CHAPTER 12. THE RED RIVER AND PETERSBURG

Epigraph. Quoted in David Lipsky, *Absolutely American: Four Years at West Point* (Boston: Houghton Mifflin, 2003), 189.

1. Waymarking.Com, American Guide Series, "Site of the Burnet House—Cincinnati, Ohio," at www.waymarking.com/waymarks/WM3NVN. Today, the PNC Annex sits on the site of the old hotel.

2. U. S. Grant, *Personal Memoirs of U. S. Grant* (New York, 1885), 364–373.

3. James G. Hollandsworth, *The Louisiana Native Guards: The Black Military Experience during the Civil War* (Baton Rouge: Louisiana State University Press, 1998), 97.

4. Technically, the 1st, 2nd, and 3rd regiments were designated Native Guards until after the capture of Port Hudson, when their named was changed to Corps d'Afrique. See BlackPast .org, Remembered & Reclaimed, "1st Louisiana Guard, USA / Corps d'Afrique (1862–1863)," at www.blackpast.org/?q=aah/1st-louisiana-native-guard-usa-corps-d-afrique-1862-1863. Incidentally, at Port Hudson, Confederate Major General Franklin Gardner had no engineers available inside the fortification, so he promoted Private Henry Glinder to lieutenant of engineers because Glinder had been a member of the Coast Survey before the war.

5. Before the war, Banks was president of the Illinois Central Railroad and served as governor of Massachusetts, Franklin graduated first in his West Point class (1843) and was a member of the Corps of Engineers, and Ransom was a civil engineer.

6. OR, ser. 1, vol. 34, pt. 1, 402–404.

7. Ibid., 402.

8. Ibid., 404.

9. Benjamin W. Bacon, *Sinews of War: How Technology, Industry, and Transportation Won the Civil War* (Novato, CA: Presidio Press, 1997), 123.

10. Technically, Robinson commanded both the 97th and 99th in what was designated the Engineer Brigade.

11. OR, ser. 1, vol. 34, pt. 1, 248–253.

12. Ibid., 255.

13. Simsport rested on the south bank of the Atchafalaya River five miles north of the town of Red River, situated on the Mississippi.

14. Wickman Hoffman, *Camp Court and Siege: A Narrative of Personal Adventure and Observation during Two Wars* (New York: Harper & Brothers, 1877), 103.

15. John W. Merwin, *Roster and Monograph, 161st Reg't, New York State Volunteer Infantry* (Elmira, NY: Gazette Print, 1902), 130.

16. Richard B. Irwin, *History of the Nineteenth Army Corps* (New York: G. P. Putman's, 1893), 346–347.

17. OR, ser. 1, vol. 34, pt. 1, 256.

18. Ibid., 253.

19. Earl J. Hess, *Trench Warfare under Grant & Lee: Field Fortifications in the Overland Campaign* (Chapel Hill: University of North Carolina Press, 2007), 12.

20. Harry L. Jackson, *First Regiment Engineer Troops P. A. C. S.: Robert E. Lee's Combat Engineers* (Louisa, VA: R. A. E. Design, 1998), 24.

21. W. Harrison Daniel, ed., "H. H. Harris' Civil War Diary," *Virginia Baptist Register* 35 (1996): 1885.

22. Philip Katcher, *Building the Victory: The Order Book of the Volunteer Engineer Brigade Army of the Potomac* (Shippensburg, PA: White Mane, 1998), 40.

23. Companies B, C, F, H, and I were capable of "throwing a bridge."

24. Jackson, 27.

25. For the roster of men in the 1st Confederate Engineers and their ranks, see Jackson, 167–185. For the 15th New York's roster, see *Roster of the 50th and 15th Regiment New York State Volunteer Engineers, Army of the Potomac* (New York: Union and Advertise Press, 1894).

26. Jackson, 164, n. 31.

27. C. B. Denson, "The Corps of Engineers and Engineer Troops," in *Histories of the Several Regiments and Battalions from North Carolina in the Great War, 1861–65*, ed. Walter Clark (Goldsboro, NC: Nash Brothers, 1901), 412–413, 426; T. M. R. Talcott, "Reminiscences of the Confederate Engineer Service," in *The Photographic History of the Civil War*, ed. Francis Trevelyan Miller (New York: Review of Reviews, 1911), 258.

28. *OR*, ser. 1, vol. 40, pt. 2, 699.

29. A. L. Rives Paper, file no. MSS1 R5247, pp. 13–18, Virginia Historical Society, Richmond, Virginia.

30. The Overland Campaign (1864) was made up of the following engagements: the Wilderness (May 4–7), Spotsylvania (May 8–21), New Market (May 15), Drewry's Bluff (May 16), North Anna River (May 23–26), and Cold Harbor (June 3).

31. The Army of the Potomac remained under the command of George Meade, but Grant attached his headquarters to the field army and ostensibly dictated strategic and tactical policy. Meade understood his relationship with the commander-in-chief, and for this reason, although Meade occasionally complained to his wife, tension between the two men was limited.

32. The other two observers sent to the Crimean were George McClellan and Alfred Mordecai. Mordecai, born and raised in North Carolina, resigned his commission in the army at the commencement of the war. His family and friends desperately wanted him to join the Confederate army, but his son had joined the Union. Rather than suffer bewilderment and guilt from making such a painful decision, he chose not to fight.

33. *OR*, ser. 1, vol. 38, pt. 1, 127–139; Letters Sent by the Chief of Engineers, RG 77, National Archives, Washington, DC.

34. Farquhar graduated in 1861, Gillespie in 1862, and Michie in 1863.

35. Mark M. Boatner III, *The Civil War Dictionary* (New York: David McKay, 1976), 45.

36. Mendell graduated from West Point in 1852, and Benham in 1837. Two reasons for Benham's assignment to the Volunteer Brigade rather than the US Engineer Battalion were possible. The first was that he was a skilled engineer and bridge builder and would see that things were done right. The second reason perhaps was Grant's attempt to deal with the army's awkward command structure. Duane, a major, was chief engineer of the Army of the Potomac, and Benham, a major of engineers (brigadier general of volunteer infantry), commanded the Volunteer Engineer Brigade. Benham was eleven years older than Duane, and he was promoted to major of engineers two years ahead of his chief. Furthermore, Mendell, a captain, commanded the more prestigious regular army Engineer Battalion, whereas Benham, a major, commanded volunteers. By sending Benham to Washington, Grant temporarily solved his chain-of-command problem.

37. Hess, *Trench Warfare*, 10.

38. John Y. Simon, ed., *The Papers of Ulysses S. Grant* (Carbondale: Southern Illinois University Press, 1967–2003), 11:19.

39. James A. Huston, "Grant's Crossing of the James with the Longest Ponton Bridge Ever Built," *Military Engineer* 45 (January–February 1953): 18; Warren T. Hannum, "The Crossing of the James River in 1864," *Military Engineer* 15 (May–June 1923): 237.

40. Huston, 19.

41. *OR*, ser. 1, vol. 40, pt. 1, 297.

42. *Atlas to Accompany the Official Records of the Union and Confederate Armies* (New York: Barnes & Noble, 2003 [1891–95]), plates 16, 17, and 92. Lieutenant Peter Michie suggested three possible sites for the James River crossing: (1) at Fort Powhatan, where the width of the river was 1,250 feet and the eastern approaches over a marsh were 1,000 yards wide; (2) a quarter of a mile north of Fort Powhatan, where the marshes were only 800 yards wide but the river was 1,570 feet wide; or (3) three-quarters of a mile north of Fort Powhatan, where the approaches required the clearing of trees and the river was 1,992 feet wide. General Weitzel, chief engineer of the Army of the James, selected the third location and Grant approved the choice on the same day, June 13. See Hannum, 234.

43. *OR*, ser. 1, vol. 40, pt. 1, 298.

44. While Colonel Spaulding and company were building the bridge at Cole's Ferry, the regular Engineer Battalion and the 15th New York Engineers were building a 2,200-foot-long bridge over the James River, using pontoons floated downriver from Fort Monroe. This is why Spaulding referred to the boats at Cole's Ferry as land pontoons.

45. *OR*, ser. 1, vol. 40, pt. 1, 298. A balk was a 27-foot-long board that served to hold two pontoons together. The balks had cleats and were locked into the gunwales of the pontoons to create a frame of timber on which the chess (flooring) was laid. Under normal circumstances the pontoons floated 20 feet apart, but in the Cole's Ferry operation the space was expanded. This required additional balk to help compensate for a frailer structure. Note that Lieutenant Colonel Spaulding's report mistakenly listed William H. Pettes as the captain of Company D.

46. *OR*, ser. 1, vol. 40, pt. 1, 298–299.

47. Huston, 21.

48. Ibid.

49. On June 12, Benham reported that although he had 155 pontoons available for use, he had enough flooring (chess) for only about 1,540 feet of bridge. He ordered troops from the 15th New York at the engineer depot in Washington to cut more chess. This was done, and the material arrived with the rest of the equipment on June 14. See *OR*, ser. 1, vol. 36, pt. 3, 772; *OR*, ser. 1, vol. 40, pt. 1, 210–211.

50. *OR*, ser. 1, vol. 40, pt. 1, 297.

51. Ibid., 297–298.

52. Horace Porter, *Campaigning with Grant* (New York: Century, 1897), 137.

53. *OR*, ser. 1, vol. 40, pt. 1, 301.

54. John Westervelt, *Diary of a Yankee Engineer: The Civil War Story of John H. Westervelt, Engineer, 1st New York Volunteer Engineer Corps*, ed. Anita Palladino (New York: Fordham University Press, 1997), 142.

55. *Harper's Weekly* 8, no. 392 (1864): 419.

56. *OR*, ser. 3, vol. 4, 796–797.

57. *OR*, ser. 1, vol. 36, pt. 1, 315–316; Westervelt, 170; *OR*, ser. 1, vol. 42, pt. 1, 661; *OR*, ser. 1, vol. 46, pt. 2, 1169; *Atlas*, plates 67 and 76.

58. Abatis consisted of felled trees stripped of their leaves and with branches sharpened, laid side by side on the ground sloping away from a ditch in front of a basic field fortification. Fraise consisted of sharpened, ten- to twelve-foot-long solid tree branches anchored into a ditch, serving as an obstacle to attacking soldiers.

59. *OR*, ser. 1, vol. 42, pt. 1, 163–165.

60. Earl J. Hess, *Into the Crater: The Mine Attack at Petersburg* (Columbia: University of South Carolina Press, 2010), 11.

61. Ibid., 14; *OR*, ser. 1, vol. 40, pt. 1, 566–567. A theodolite is an optical instrument consisting of a small mounted telescope that can rotate in both the horizontal and vertical planes, used to measure angles in surveying.

62. *OR*, ser. 1, vol. 40, pt. 1, 557–558.

63. Kevin M. Levin, "'The Earth Seemed to Tremble,'" *America's Civil War*, May 2006, 24.

64. Ibid., 25.

65. *OR*, ser. 1, vol. 42, pt. 2, 1161–1163.

66. James L. Nichols, *Confederate Centennial Studies Number Five: Confederate Engineers* (Tuscaloosa, AL: Confederate Publishing, 1957), 108.

67. Levin, 28.

68. Abraham Lincoln, "Memorandum on Probable Failure of Re-election, Washington, DC, August 23, 1864," in *Lincoln: Speeches, Letters, Miscellaneous Writings, Presidential Messages and Proclamations, 1859–1865*, ed. Don E. Fehrenbacher (New York: Library of America, 1989), 624.

CHAPTER 13. ATLANTA AND THE CAROLINA CAMPAIGNS

1. D. C. McCallum, *United States Military Railroads: Report of Bvt. Brig. Gen. D. C. McCallum, Director and General Manager, from 1861 to 1866* (ULAN Press, 2013 [n.d.]), 15. See also *OR*, ser. 3, vol. 5, 987–990. The USMRR was responsible for a much more extended supply route than the one from Nashville to the front lines in northern Georgia. The focal point of army supplies in the western theater was Louisville. All the war materials from the agricultural areas of the Midwest and the manufacturing centers of the East came to the city by railroad or the Ohio River. Then the Louisville & Nashville Railroad transported provisions to Nashville, the main supply depot for the Military Division of the Mississippi. In Nashville, the engineers had built warehouses that covered entire city blocks, as well as several acres of corrals and stables. From Nashville supplies traveled along the Nashville & Chattanooga Railroad or Cumberland River to Chattanooga, and then on the Western & Atlantic Railroad. By early July 1864, Sherman's engineers were building an advanced depot at Ringgold, Georgia. Along the railroads, detachments stockpiled repair equipment: spikes, bridge timbers, crossties, and rails.

2. McCallum, 18.

3. Ibid.

4. Ibid., 21.

5. Report of Adna Anderson, General Superintendent, U.S. Government Railroad, Military Division of the Mississippi for the Year Ending June 30, 1864, in RG 92, Office of the Quartermaster General, National Archives, Washington, DC; Military Railroads, Unclassified Correspondence, box 10, entry 1525, National Archives, Washington, DC. See also McCallum,

22; John E. Clark Jr., *Railroads in the Civil War: The Impact of Management on Victory and Defeat* (Baton Rouge: Louisiana State University Press, 2001), 210–211; OR, ser. 3, vol. 5, 996–998.

6. William Tecumseh Sherman, *Memoirs of General William T. Sherman*, introduction by William S. McFeely (New York: Da Capo Press, 1984), 2:11.

7. Sidings were short stretches of track used for parking rolling stock or for enabling trains on the same line to pass.

8. McCallum, 24.

9. Ibid., 25.

10. OR, ser. 1, vol. 32, pt. 3, 434.

11. The operations, called the Atlanta Campaign, lasted 117 days and included seventeen battles. The battles are divided into two categories: Sherman versus Johnston and Sherman versus John Bell Hood. Battles fought with Johnston were the following: Rocky Face Ridge (May 7–13), Resaca (May 13–15), Adairsville (May 17), New Hope Church (May 25–26), Dallas (May 26–June 1), Pickett's Mill (May 27), operations around Marietta (June 9–July 3), Kolb's Farm (June 22), Kennesaw Mountain (June 27), and Pace's Ferry (July 5). The following battles were fought with Hood, in command of the Army of Tennessee: Peachtree Creek (July 20), Atlanta (July 22), Ezra Church (July 28), Utoy Creek (August 5–7), Dalton (August 14–15), Lovejoy's Station (August 20), and Jonesborough (August 31–September 1).

12. Paul Taylor, *Orlando M. Poe: Civil War General and Great Lakes Engineer* (Kent, OH: Kent State University Press, 2009), 26.

13. William Merrill, "The Engineer Service in the Army of the Cumberland," in *History of the Army of the Cumberland: Its Organization, Campaigns, and Battles*, ed. Thomas B. Van Horne (Cincinnati: Robert Clarke, 1885), 440–444. See also OR, ser. 1, vol. 38, pt. 4, 639–640.

14. The Volunteer Regiment was known by several names, including the United States Veteran Volunteers, the 1st Regiment Engineers, and if you were from Ohio, the 1st Ohio Veteran Engineers. Major Patrick O'Connell commanded the Ohio battalion within the regiment. See Mark Hoffman, *"My Brave Mechanics": The First Michigan Engineers and Their Civil War* (Detroit: Wayne State University Press, 2007), 214; Civil War Index, "Military Records: 1st Ohio Veteran Engineers," at www.civilwarindex.com/armyoh/1st_oh_veteran_engineers. html. See also OR, ser. 1, vol. 31, pt. 2, 13, 21; OR, ser. 1, vol. 32, pt. 1, 40.

15. John J. Hight, *History of the Fifty-eighth Regiment Indiana Volunteer Infantry* (Princeton: Press of Clarion, 1895), 287, 289.

16. OR, ser. 1, vol. 38, pt. 1, 128. Captain Charles E. McAlester from Flint, Michigan, commanded the battalion. Before this, McAlester was a company commander in the 23rd Michigan Infantry.

17. Taylor, 149. The nine regular engineer officers under Poe's command were the following: captains Merrill, Barlow, McAlester, Reese, and Twining, and lieutenants Ernest, Ludlow, Damrell, and Wharton. Both Merrill and Wharton served with the Veteran Volunteer Engineers and, consequently, Merrill was commissioned as colonel of volunteers and Wharton as lieutenant colonel of volunteers. As was the tradition, when their command of the volunteers ended, their ranks reverted to those they held in the regular army.

18. Hight, 290, 293–296.

19. Albert Castel, *Decision in the West: The Atlanta Campaign of 1864* (Lawrence: University Press of Kansas, 1992), 162.

20. OR, ser. 1, vol. 38, pt. 2, 515–516.

21. OR, ser. 1, vol. 38, pt. 1, 130.

22. Taylor, 183.

23. *OR*, ser. 1, vol. 38, pt. 1, 137.

24. Robert J. Fryman, "Fortifying the Landscape: An Archaeological Study of Military Engineering and the Atlanta Campaign," in *Archaeological Perspectives on the Civil War*, ed. Clarence R. Geier and Stephen R. Potter (Gainesville: University Press of Florida, 2000), 50–51.

25. Ibid., 54–55.

26. Ibid., 55. The Army of Tennessee did have rifled artillery pieces. The Federal Ordnance Corps reported the capture of 3.80-inch James rifles and 12-pounder Parrotts, although most of the artillery captured was smoothbores. See *OR*, ser. 1, vol. 38, pt. 1, 124–125.

27. Jeffrey N. Nash, *Destroyer of the Iron Horse: General Joseph E. Johnston and Confederate Rail Transport, 1861–1865* (Kent, OH: Kent State University Press, 1991), 134–135, 138–145.

28. Hosea Whitford Rood, *Story of the Service of Company E, and the Twelfth Wisconsin Regiment, Veteran Volunteer Infantry in the War of the Rebellion* (Milwaukee: Swain and Tate, 1893), 280; J. W. Gaskill, *Footprints through Dixie: Everyday Life of the Man under a Musket on the Firing Line and in the Trenches, 1862–1865* (Alliance, OH: privately published, 1919), 108. Sherman also recorded a similar story about carrying extra tunnels in his *Memoirs*, 2:151.

29. *OR*, ser. 3, vol. 5, 996–998.

30. In his *Memoirs*, Sherman gives credit to Wright for the bridges over the Oostanaula and Etowah rivers and does not mention Smeed. McCallum also fails to mention Smeed in any of his reports, perhaps because he associated him with Haupt. In McCallum's final report of USMRR operations, *United States Military Railroads*, he never once mentioned Haupt. For his part, Haupt did mention Smeed's work during the Atlanta Campaign, in glowing terms, because of his reading of an unfinished letter from Smeed, sent to Haupt by Smeed's daughter, Mrs. Kate Smeed Cress. See Herman Haupt, *Reminiscences of General Herman Haupt* (Milwaukee: Wright & Joys, 1901), 51, 289–296.

31. Haupt, 296, 317.

32. McCallum, 44.

33. *OR*, ser. 1, vol. 44, 56.

34. Samuel Carter III, *The Siege of Atlanta, 1864* (New York: St. Martin's Press, 1973), 370.

35. *OR*, ser. 1, vol. 47, pt. 1, 173.

36. Ibid., 169. The left wing of the army was originally part of the Army of the Cumberland. In the winter of 1865, Sherman divided the army, sending the IV and XXIII Corps with George Thomas to destroy Hood's Confederate army, while the remaining two corps, the XIV and XX, accompanied Sherman's march and were redesignated the Army of Georgia, under the command of Henry Slocum.

37. Ibid. Colonel Reese was chief engineer for the army's right wing (the Army of the Tennessee), and his assistant was Captain Amos Stickney. Major William Ludlow was chief engineer of the Army of Georgia; Captain William Kossak, chief engineer for the XVII Corps; and Captain Klostermann, chief engineer for the XV Corps.

38. Ibid., 386.

39. Ibid., 389.

40. Ibid.

41. Benjamin W. Bacon, *Sinews of War: How Technology, Industry, and Transportation Won the Civil War* (Novato, CA: Presidio Press, 1997), 212.

42. *OR*, ser. 1, vol. 47, pt. 1, 171.

43. Ibid., 427.

44. Ibid.

45. "Rheumatism" in the nineteenth century could describe a number of ailments, but more than likely Moore suffered from a painful bursitis or tendonitis. There is no longer any recognized disorder called "rheumatism."

46. Captain J. C. Duane, *Manual for Engineer Troops* (New York: D. Van Nostrand, 1862), 24, 45–46.

47. Jack K. Bauer, *Soldiering: The Civil War Diary of Rice C. Bull* (San Rafael, CA: Presidio Press, 1997), 222.

48. *OR*, ser. 1, vol. 47, pt. 1, 173–174.

49. Ibid., 384.

50. Jacob D. Cox, *Military Reminiscences of the Civil War* (New York: Charles Scribner's, 1900), 2:531–532; Richard Harwell and Philip N. Racine, eds., *The Fiery Trial: A Union Officer's Account of Sherman's Last Campaign* (Knoxville: University of Tennessee Press, 1986), 213.

51. The permanent bridge was built across the James River near Varina, where the Varina Road ended at the river. The south side of the bridge connected to Curl's Neck.

52. *OR*, ser. 3, vol. 5, 190–192.

53. John H. Westervelt, *Diary of a Yankee Engineer: The Civil War Story of John H. Westervelt, Engineer, 1st New York Volunteer Engineer Corps*, ed. Anita Palladino (New York: Fordham University Press, 1997), 199, 206.

54. *OR*, ser. 3, vol. 5, 191–193.

55. *OR*, ser. 1, vol. 46, pt. 1, 160. Fort Siebert was located approximately five miles northeast of the Hatcher's Run crossing, and the road connecting the two points was an important supply link.

56. *OR*, ser. 1, vol. 46, pt. 1, 163–164.

57. Sheridan's Union cavalry had ridden eight to ten miles south of Warren and Ord's lines and approached Dinwiddie Court House on the Boydton Plank Road. At the tip of a V, Sheridan was equidistant from Warren's lines to the northeast and Confederate General George Pickett's soldiers to the northwest near Five Forks. Between Warren's left flank and Five Forks there was a four-mile gap unoccupied by either army.

58. *OR*, ser. 1, vol. 46, pt. 1, 643.

59. Ibid.

60. Ibid.

61. *OR*, ser. 3, vol. 5, 193.

62. The village of Blacks & Whites is today known as the city of Blackstone, Virginia.

63. Most historians identify this battle as Sayler's Creek. This is the modern spelling; Virginia county maps printed in 1865 designate the site as Sailor's Creek.

64. *OR*, ser. 3, vol. 5, 194.

65. Abraham Lincoln, *Lincoln: Speeches, Letters, Miscellaneous Writings, Presidential Messages, and Proclamations, 1859–1865*, ed. Don E. Fehrenbacher (New York: Library of America, 1989), 696.

66. Harry L. Jackson, *First Regiment Engineer Troops, P. A. C. S.: Robert E. Lee's Combat Engineers* (Louisa, VA: R. A. E. Design, 1998), 132. The Staunton River runs in a westerly direction approximately forty miles south of the Appomattox River. The Dan River runs east to west along the Virginia–North Carolina border connecting with Danville. The Staunton River forks with the Dan thirty miles below Marysville Court House.

67. *OR*, ser. 1, vol. 46, pt. 3, 1369. See also Jackson, 135.

68. Jackson, 140. It is possible that Lee did not want his men encumbered by a slow-moving pontoon train. They were in a race with Grant's army, although without a pontoon bridge their movement options were limited.

69. See *Atlas to Accompany the Official Records of the Union and Confederate Armies* (New York: Barnes & Noble, 2003 [1891–95]), plate 137

70. Jackson, 147.

71. Robert Hendrickson, *The Road to Appomattox* (New York: John Wiley, 1998), 176.

72. Shelby Foote, *The Civil War, a Narrative: Red River to Appomattox* (New York: Random House, 1974), 922.

73. Hendrickson, 198.

74. Arthur Parker, *The Life of General Ely S. Parker* (Buffalo: Buffalo Historical Society, 2005 [1919]), 102–103.

CONCLUSION. KNOW-HOW TRIUMPHANT

1. Douglas Southall Freeman, *R. E. Lee: A Biography* (New York: Scribner's, 1935), 4:154–155. See also Elizabeth R. Varon, *Appomattox: Victory, Defeat, and Freedom at the End of the Civil War* (Oxford: Oxford University Press, 2014). Varon skillfully argues that Lee's comments regarding "overwhelming resources" were not entirely true and were not just commentary on why the South lost the war. Instead, Lee consciously established the postwar theme of "might beat right," highlighting the Confederacy as a noble experiment defeated by overwhelming odds. This set the Southern political agenda in the postwar period, which worked to frame, celebrate, and maintain the principal values of the old Confederacy.

2. The ten costliest battles of the Civil War in terms of casualties (killed, wounded, captured, and missing): Gettysburg (51,112), Chickamauga (34,624), Spotsylvania (31,086), Chancellorsville (30,099), Antietam (26,134), the Wilderness (25,416), Second Manassas (25,251), Stone's River (24,645), Shiloh (23,741), and Fredericksburg (18,030).

3. Richard N. Current, "God and the Strongest Battalions," in *Why the North Won the Civil War*, ed. David Herbert Donald (Baton Rouge: Louisiana State University Press, 1960).

4. John Keegan, *The American Civil War: A Military History* (New York: Alfred A. Knopf, 2009), 70.

5. Richard E. Beringer, Herman Hattaway, Archer Jones, and William N. Still Jr., *The Elements of Confederate Defeat: Nationalism, War Aims, and Religion* (Athens: University of Georgia Press, 1988), 187; Pierre G. T. Beauregard, "The First Battle of Bull Run," in *Battles and Leaders of the Civil War*, ed. Robert U. Johnson and Clarence C. Buel (New York: Century Company, 1887), 222.

6. David M. Potter, "Jefferson Davis and the Political Factors in Confederate Defeat," in Donald, 95–101.

7. Channing M. Bolton, "With General Lee's Engineers," *Confederate Veterans Magazine* 30, no. 8 (August 1922): 300.

8. Herbert T. Coleman, "The Status of Education in the South Prior to the War between the States," *Confederate Veteran Magazine* 25, no. 10 (October 1907): 441.

9. Ibid., 444.

10. *OR*, ser. 1, vol. 46, pt. 3, 1391–1392.

11. Malcom Gladwell, *The Tipping Point: How Little Things Can Make a Big Difference* (New York: Little, Brown, 2000).

12. *OR*, ser. 4, vol. 3, 1022–1023.

13. *OR*, ser. 1, vol. 38, pt. 5, 961.

14. Confederate engineer regiments identified men with some mechanical or carpentry skills as artificers. Of the eight companies in the 3rd Regiment, five had a sufficient number of artificers: A (36), B (65), C (50), F (54), and G (60). The other companies had very few: D (1), E (5), and H (11)—not enough skilled men to be responsible for pontoon bridge building. Company K was never formed.

15. James L. Nichols, *Confederate Centennial Studies Number Five: Confederate Engineers* (Tuscaloosa, AL: Confederate Publishing, 1957), 100.

16. "Co. E, 3rd Regiment Confederate Engineers," at hartsengineers.com. See also *OR*, ser. 1, vol. 49, 1022; *OR*, ser. 1, vol. 43, pt. 2, 945.

17. *OR*, ser. 1, vol. 47, pt. 2, 1210–1230.

18. Henry Herbert Harris, "H. H. Harris' Civil War Diary (1863–1865) Part II," ed. W. Harrison Daniel, *Virginia Baptist Register* 36 (1997): 1855.

19. *OR*, ser. 1, vol. 38, pt. 2, 844–845, 855–856, 903.

20. Harry L. Jackson, *First Regiment Troops P. A. C. S.: Robert E. Lee's Combat Engineers* (Louisa, VA: R. A. E. Design, 1998), 121.

21. *Daily Richmond Examiner*, January 3 and January 31, 1865.

22. *Journal of the Confederate States Congress* 7 (1865): 584–586.

23. Bruce Catton, *Never Call Retreat* (New York: Doubleday, 1965), 256–257; "Men at War: An Interview with Shelby Foote," in *The Civil War: An Illustrated History*, ed. Geoffrey C. Ward with Ric Burns and Ken Burns (New York: Alfred A. Knopf, 1990), 272.

24. James A. Huston, *The Sinews of War: Army Logistics 1775–1953* (Washington, DC: Office of the Chief of Military History, United States Army, 1966), 210–211.

25. D. C. McCallum, *United States Military Railroads: Report of Bvt. Brig. Gen. D. C. McCallum, Director and General Manager, from 1861 to 1866* (ULAN Press, 2013 [n.d.]), 42, 43.

26. Ibid., 34.

27. Ibid., 10, 31, 45. In the Virginia, Pennsylvania, and Maryland sectors, the average number of persons employed monthly rose over the four years of the war. In 1862, the average was 750 men; in 1863, 1,974 men; in 1864, 2,378 men; and in 1865, 3,060 men. In the Division of the Mississippi, the monthly average went from 11,580 men in 1864 to 10,061 men in 1865. For North Carolina, only the greatest number of persons employed monthly in 1865 was recorded: 3,387.

28. Record of the Office of the Quartermaster General: Transportation, 1834–1917, Printed and Manuscript Orders, Circulars, and Letters of Instruction Relating to the U.S. Military Railroad, 1861–1864, box no. 1, National Archives, Washington, DC.

29. McCallum, 41.

30. *OR*, ser. 1, vol. 46, pt. 1, 649–650.

31. Abraham Lincoln, "Speech at New Haven, Connecticut," in *Lincoln: Speeches, Letters, Miscellaneous Writings, Presidential Messages and Proclamations, 1859–1865*, ed. Don E. Fehrenbacher (New York: Library of America, 1989), 144.

Essay on Sources

The literature on Civil War engineering is meager. Since 1865, the focus of Civil War scholarship has been aimed at every imaginable aspect of the war except engineering. The few books that do exist, however, make a rich historiographic contribution to our understanding of what Shelby Foote described as the seminal event in American history.

Mr. Lincoln's Bridge Builders: The Right Hand of American Genius (2000) by Phillip M. Thienel and *"My Brave Mechanics": The First Michigan Engineers and Their Civil War* (2007) by Mark Hoffman are two excellent works about Union engineering and a good place to start. Several other fine books and essays describe Union engineering operations during specific campaigns, such as Benjamin Franklin Cooling's *Forts Henry and Donelson: The Key to the Confederate Heartland* (1987); Larry J. Daniel and Lynn N. Bock's *Island No. 10: Struggle for the Mississippi Valley* (1996); William J. Miller's "I Only Wait for the River: McClellan and His Engineers on the Chickahominy," in *The Richmond Campaign of 1862: The Peninsula and the Seven Days* (2000), edited by Gary W. Gallagher; Edward Hagerman's *The American Civil War and the Origins of Modern Warfare: Ideas, Organization, and Field Command* (1988); Edwin C. Bearss's three-volume *The Campaign for Vicksburg* (1985); Warren E. Grabau's *Ninety-eight Days: A Geographer's View of the Vicksburg Campaign* (2000); Albert Castel's *Decision in the West: The Atlanta Campaign of 1864* (1992); and Earl J. Hess's *The Civil War in the West: Victory and Defeat from the Appalachians to the Mississippi* (2012) and *Field Armies and Fortifications in the Civil War: The Eastern Campaigns, 1861–1864* (2005).

Philip L. Shiman's essay "Engineering and Command: The Case of General William S. Rosecrans, 1862–1863," in *The Art of Command in the Civil War* (1998), edited by Steven E. Woodworth, is a superb account of the development of the Pioneer Corps in the Army of the Cumberland. For information about the Corps d'Afrique and the role African Americans played as engineers, especially during the Red River Campaign, see James G. Hollandsworth's *The Louisiana Native Guards: The Black Military Experience during the Civil War* (1998). Earl B. McElfresh's *Maps and Mapmakers of the Civil War* (1999) provides biographical information on some of the best-known topographers on both sides and describes in detail the art and science of their craft.

Memoirs, journals, and order books serve as a valuable resource for readers wishing to learn more about the personal experiences of those who served as engineers in the Union army, including men from infantry regiments. *Bridge Building in Wartime: Colonel Wesley Brainerd's Memoir of the 50th New York Volunteer Engineers* (1997), edited by Ed

Malles, tells the story of the regiment from its recruitment to its arrival in the trenches of Petersburg in 1864. In addition to Brainerd's memoir there are the following: *Building the Victory: The Order Book of the Volunteer Engineer Brigade Army of the Potomac* (1998), edited by Philip Katcher; *Diary of a Yankee Engineer: The Civil War Story of John H. Westervelt, Engineer, 1st New York Volunteer Engineer Corps* (1997), edited by Anita Palladino; William W. Belknap's *History of the Fifteenth Regiment Iowa Veteran Volunteer Infantry* (1887); John J. Hight's *History of the Fifty-eighth Regiment Indiana Volunteer Infantry* (1895); W. A. Neal's *An Illustrated History of the Missouri Engineers and the Twenty-fifth Infantry Regiment* (1889); James R. Willett's *Rambling Recollections of a Military Engineer* (1888); and *Dear Friends at Home: The Letters and Diary of Thomas James Owen, 50th New York Volunteer Engineer Regiment* (1985), edited by Dale E. Floyd.

There are only a few books that provide specific information on Confederate engineering. James Lynn Nichols's *Confederate Engineers* (1957), Harry L. Jackson's *First Regiment Engineer Troops P. A. C. S.: Robert E. Lee's Combat Engineers* (1998), and George G. Kundahl's *Confederate Engineer: Training and Campaigning with John Morris Wampler* (2000) are the best. Other books that investigate Confederate staff operations during the war and briefly describe the work of engineers include McHenry Howard's *Recollections of a Maryland Confederate Soldier and Staff Officer* (1914); Richard Goff's *Confederate Supply* (1969); William Willis Blackford's *War Years with Jeb Stuart* (1945); Edward Porter Alexander's *Fighting for the Confederacy: The Personal Recollections of General Edward Porter Alexander*, edited by Gary W. Gallagher (1989); Frank E. Vandiver's *Plowshares into Swords: Josiah Gorgas and Confederate Ordnance* (1952) and *Rebel Brass: The Confederate Command System* (1956); and J. Boone Bartholomees Jr.'s *Buff Facings and Gilt Buttons: Staff and Headquarters Operations in the Army of Northern Virginia, 1861–1865* (1998).

Military engineering is a highly technical subject, which helps explain why so little has been written about it except in professional journals and manuals. For example, J. C. Duane's *Manual for Engineer Troops* (1862), Henry D. Grafton's *A Treatise on the Camp and March, with Which Is Connected the Construction of Field Works and Military Bridges* (1861), Dennis Hart Mahan's *A Treatise on Field Fortification* (1863), Henry Wager Halleck's *Elements of Military Art and Science* (1846), and Herman Haupt's *Military Bridges: With Suggestions of New Expedients and Constructions for Crossing Streams and Chasms* (1864) are rich in detail about the problems faced and solutions developed by engineers in mid-nineteenth-century warfare.

The interested reader who is hunting for less technical information about engineering and for broader studies of the historical development of military engineering should consider Bernard Brodie and Fawn M. Brodie's *From Crossbow to H-bomb: The Evolution of Weapons and Tactics of Warfare* (1973) and William H. McNeill's *The Pursuit of Power: Technology, Armed Force, and Society from A. D. 1000 to the Present* (1982). Henry Guerlac's essay "Vauban: The Impact of Science on War," in *Makers of Modern Strategy: From Machiavelli to the Nuclear Age* (1986), edited by Peter Paret, offers insight on the intersection of the Enlightenment with the increasing professionalization of engineers. This growing professionalism is explored further in Janis Langins's *Conserving the Enlightenment: French Military Engineering from Vauban to the Revolution* (2003).

The best book on the relationship between science and warfare in the United States is *Military Enterprise and Technological Change: Perspectives on the American Experience* (1985), edited by Merritt Roe Smith. Those interested in pursuing a better understanding of engineering within the larger framework of army logistics should read James A. Huston's *The Sinews of War: Army Logistics, 1775–1953* (1977) and Martin L. van Creveld's *Supplying War: Logistics from Wallenstein to Patton* (1977).

Those fascinated by the development and work of the United States Corps of Engineers in the prewar years should look at the following: Paul K. Walker's *Engineers of Independence: A Documentary History of the Army Engineers in the American Revolution, 1775–1783* (1981); James L. Morrison Jr.'s *"The Best School in the World": West Point, the Pre–Civil War Years, 1833–1866* (1986); William B. Skelton's *An American Profession of Arms: The Army Officer Corps, 1784–1861* (1992); Russell F. Weigley's *History of the United States Army* (1984); John Lauritz Larson's *Internal Improvement: National Public Works and the Promise of Popular Government in the Early United States* (2001); and Mark A. Smith's *Engineering Security: The Corps of Engineers and the Third System Defense Policy, 1815–1861* (2009).

Although secondary sources exploring Civil War engineering or development of the United States Army's Corps of Engineers are relatively few in number, there are many more works to consider when it comes to the growth of American railroads. During the antebellum period, railroads brought about significant changes in the transportation of goods and services, in manufacturing and tool making, in the development of modern management systems, and in the labor force. For the impact of railroads on transportation, manufacturing, and labor, see Charles B. Dew's *Ironmaker to the Confederacy: Joseph R. Anderson and the Tredegar Iron Works* (1966); Albert Fishlow's *American Railroads and the Transformation of the Antebellum Economy* (1965); Allen W. Trelease's *The North Carolina Railroad, 1849–1871, and the Modernization of North Carolina* (1991); George Rogers Taylor's *The Transportation Revolution, 1815–1860* (1951); David Walker Howe's *What Hath God Wrought: The Transformation of America, 1815–1848*; Alfred D. Chandler Jr.'s *The Visible Hand: The Managerial Revolution in American Business* (1977); and *The Railroads: Pioneers in Modern Management* (1979), edited by Chandler.

The new management models described by Chandler become the operational structure adopted by the highly efficient United States Military Railroad during the war. Works on the military use of railroads between 1861 and 1865 include the classic *Victory Rode the Rails: The Strategic Place of Railroads in the Civil War* (1953) by George Edgar Turner and *The Northern Railroad in the Civil War, 1861–1865* (1952) by Thomas Weber. Other excellent studies are Roger Pickenpaugh's *Rescue by Rail: Troop Transfer and the Civil War in the West, 1863* (1998); Jeffrey N. Lash's *Destroyer of the Iron Horse: General Joseph E. Johnston and Confederate Rail Transport, 1861–1865* (1991); John E. Clark Jr.'s *Railroads in the Civil War: The Impact of Management on Victory and Defeat* (2001); and Robert C. Black III's *The Railroads of the Confederacy* (1998). Railroad operations during the war also required the effective use of the telegraph. The two best books on this subject are William Rattle Plum's *The Military Telegraph during the Civil War in the United States* (1974) and Rebecca Robins Raines's *Getting the Message Through: A Branch History of the United States Army Signal Corps* (1996).

For learning more about early American education in general and the common school movement in particular, several excellent studies are available. Rush Welter's

American Writings on Popular Education (1971) and Carl F. Kaestle's *Pillars of the Republic: Common Schools and American Society, 1780–1860* (1983) offer valuable insight into the central questions of the day, such as how schools, including Catholic schools, should be financed, whether teachers needed to be trained, and what role women should play in American education. Other highly recommended works are Christopher Collier's *Connecticut's Public Schools: A History, 1650–2000* (2009) and Lee Soltow and Edward Stevens's *The Rise of Literacy and the Common Schools in the United States: A Socioeconomic Analysis to 1870* (1981).

For works on science and engineering in the United States during the first half of the nineteenth century, I recommend the following: A. J. Angulo's *William Barton Rogers and the Idea of MIT* (2009); Keith Hoskin and Richard Macve's "The Genesis of Accountability: The West Point Connection," in the journal *Accounting Organization and Society* (1988); Samuel Rezneck's *Education for a Technological Society: A Sesquicentennial History of Rensselaer Polytechnic Institute* (1968); and Dirk J. Struik's *Yankee Science in the Making: Science and Engineering in New England from Colonial Times to the Civil War* (1991).

There is a wealth of material on the expansion of nineteenth-century American manufacturing and industrialization. These works include "Manufacturing: Reports filed by Joseph Whitworth to the House of Commons," in *The American System of Manufactures* (1969), edited by Nathan Rosenberg; Eugene S. Ferguson's "On the Origin and Development of American Mechanical 'Know-How,'" in *Midcontinent American Studies Journal* (1962); Stuart Bruchey's *The Roots of American Economic Growth, 1607–1861: An Essay in Social Causation* (1988); Charles B. Dew's *Bond of Iron: Master and Slave at Buffalo Forge* (1994); John Majewski's *Modernizing a Slave Economy: The Economic Vision of the Confederate Nation* (2009); Merritt Roe Smith's *Harper's Ferry and the New Technology: The Challenge of Change* (1977); David Hounshell's *From the American System to Mass Production, 1800–1932* (1984); and Fred Bateman and Thomas Weiss's *A Deplorable Scarcity: The Failure of Industrialization in the Slave Economy* (1981).

These sources should provide the reader with fodder for further study. For additional primary and secondary sources, please consult my endnotes.

Index

abatis: definition of, 349n58

Accokeek Creek railroad bridge. *See* Haupt, Herman

African Americans: as craftsmen, 297; as engineers, 355; and industrialization, 53, 313n8, 319n29; as slave labor, 38–39, 41, 89, 94, 297; Union army's treatment of, 164

Aiken's Landing. *See* 1st New York Volunteer Engineers

Alexander, Barton A.: as chief engineer, 118; at Sumner's bridge, 125; on volunteer engineers, 76–79

Alexander, Edward P., 248

American India-rubber pontoons. *See* India-rubber pontoons

American Institute of New York City, 33

American Railroad Journal, 58, 61

Ames Manufacturing Company (Chicopee Falls, MA), 56

Amoskeag Machine Shop (New Hampshire), 47

Anderson, Adna, 239–40; Atlanta Campaign, 265; Gettysburg Campaign, 218; hired by Tennessee governor, 90; hired by Union army, 94; United States Military Railroad Construction Corps, 333n35

Anderson, John Byers, 112, 113, 114, 224, 235; construction corps, 162; relieved of command, 239

Anderson, Robert, 83–84

Annapolis & Elkridge Railroad, 134

Antietam, battle of, 8, 153–54

Appomattox Court House, 7

Army of Mississippi, 147

Army of Northern Virginia, 7, 147, 149, 154, 157, 165, 248, 282; transportation problems (spring 1863), 210

Army of Tennessee, 89, 151, 212

Army of the Cumberland, 161–62, 165, 220–21, 223, 227, 230, 238; engineer organization for Atlanta Campaign, 269–70

Army of the Gulf, 179

Army of the James, 251, 255

Army of the Ohio, 159–60, 238; Atlanta Campaign, 269–70

Army of the Potomac, 206, 208; engineer organization, 209; school for engineers, 248–49

Army of the Tennessee, 1, 206, 220, 238; pioneer force, 271–72

Army of Virginia, 141

Asboth, Alexander, 82

Ashe, William S., 60, 62; Florida Railroad, 61; railroad transportation in Virginia, 144

Atchafalaya River bridge. *See* Bailey, Joseph

Bache, Alexander Dallas, 25, 118

Bailey, Joseph: Atchafalaya River bridge, 246–47; dam project, 244; official report, 245

Baltimore & Ohio Railroad, 55, 134; management model, 57

Balz's Company Sappers and Miners, 323n48, 323n91

Banks, Nathaniel, 173; and Red River Campaign, 241–43

Barnard, Henry, 21–22

Barnard, John G.: brevet rank, 322n24; evaluation of pontoon equipment, 131–32; named chief engineer, 251; promotion of, 76; report on Peninsula Campaign engineering operation, 122, 130–31

Battle Creek Corps (Engineers), 86–87

Bayou Baxter, 183

Bayou Macon, 184